Student Solutions Manual

Fundamentals of Mathematics

TENTH EDITION

James Van Dyke
James Rogers
Hollis Adams
Portland Community College

Prepared by

Michael G. Welden
Mt. San Jacinto College

Australia • Brazil • Japan • Korea • Mexico • Singapore • Spain • United Kingdom • United States

For product information and technology assistance, contact us at
Cengage Learning Customer & Sales Support,
1-800-354-9706

For permission to use material from this text or product, submit all requests online at **www.cengage.com/permissions**
Further permissions questions can be emailed to
permissionrequest@cengage.com

ISBN-13: 978-1-111-42948-5
ISBN-10: 1-111-42948-0

Brooks/Cole
20 Davis Drive
Belmont, CA 94002-3098
USA

Cengage Learning is a leading provider of customized learning solutions with office locations around the globe, including Singapore, the United Kingdom, Australia, Mexico, Brazil, and Japan. Locate your local office at: **www.cengage.com/global**

Cengage Learning products are represented in Canada by Nelson Education, Ltd.

To learn more about Brooks/Cole, visit **www.cengage.com/brookscole**

Purchase any of our products at your local college store or at our preferred online store **www.cengagebrain.com**

Printed in the United States of America
1 2 3 4 5 15 14 13 12 11

FD276

Contents

Preface

This manual contains detailed solutions to all of the odd numbered section exercises of the text *Fundamentals of Mathematics*, tenth edition, by James Van Dyke, James Rogers, and Hollis Adams. It also contains solutions to all problems in the chapter review and chapter test sections of the text.

Many of the exercises in the text may be solved using more than one method, but it is not feasible to list all possible solutions in this manual. Also, some of the exercises may have been solved in this manual using a method that differs slightly from that presented in the text. There are a few exercises in the text whose solutions may vary from person to person. Some of these solutions may not have been included in this manual. For the solution to an exercise like this, the notation "answers may vary" has been included.

Please remember that only reading a solution does not teach you how to solve a problem. To repeat a commonly used phrase, mathematics is not a spectator sport. You MUST make an honest attempt to solve each exercise in the text without using this manual first. This manual should be viewed more or less as a last resort. Above all, DO NOT simply copy the solution from this manual onto your own paper. Doing so will not help you learn how to do the exercise, nor will it help you to do better on quizzes or tests.

I would like to thank Paul McCombs from Rock Valley College and Shaun Williams and Jennifer Cordoba of Brooks/Cole Publishing Company for their help and support. This solutions manual was prepared using EXP 5.1, a scientific word processor.

This book is dedicated to Jack Simon in appreciation for his guidance and friendship.

May your study of this material be successful and rewarding.

Michael G. Welden

SECTION 1.1

Exercises 1.1 (page 10)

1. 8 hundreds + 4 tens + 3 ones; eight hundred forty-three

3. 4 hundreds + 6 tens + 0 ones; four hundred sixty

5. 7 thousands + 0 hundreds + 2 tens + 0 ones; seven thousand, twenty

7. eighty-seven = 8 tens + 7 ones = 87

9. nine thousand, five hundred = 9 thousands + 500 ones = 9500

11. 101 million = 101 millions + 0 thousands + 0 ones = 101,000,000

13. 27,680 = 27 thousands + 680 ones = twenty-seven thousand, six hundred eighty

15. 207,690 = 207 thousands + 690 ones = two hundred seven thousand, six hundred ninety

17. 54,000,000 = 54 millions + 0 thousands + 0 ones = fifty-four million

19. two hundred forty-three thousand, seven hundred = 243 thousands + 700 ones = 243,700

21. twenty-two thousand, five hundred seventy = 22 thousands + 570 ones = 22,570

23. nineteen billion = 19 billions + 0 millions + 0 thousands + 0 ones = 19,000,000,000

25. 12 is to the left of 22, so $12 < 22$.

27. 61 is to the right of 54, so $61 > 54$..

29. 246 is to the left of 251, so $246 < 251$.

31. 7470 is to the left of 7850, so $7470 < 7850$.

33. The digit in the tens position is 4. Since the digit in the ones position is 2, and 2 is less than 5, round down. Leave the 4 in the tens position, and replace the 2 in the ones position with 0. The rounded number is 740.

35. The digit in the hundreds position is 6. Since the digit in the tens position is 5, round up. Change the 6 in the hundreds position to 7, and replace the digits in the tens and ones positions with 0. The rounded number is 2700.

37. **TEN**
The digit in the tens position is 4. Since the digit in the ones position is 6, and 6 is more than 5, round up. Change the 4 in the tens position to 5, and replace the 6 in the ones position with 0. The rounded number is 607,550.

HUNDRED
The digit in the hundreds position is 5. Since the digit in the tens position is 4, and 4 is less than 5, round down. Leave the 5 in the hundreds position, and replace the digits in the tens and ones positions with 0. The rounded number is 607,500.

continued on next page...

37. continued

THOUSAND

The digit in the thousands position is 7. Since the digit in the hundreds position is 5, round up. Change the 7 in the thousands position to 8, and replace the digits in the hundreds, tens and ones positions with 0. The rounded number is 608,000.

TEN THOUSAND

The digit in the ten thousands position is 0. Since the digit in the thousands position is 7, and 7 is more than 5, round up. Change the 0 in the ten thousands position to 1, and replace the digits in the thousands, hundreds, tens and ones positions with 0. The rounded number is 610,000.

39. TEN

The digit in the tens position is 4. Since the digit in the ones position is 2, and 2 is less than 5, round down. Leave the 4 in the tens position, and replace the 2 in the ones position with 0. The rounded number is 6,545,740.

HUNDRED

The digit in the hundreds position is 7. Since the digit in the tens position is 4, and 4 is less than 5, round down. Leave the 7 in the hundreds position, and replace the digits in the tens and ones positions with 0. The rounded number is 6,545,700.

THOUSAND

The digit in the thousands position is 5. Since the digit in the hundreds position is 7, and 7 is greater than 5, round up. Change the 5 in the thousands position to 6, and replace the digits in the hundreds, tens and ones positions with 0. The rounded number is 6,546,000.

TEN THOUSAND

The digit in the ten thousands position is 4. Since the digit in the thousands position is 5, round up. Replace the 4 in the ten thousands position with 5, and replace the digits in the thousands, hundreds, tens and ones positions with 0. The rounded number is 6,550,000.

41. Find the row labeled "$15 - $24,999" and the column labeled "Does Exercise." The value at the intersection is 40%, so the percent who exercise regularly is 40%.

43. Find the column labeled "Does Not Exercise." Look for the row with the lowest entry in this column. The lowest entry is 48%, in the row labeled "Over $50,000." The over $50,000 category has the lowest percentage of non-exercisers.

45. Look at the entries in the column labeled "Does Exercise." As you move down the column, the percentages increase. Thus, as the income level goes up, so does the percentage of regular exercisers.

47. Look at each row and list the cities where the number increased from the "2005" column to the "2007 column. There was an increase in Seattle. Look at each row and list the cities where the number decreased from the "2005" column to the "2007 column. There were decreases in Boston, Chicago, Minneapolis, New Orleans, and San Diego. There was a slight increase in Atlanta, but the value is the same when rounded.

49. The number of homeless in Chicago in 2007 was 5979, while the number in Atlanta was 6840.
$6840 > 5979$

SECTION 1.1

51. six hundred fifty-six million, seven hundred thirty-two thousand, four hundred ten = 656 millions + 732 thousands + 410 ones = 656,732,410

53. 18,503 = 18 thousands + 503 ones = eighteen thousand, five hundred three dollars

55. 4553 is to the right of 4525, so $4553 > 4525$.

57. The largest 4-digit number is 9999.

59. The digit in the hundred thousands position is 5. Since the digit in the ten thousands position is 7, and 7 is more than 5, round up. Change the 5 in the hundred thousands position to 6, and replace the digits in the ten thousands,thousands, hundreds, tens, and ones positions with 0. The rounded number is 74,600,000.

61. Rounding 63,749 to the nearest hundred results in 63,700, since 4 is less than 5. Rounding 63,749 to the nearest ten results in 63,750, since 9 is greater than 5. Rounding 63,750 to the nearest hundred results in 63,800. The second result of rounding to the nearest hundred is different because rounding to the nearest ten first changes the tens digit. The first result is correct, because 63,749 is closer to 63,700 than to 63,800.

63. 11,475 = 11 thousands + 475 ones = eleven thousand, four hundred seventy-five. Kimo wrote "eleven thousand, four hundred seventy-five" on the check.

65. $25{,}400 < 230{,}000$

67. thirty-six thousand, four hundred seven = 36 thousands + 407 ones = 36,407. The bid is $36,407.

69. The digit in the thousands position is 5. Since the digit in the hundreds position is 2, and 2 is less than 5, round down. Leave the 5 in the thousands position, and replace the digits in the hundreds, tens, and ones positions with 0. The rounded number is 185,000. To the nearest thousand dollars, the value of the shares is $185,000.

71. The number in the millions position is 0. Since the digit in the hundred thousands position is 6, and 6 is more than 5, round up. Change the 0 in the millions position to 1, and change the digits in the hundred thousands, ten thousands, thousands, hundreds, tens, and ones positions to 0. There were about 101,000,000 metric tons of carbon emissions in 2000.

73. 32,095 = 32 thousands + 95 ones = thirty-two thousand, ninety-five dollars for Maine

75. The smallest per capita income is $32,095, in Maine.

77. The digit in the millions position is 2. Since the digit in the hundred thousands position is 8, and 8 is greater than 5, round up. Change the 2 in the millions position to 3, and replace the digits in the hundred thousands, ten thousands, thousands, hundreds, tens, and ones positions with 0.
The rounded number is 93,000,000 miles.

79.

Month	Number of Marriages	Month	Number of Marriages
June	242,000	December	184,000
May	241,000	April	172,000
August	239,000	November	171,000
July	235,000	February	155,000
October	231,000	March	118,000
September	225,000	January	110,000

81. The rivers in increasing length are: Yukon, Colorado, Arkansas, Rio Grande, Missouri, Mississippi.

83. The estimate is accurate, since 2,376,499 rounded to the nearest thousand is 2,376,000.

85. Jif has fewer of the following nutrients: fat (and saturated fat), sodium, carbohydrates, and sugars.

87. $533{,}184{,}219 = 533$ millions $+$ 184 thousands $+$ 219 ones; *The Dark Knight* took in five hundred thirty-three million, one hundred eighty-four thousand, two hundred nineteen dollars.

89. The digit in the millions position is 7. Since the digit in the hundred thousands position is 2, and 2 is less than 5, round down. Leave the 7 in the millions position, and change all digits to the right to 0. *Shrek 2* took in about $437,000,000.

91. "Base ten" is a good name for our number system because each place value in the system is 10 times the previous place and one-tenth the succeeding place.

93. Rounding a number is a method of obtaining an approximation of that number. The purpose is to gain an idea of the number without listing digits that do not add to our understanding. The number 87,452 rounds to 87,000 to the nearest thousand because 452 is less than halfway between 0 and 1000. To the nearest hundred, the number is 87,500, because 52 is more than halfway between 0 and 100.

95. five trillion, three hundred twenty-six billion, nine hundred one million, five hundred seventy thousand

97. The value $X = 7$ makes the statement false.

99. The digit in the hundred thousands position is 0. Since the digit in the ten thousands position is 4, and 4 is less than 5, round down. Leave the 0 in the hundred thousands position, and replace the digits in the ten thousands, thousands, hundreds, tens and ones positions with 0. The rounded number is 0.

Exercises 1.2 (page 24)

1.
$$\begin{array}{r} {\scriptstyle 1} \\ 75 \\ +\ 38 \\ \hline 113 \end{array}$$

3.
$$\begin{array}{r} {\scriptstyle 1} \\ 724 \\ +\ 218 \\ \hline 942 \end{array}$$

5.
$$\begin{array}{r} {\scriptstyle 1} \\ 212 \\ +\ 495 \\ \hline 707 \end{array}$$

SECTION 1.2

7. 1

9.
$$\begin{array}{rrrrr} & {\scriptstyle 1} & {\scriptstyle 1} & {\scriptstyle 2} & \\ & & 5 & 1 & 5 \\ & 2 & 9 & 0 & 8 \\ + & & 3 & 8 & 7 \\ \hline & 3 & 8 & 1 & 0 \end{array}$$

11.
$$\begin{array}{rrrrr} & & {\scriptstyle 1} & {\scriptstyle 1} & \\ & & & & 7 \\ & & & 8 & 5 \\ & & 6 & 0 & 7 \\ + & 5 & 0 & 9 & 0 \\ \hline & 5 & 7 & 8 & 9 \end{array}$$

13.
$$\begin{array}{rrrrrr} & & {\scriptstyle 1} & {\scriptstyle 1} & {\scriptstyle 1} & \\ & & 2 & 7 & 9 & 5 \\ & & 3 & 6 & 4 & 3 \\ & & 7 & 0 & 5 & 5 \\ + & & 4 & 0 & 0 & 4 \\ \hline & 1 & 7 & 4 & 9 & 7 \end{array}$$

$17{,}497 \approx 17{,}500$

15.
$$\begin{array}{rcccccc} & \text{8 hundreds} & + & \text{7 tens} & + & \text{4 ones} \\ - & \text{5 hundreds} & + & \text{7 tens} & + & \text{2 ones} \\ \hline & \text{3 hundreds} & + & \text{0 tens} & + & \text{2 ones} \end{array}$$

3 hundreds + 0 tens + 2 ones = 302

17.
$$\begin{array}{rrrr} & {\scriptstyle 3} & {\scriptstyle 10} & \\ & \not{4} & \not{0} & 6 \\ - & & 7 & 2 \\ \hline & 3 & 3 & 4 \end{array}$$

19.
$$\begin{array}{rrrr} & 8 & 7 & 6 \\ - & 3 & 4 & 5 \\ \hline & 5 & 3 & 1 \end{array}$$

21. 10

23.
$$\begin{array}{rrrr} & {\scriptstyle 8} & {\scriptstyle 13} & \\ & & \not{3} & {\scriptstyle 14} \\ & \not{9} & \not{4} & \not{4} \\ - & 4 & 5 & 8 \\ \hline & 4 & 8 & 6 \end{array}$$

25.
$$\begin{array}{rrrr} & {\scriptstyle 2} & {\scriptstyle 9} & \\ & & \not{1}\not{0} & {\scriptstyle 10} \\ & \not{3} & \not{0} & \not{0} \\ - & 1 & 6 & 4 \\ \hline & 1 & 3 & 6 \end{array}$$

27.
$$\begin{array}{rrrrr} & & {\scriptstyle 6} & {\scriptstyle 16} & \\ & 8 & \not{7} & \not{0} & 9 \\ - & 4 & 0 & 7 & 3 \\ \hline & 4 & 6 & 9 & 6 \end{array}$$

$4696 \approx 4700$

29.
$$\begin{array}{rrrr} & 5 & 4 & 6 \\ + & 5 & 7 & 7 \\ \hline \end{array} \Rightarrow \begin{array}{rrrrr} & & 5 & 0 & 0 \\ + & & 6 & 0 & 0 \\ \hline & 1 & 1 & 0 & 0 \end{array}$$

31.
$$\begin{array}{rrrrr} & 2 & 0 & 4 & 4 \\ & 4 & 5 & 5 & 0 \\ + & 3 & 4 & 4 & 9 \\ \hline \end{array} \Rightarrow \begin{array}{rrrrrr} & & {\scriptstyle 1} & & & \\ & & 2 & 0 & 0 & 0 \\ & & 4 & 6 & 0 & 0 \\ + & & 3 & 4 & 0 & 0 \\ \hline & 1 & 0 & 0 & 0 & 0 \end{array}$$

33.
$$\begin{array}{rrrr} & 6 & 7 & 5 \\ - & 3 & 4 & 9 \\ \hline \end{array} \Rightarrow \begin{array}{rrrr} & 7 & 0 & 0 \\ - & 3 & 0 & 0 \\ \hline & 4 & 0 & 0 \end{array}$$

35.
$$\begin{array}{rrrrr} & 9 & 7 & 6 & 5 \\ - & 4 & 7 & 6 & 6 \\ \hline \end{array} \Rightarrow \begin{array}{rrrrr} & 9 & 8 & 0 & 0 \\ - & 4 & 8 & 0 & 0 \\ \hline & 5 & 0 & 0 & 0 \end{array}$$

37.
$$\begin{array}{rrrrr} & 3 & 2 & 0 & 9 \\ & 7 & 0 & 9 & 5 \\ & 4 & 4 & 4 & 4 \\ & 2 & 0 & 0 & 4 \\ + & 3 & 1 & 6 & 6 \\ \hline \end{array} \Rightarrow \begin{array}{rrrrrr} & & 3 & 0 & 0 & 0 \\ & & 7 & 0 & 0 & 0 \\ & & 4 & 0 & 0 & 0 \\ & & 2 & 0 & 0 & 0 \\ + & & 3 & 0 & 0 & 0 \\ \hline & 1 & 9 & 0 & 0 & 0 \end{array}$$

39.

$$\begin{array}{rrrrrr} & 4 & 5 & 9 & 0 & 2 \\ & 3 & 3 & 3 & 3 & 3 \\ & 5 & 7 & 7 & 0 & 0 \\ + & 2 & 3 & 6 & 5 & 3 \\ \hline \end{array} \Rightarrow \begin{array}{rrrrrr} & & 2 & & & \\ & & 4 & 6 & 0 & 0 & 0 \\ & & 3 & 3 & 0 & 0 & 0 \\ & & 5 & 8 & 0 & 0 & 0 \\ + & & 2 & 4 & 0 & 0 & 0 \\ \hline & 1 & 6 & 1 & 0 & 0 & 0 \end{array}$$

41.

$$\begin{array}{rrrrr} & 7 & 8 & 2 & 2 \\ - & 3 & 0 & 9 & 8 \\ \hline \end{array} \Rightarrow \begin{array}{rrrrr} & 8 & 0 & 0 & 0 \\ - & 3 & 0 & 0 & 0 \\ \hline & 5 & 0 & 0 & 0 \end{array}$$

43.

$$\begin{array}{rrrrrr} & 6 & 5 & 8 & 0 & 8 \\ - & 3 & 2 & 1 & 7 & 5 \\ \hline \end{array} \Rightarrow \begin{array}{rrrrrr} & 6 & 6 & 0 & 0 & 0 \\ - & 3 & 2 & 0 & 0 & 0 \\ \hline & 3 & 4 & 0 & 0 & 0 \end{array}$$

45. 11 cm + 11 cm + 11 cm + 11 cm = 44 cm

47. 40 in. + 56 in. + 36 in. + 89 in. = 221 in.

49. 7 ft + 7 ft + 7 ft + 7 ft + 7 ft + 7 ft + 7 ft + 7 ft = 56 ft

51. 10 in. + 40 in. + 4 in. + 18 in. + 18 in. + 4 in. + 40 in. = 134 in.

53. Fords + Toyotas + Lexuses
= 1837 + 1483 + 241

$$\begin{array}{rrrrr} & 1 & 1 & 1 & \\ & 1 & 8 & 3 & 7 \\ & 1 & 4 & 8 & 3 \\ + & & 2 & 4 & 1 \\ \hline & 3 & 5 & 6 & 1 \end{array}$$

The total number of Fords, Toyotas, and Lexuses sold is 3561.

55. Hondas − Fords = 2000 − 1837

$$\begin{array}{rrrrr} & & 9 & 9 & 10 \\ & 1 & \not{10} & \not{10} & \\ & \not{2} & \not{0} & \not{0} & \not{0} \\ - & 1 & 8 & 3 & 7 \\ \hline & & 1 & 6 & 3 \end{array}$$

The are 163 more Hondas sold than Fords.

57. Hondas + Fords + Toyotas
= 2000 + 1837 + 1483

$$\begin{array}{rrrrr} & 1 & 1 & 1 & \\ & 2 & 0 & 0 & 0 \\ & 1 & 8 & 3 & 7 \\ + & 1 & 4 & 8 & 3 \\ \hline & 5 & 3 & 2 & 0 \end{array}$$

The total number of the three best-selling cars is 5320.

59.

$$\begin{array}{rrrrr} & 3 & 2 & 2 & \\ & 1 & 0 & 4 & 6 \\ & & 8 & 7 & 3 \\ & & 4 & 5 & 4 \\ & 1 & 1 & 5 & 6 \\ & & 6 & 0 & 7 \\ & & 5 & 4 & 1 \\ + & & 8 & 1 & 0 \\ \hline & 5 & 4 & 8 & 7 \end{array} \qquad \begin{array}{rrrr} & 8 & 7 & 3 \\ - & 5 & 4 & 1 \\ \hline & 3 & 3 & 2 \end{array}$$

The total count for the week was 5487 salmon. Tuesday's count was 332 more than Saturday's.

SECTION 1.2

61.

$$\begin{array}{crrrrr} & 1 & 0 & 0 & 3 & 4 \\ - & & 7 & 9 & 5 & 9 \\ \hline \end{array} \Rightarrow \begin{array}{crrrrr} & 1 & 0 & 0 & 0 & 0 \\ - & & 8 & 0 & 0 & 0 \\ \hline & & 2 & 0 & 0 & 0 \end{array}$$

Ralph's answer is reasonable.

63.

$$\begin{array}{crrr} & 3 & 7 & 6 \\ & 4 & 8 & 2 \\ & 2 & 8 & 9 \\ + & 1 & 4 & 8 \\ \hline \end{array} \Rightarrow \begin{array}{crrrr} & & 4 & 0 & 0 \\ & & 5 & 0 & 0 \\ & & 3 & 0 & 0 \\ + & & 1 & 0 & 0 \\ \hline & 1 & 3 & 0 & 0 \end{array}$$

The total estimated cost is $1300.

65.

$$\begin{array}{crrrrr} & & ^{2} & ^{3} & ^{1} & \\ & & & 4 & 8 & 2 \\ & 1 & 2 & 4 & 8 & 0 \\ & & 3 & 8 & 7 & 6 \\ + & & & 1 & 7 & 6 \\ \hline & 1 & 7 & 0 & 1 & 4 \end{array}$$

The total number of property crimes is 17,014.

67.

$$\begin{array}{crrr} & ^{1} & & \\ & 2 & 0 & 0 \\ & 2 & 0 & 0 \\ & & 6 & 0 \\ + & & 6 & 0 \\ \hline & 5 & 2 & 0 \end{array}$$

Sasha consumes 520 calories.

69. From #67 and #68, Sasha consumes 1070 calories at dinner.

$$\begin{array}{crrrr} & & ^{1} & ^{15} & \\ & 2 & \not{2} & \not{5} & 0 \\ - & 1 & 0 & 7 & 0 \\ \hline & 1 & 1 & 8 & 0 \end{array}$$

She could have eaten 1180 calories at breakfast and lunch.

71.

$$\begin{array}{crrrrr} & ^{2} & ^{10} & & & \\ & & \not{0} & ^{14} & & \\ & \not{3} & \not{1} & \not{4} & 7 & 8 \\ - & & 8 & 5 & 4 & 3 \\ \hline & 2 & 2 & 9 & 3 & 5 \end{array}$$

A total of 22,935 trees can be harvested.

73.

$$\begin{array}{crrrrr} & ^{1} & ^{2} & ^{2} & & \\ & & 1 & 6 & 7 & 0 \\ & & 3 & 6 & 7 & 0 \\ & & & 6 & 7 & 0 \\ + & 1 & 5 & 6 & 7 & 0 \\ \hline & 2 & 1 & 6 & 8 & 0 \end{array} \qquad \begin{array}{crrrrr} & & ^{5} & ^{13} & & \\ & & & \not{3} & ^{15} & \\ & 3 & \not{6} & \not{4} & \not{5} & 0 \\ - & 2 & 1 & 6 & 8 & 0 \\ \hline & 1 & 4 & 7 & 7 & 0 \end{array}$$

$14{,}770 \approx 14{,}800$

Fong's grocery still owes about $14,800.

75. The median incomes were probably rounded to the nearest thousand.

$$\begin{array}{crrrrr} & 9 & 5 & 0 & 0 & 0 \\ - & 7 & 2 & 0 & 0 & 0 \\ \hline & 2 & 3 & 0 & 0 & 0 \end{array}$$

San Francisco's median income was $23,000 higher than Seattle's.

77.

$$\begin{array}{crrr} & ^{8} & ^{11} & \\ & \not{9} & \not{1} & 3 \\ - & 3 & 8 & 3 \\ \hline & 5 & 3 & 0 \end{array}$$

530 more husbands killed their wives than wives killed their husbands.

SECTION 1.2

79.

$$\begin{array}{r} {\scriptstyle 1} \\ 167 \\ +\ 42 \\ \hline 209 \end{array}$$

209 people killed a sibling.

81.

$$\begin{array}{rccccc} & & & {\scriptstyle 8} & {\scriptstyle 14} & \\ & 3 & 5 & \cancel{9} & \cancel{4} & 0 \\ - & 1 & 2 & 1 & 8 & 0 \\ \hline & 2 & 3 & 7 & 6 & 0 \end{array}$$

There were 23,760 more cars sold than sport utility vehicles.

83.

$$\begin{array}{rccccc} & 2 & 4 & 3 & 1 & 0 \\ + & 1 & 2 & 1 & 8 & 0 \\ \hline & 3 & 6 & 4 & 9 & 0 \end{array}$$

$$\begin{array}{rccccc} & & {\scriptstyle 5} & {\scriptstyle 14} & & \\ & 3 & \cancel{6} & \cancel{4} & 9 & 0 \\ - & 3 & 5 & 9 & 4 & 0 \\ \hline & & & 5 & 5 & 0 \end{array}$$

There were 550 more utility vehicles sold than cars.

85.

$$\begin{array}{rcccccccc} & & & {\scriptstyle 1} & {\scriptstyle 10} & & & & \\ & 2 & 2 & \cancel{0} & 0 & 0 & 0 & 0 & 0 \\ - & & 1 & 8 & 0 & 0 & 0 & 0 & 0 \\ \hline & 2 & 0 & 2 & 0 & 0 & 0 & 0 & 0 \end{array}$$

There are 20,200,000 adults with HIV.

87. 62 ft + 38 ft + 62 ft + 38 ft = 200 ft
The perimeter of the house is 200 ft.

89. 64 in. + 64 in. + 48 in. + 48 in. = 224 in.
Blanche needs 224 in. of lace.

91. You could explain that $15 - 9 = 6$ by showing what happens when 9 circles are removed from a group of 15 circles.

93. A sum is the result of adding numbers. The sum of 2 and 3 is 5, indicated by $2 + 3 = 5$.

95.

$$\begin{array}{rcccccc} & {\scriptstyle 2} & {\scriptstyle 1} & {\scriptstyle 1} & {\scriptstyle 1} & & \\ & & & & 1 & 6 & 0 \\ & & & 8 & 0 & 0 & 0 \\ & & & & 3 & 1 & 2 \\ & & 4 & 7 & 2 & 0 & 0 \\ & & & & 9 & 5 & 2 \\ + & 1 & 4 & 7 & 5 & 2 & 3 \\ \hline & 7 & 0 & 0 & 9 & 4 & 7 \end{array}$$

$700{,}947 \approx 700{,}900$;
seven hundred thousand, nine hundred

97. Civics

$$\begin{array}{rrrrrr} & 1 & 1 & 2 & 2 & \\ & 1 & 5 & 4 & 8 & 8 \\ & 1 & 5 & 4 & 8 & 8 \\ + & 1 & 5 & 4 & 8 & 8 \\ \hline & 4 & 6 & 4 & 6 & 4 \end{array}$$

Accords

$$\begin{array}{rrrrrr} & 3 & 3 & 3 & 2 & \\ & 1 & 8 & 9 & 8 & 5 \\ & 1 & 8 & 9 & 8 & 5 \\ & 1 & 8 & 9 & 8 & 5 \\ + & 1 & 8 & 9 & 8 & 5 \\ \hline & 7 & 5 & 9 & 4 & 0 \end{array}$$

Acuras

$$\begin{array}{rrrrrr} & & 1 & 1 & 1 & \\ & 3 & 0 & 7 & 9 & 8 \\ + & 3 & 0 & 7 & 9 & 8 \\ \hline & 6 & 1 & 5 & 9 & 6 \end{array}$$

Total

$$\begin{array}{rrrrrrr} & & 1 & 2 & 2 & 1 & \\ & & 4 & 6 & 4 & 6 & 4 \\ & & 7 & 5 & 9 & 4 & 0 \\ + & & 6 & 1 & 5 & 9 & 6 \\ \hline & 1 & 8 & 4 & 0 & 0 & 0 \end{array}$$

Accords − Civics

$$\begin{array}{rrrrrr} & 6 & 15 & 8 & 13 & \\ & & & & \not 3 & 10 \\ & \not 7 & \not 5 & \not 9 & \not 4 & \not 0 \\ - & 4 & 6 & 4 & 6 & 4 \\ \hline & 2 & 9 & 4 & 7 & 6 \end{array}$$

The total sales were \$184,000.
The sales for the Accord were \$29,476 more than the sales for the Civic.

99.

$$\begin{array}{rrrrr} & 4 & \text{A} & 6 & \text{B} \\ - & \text{C} & 2 & 5 & 1 \\ \hline & 1 & 5 & \text{D} & 1 \end{array}$$

$\text{B} - 1 = 1 \Rightarrow \text{B} = 2$. $\text{D} = 6 - 5 = 1$.
$\text{A} - 2 = 5 \Rightarrow \text{A} = 7$. $4 - \text{C} = 1 \Rightarrow \text{C} = 3$.

Getting Ready for Algebra (page 34)

1.
$$\begin{aligned} x + 12 &= 24 \\ x + 12 - 12 &= 24 - 12 \\ x &= 12 \end{aligned}$$
Check:
$$\begin{aligned} x + 12 &= 24 \\ 12 + 12 &\stackrel{?}{=} 24 \\ 24 &= 24 \end{aligned}$$

3.
$$\begin{aligned} x - 6 &= 17 \\ x - 6 + 6 &= 17 + 6 \\ x &= 23 \end{aligned}$$
Check:
$$\begin{aligned} x - 6 &= 17 \\ 23 - 6 &\stackrel{?}{=} 17 \\ 17 &= 17 \end{aligned}$$

5.
$$\begin{aligned} z + 13 &= 27 \\ z + 13 - 13 &= 27 - 13 \\ z &= 14 \end{aligned}$$
Check:
$$\begin{aligned} z + 13 &= 27 \\ 14 + 13 &\stackrel{?}{=} 27 \\ 27 &= 27 \end{aligned}$$

7.
$$\begin{aligned} c + 24 &= 63 \\ c + 24 - 24 &= 63 - 24 \\ c &= 39 \end{aligned}$$
Check:
$$\begin{aligned} c + 24 &= 63 \\ 39 + 24 &\stackrel{?}{=} 63 \\ 63 &= 63 \end{aligned}$$

9.
$$\begin{aligned} a - 40 &= 111 \\ a - 40 + 40 &= 111 + 40 \\ a &= 151 \end{aligned}$$
Check:
$$\begin{aligned} a - 40 &= 111 \\ 151 - 40 &\stackrel{?}{=} 111 \\ 111 &= 111 \end{aligned}$$

11.
$$\begin{aligned} x + 91 &= 105 \\ x + 91 - 91 &= 105 - 91 \\ x &= 14 \end{aligned}$$
Check:
$$\begin{aligned} x + 91 &= 105 \\ 14 + 91 &\stackrel{?}{=} 105 \\ 105 &= 105 \end{aligned}$$

13.
$$\begin{aligned} y + 67 &= 125 \\ y + 67 - 67 &= 125 - 67 \\ y &= 58 \end{aligned}$$
Check:
$$\begin{aligned} y + 67 &= 125 \\ 58 + 67 &\stackrel{?}{=} 125 \\ 125 &= 125 \end{aligned}$$

15.
$$\begin{aligned} k - 56 &= 112 \\ k - 56 + 56 &= 112 + 56 \\ k &= 168 \end{aligned}$$
Check:
$$\begin{aligned} k - 56 &= 112 \\ 168 - 56 &\stackrel{?}{=} 112 \\ 112 &= 112 \end{aligned}$$

17.
$$\begin{aligned} 73 &= x + 62 \\ 73 - 62 &= x + 62 - 62 \\ 11 &= x \end{aligned}$$
Check:
$$\begin{aligned} 73 &= x + 62 \\ 73 &\stackrel{?}{=} 11 + 62 \\ 73 &= 73 \end{aligned}$$

19.
$$\begin{aligned} 87 &= w - 29 \\ 87 + 29 &= w - 29 + 29 \\ 116 &= w \end{aligned}$$
Check:
$$\begin{aligned} 87 &= w - 29 \\ 87 &\stackrel{?}{=} 116 - 29 \\ 87 &= 87 \end{aligned}$$

21.
$$\boxed{\text{Cost}} + \boxed{\text{Markup}} = \boxed{\text{Selling price}}$$
$$\begin{aligned} C + M &= S \\ 917 + M &= 1265 \\ 917 - 917 + M &= 1265 - 917 \\ M &= 348 \end{aligned}$$
The markup is \$348.

23.
$$\boxed{\text{Length}} = \boxed{\text{Width}} + 2$$
$$\begin{aligned} L &= W + 2 \\ L &= 7 + 2 \\ L &= 9 \end{aligned}$$
The length of the garage is 9 meters.

25. Let $S =$ the rating of the Saturn and let $I =$ the rating of the Impreza.
$$\begin{aligned} I + 5 &= S \\ I + 5 &= 35 \\ I + 5 - 5 &= 35 - 5 \\ I &= 30 \end{aligned}$$
The rating for the Impreza is 30 mpg.

27. Let $B =$ the total amount budgeted in a category. Let $S =$ the amount spent in a category. Let $R =$ the amount not yet spent in a category.
$$S + R = B$$

Exercises 1.3 (page 40)

1.
$$\begin{array}{r} {\scriptstyle 2} \\ 83 \\ \times\quad 7 \\ \hline 581 \end{array}$$

3.
$$\begin{array}{r} {\scriptstyle 2} \\ 97 \\ \times\quad 3 \\ \hline 291 \end{array}$$

5.
$$\begin{array}{r} {\scriptstyle 2} \\ 76 \\ \times\quad 4 \\ \hline 304 \end{array}$$

7.
$$\begin{array}{r} {\scriptstyle 2} \\ 93 \\ \times\quad 7 \\ \hline 651 \end{array}$$

9.
$$\begin{array}{r} {\scriptstyle 5} \\ 37 \\ \times\quad 8 \\ \hline 296 \end{array}$$

11.
$$\begin{array}{r} 239 \\ \times\quad 0 \\ \hline 0 \end{array}$$

SECTION 1.3

13.
$$\begin{array}{crrrr} & & & {\scriptstyle 2} & \\ & & & 7 & 6 \\ \times & & & 4 & 0 \\ \hline & & & 0 & 0 \\ & 3 & 0 & 4 & 0 \\ \hline & 3 & 0 & 4 & 0 \end{array}$$

15. thousands

17.
$$\begin{array}{crrrr} & & {\scriptstyle 3} & {\scriptstyle 4} & \\ & & 6 & 4 & 6 \\ \times & & & & 7 \\ \hline & 4 & 5 & 2 & 2 \end{array}$$

19.
$$\begin{array}{crrrr} & & & {\scriptstyle 2} & \\ & & 8 & 0 & 4 \\ \times & & & & 7 \\ \hline & 5 & 6 & 2 & 8 \end{array}$$

21.
$$\begin{array}{crrrr} & & & {\scriptstyle 1} & \\ & & & \cancel{2} & \\ & & & 5 & 3 \\ \times & & & 6 & 7 \\ \hline & & 3 & 7 & 1 \\ & 3 & 1 & 8 & 0 \\ \hline & 3 & 5 & 5 & 1 \end{array}$$

23.
$$\begin{array}{crrrr} & & & {\scriptstyle 1} & \\ & & & \cancel{2} & \\ & & & 9 & 4 \\ \times & & & 3 & 7 \\ \hline & & 6 & 5 & 8 \\ & 2 & 8 & 2 & 0 \\ \hline & 3 & 4 & 7 & 8 \end{array}$$

25.
$$\begin{array}{crrrrrr} & & & & {\scriptstyle 1} & \\ & & & 4 & 1 & 6 \\ \times & & & 3 & 0 & 0 \\ \hline & & & 0 & 0 & 0 \\ & & 0 & 0 & 0 & 0 \\ & 1 & 2 & 4 & 8 & 0 & 0 \\ \hline & 1 & 2 & 4 & 8 & 0 & 0 \end{array}$$

27.
$$\begin{array}{crrrrr} & & & & {\scriptstyle 2} & \\ & & & & \cancel{1} & \\ & & & 9 & 0 & 4 \\ \times & & & & 7 & 4 \\ \hline & & 3 & 6 & 1 & 6 \\ & 6 & 3 & 2 & 8 & 0 \\ \hline & 6 & 6 & 8 & 9 & 6 \end{array}$$

29.
$$\begin{array}{crrrrr} & & & {\scriptstyle 1} & {\scriptstyle 2} & \\ & & & \cancel{3} & \cancel{5} & \\ & & & 7 & 4 & 7 \\ \times & & & & 4 & 8 \\ \hline & & 5 & 9 & 7 & 6 \\ & 2 & 9 & 8 & 8 & 0 \\ \hline & 3 & 5 & 8 & 5 & 6 \end{array}$$

31.
$$\begin{array}{crrrrr} & & & {\scriptstyle 4} & {\scriptstyle 1} & \\ & & & \cancel{3} & \cancel{1} & \\ & & & 2 & 5 & 2 \\ \times & & & & 8 & 7 \\ \hline & & 1 & 7 & 6 & 4 \\ & 2 & 0 & 1 & 6 & 0 \\ \hline & 2 & 1 & 9 & 2 & 4 \end{array}$$

21,924 $\approx$ 21,900

33.
$$\begin{array}{crrrrr} & & & & {\scriptstyle 1} & \\ & & & 3 & 1 & 2 \\ \times & & & & 5 & 0 \\ \hline & & & 0 & 0 & 0 \\ & 1 & 5 & 6 & 0 & 0 \\ \hline & 1 & 5 & 6 & 0 & 0 \end{array}$$

35.
$$\begin{array}{crrrrr} & & & {\scriptstyle 1} & {\scriptstyle 4} & \\ & & & & \cancel{2} & \\ & & & 5 & 2 & 7 \\ \times & & & & 7 & 3 \\ \hline & & 1 & 5 & 8 & 1 \\ & 3 & 6 & 8 & 9 & 0 \\ \hline & 3 & 8 & 4 & 7 & 1 \end{array}$$

37.
$$\begin{array}{crrrrr} & & & {\scriptstyle 1} & {\scriptstyle 3} & \\ & & & \cancel{2} & \cancel{5} & \\ & & & 7 & 3 & 8 \\ \times & & & & 4 & 7 \\ \hline & & 5 & 1 & 6 & 6 \\ & 2 & 9 & 5 & 2 & 0 \\ \hline & 3 & 4 & 6 & 8 & 6 \end{array}$$

39.
$$\begin{array}{crrrrrr} & & & {\scriptstyle 2} & {\scriptstyle 1} & & \\ & & & \cancel{1} & \cancel{1} & & \\ & & & 4 & 3 & 2 & 1 \\ \times & & & & & 7 & 6 \\ \hline & & 2 & 5 & 9 & 2 & 6 \\ & 3 & 0 & 2 & 4 & 7 & 0 \\ \hline & 3 & 2 & 8 & 3 & 9 & 6 \end{array}$$

41.
$$\begin{array}{crr} & 4 & 3 \\ \times & 8 & 4 \\ \hline \end{array} \Rightarrow \begin{array}{crrrr} & & & 4 & 0 \\ \times & & & 8 & 0 \\ \hline & & 0 & 0 & 0 \\ & 3 & 2 & 0 & 0 \\ \hline & 3 & 2 & 0 & 0 \end{array}$$

43.
$$\begin{array}{crrr} & 5 & 2 & 8 \\ \times & & 4 & 8 \\ \hline \end{array} \Rightarrow \begin{array}{crrrrr} & & & 5 & 0 & 0 \\ \times & & & & 5 & 0 \\ \hline & & & 0 & 0 & 0 \\ & 2 & 5 & 0 & 0 & 0 \\ \hline & 2 & 5 & 0 & 0 & 0 \end{array}$$

45.
$$\begin{array}{r} 4510 \\ \times \quad 53 \\ \hline \end{array} \Rightarrow \begin{array}{r} 5000 \\ \times \quad 50 \\ \hline 0000 \\ 250000 \\ \hline 250000 \end{array}$$

47.
$$\begin{array}{r} 3046 \\ \times \quad 83 \\ \hline \end{array} \Rightarrow \begin{array}{r} 3000 \\ \times \quad 80 \\ \hline 0000 \\ 240000 \\ \hline 240000 \end{array}$$

49.
$$\begin{array}{r} 17121 \\ \times \quad 39 \\ \hline \end{array} \Rightarrow \begin{array}{r} 20000 \\ \times \quad 40 \\ \hline 00000 \\ 800000 \\ \hline 800000 \end{array}$$

51.
$$\begin{array}{r} 34560 \\ \times \quad 610 \\ \hline \end{array} \Rightarrow \begin{array}{r} 30000 \\ \times \quad 600 \\ \hline 00000 \\ 000000 \\ 18000000 \\ \hline 18000000 \end{array}$$

53. Area = length · width
= 38 · 11 = 418 square yd

55. Area = length · width
= 31 · 31 = 961 square ft

57. Area = length · width
= 36 · 54 = 1944 square cm

59. Area = length · width
= 17 · 6 = 102 square ft

61. Area = length · width
= 512 · 102 = 52,224 square cm

63. Area = length · width + length · width
= 26 · 9 + 26 · 9
= 234 + 234 = 468 square m

65. Area = length · width + length · width + length · width
= 14 · 7 + 14 · 7 + 14 · 7
= 98 + 98 + 98 = 294 square ft

67. Area = length · width + length · width + length · width
= 6 · 11 + 6 · 11 + 6 · 11
= 66 + 66 + 66 = 198 square ft

69.

$$\begin{array}{cccccc} & & & & 3 & \\ & & & & \not{3} & \\ & & & & \not{1} & \\ & & & 5 & 0 & 5 \\ \times & & & 7 & 7 & 3 \\ \hline & & 1 & 5 & 1 & 5 \\ & 3 & 5 & 3 & 5 & 0 \\ 3 & 5 & 3 & 5 & 0 & 0 \\ \hline 3 & 9 & 0 & 3 & 6 & 5 \end{array}$$

71.

$$\begin{array}{ccccccc} & & & 1 & 6 & 2 & \\ & & & & \not{3} & \not{1} & \\ & & & & \not{3} & \not{1} & \\ & & & 3 & 1 & 9 & 3 \\ \times & & & & 7 & 4 & 4 \\ \hline & & 1 & 2 & 7 & 7 & 2 \\ & 1 & 2 & 7 & 7 & 2 & 0 \\ 2 & 2 & 3 & 5 & 1 & 0 & 0 \\ \hline 2 & 3 & 7 & 5 & 5 & 9 & 2 \end{array}$$

$2{,}375{,}592 \approx 2{,}376{,}000$

73.

$$\begin{array}{r} 482 \\ \times\quad 59 \\ \hline \end{array} \Rightarrow \begin{array}{r} 500 \\ \times\quad 60 \\ \hline 000 \\ 30000 \\ \hline 30000 \end{array}$$

The estimate is 30,000, which is close to Maria's answer of 28,438. Her answer is reasonable.

75. $341 \cdot 15 = 5115$
The Rotary club estimates that it will sell 5115 dozen roses.

77. $17{,}929 \cdot 31 = 555{,}799$
The gross receipts from the sale of the Prius are \$555,799.

79. Civics: $15{,}844 \cdot 23 = 364{,}412$; Prius: $17{,}929 \cdot 31 = 555{,}799$
Smart Cars: $15{,}237 \cdot 18 = 274{,}266$; Total: $364{,}412 + 555{,}799 + 274{,}266 = 1{,}194{,}477 \approx 1{,}194{,}000$
The total gross receipts are about \$1,194,000.

81. $1874 \cdot 12 = 22{,}488$
The population grew by 22,488 people.

83. Cost of stock $= 1295 \cdot 16 = 20{,}720$
Selling price $= 1295 \cdot 51 = 66{,}045$
Profit $= 66{,}045 - 20{,}720 - 1050 = 44{,}275$
She realized \$44,275 from the sale.

85.

$$\begin{array}{r} 38.35 \\ \times\quad 12 \\ \hline \end{array} \Rightarrow \begin{array}{r} 40 \\ \times\quad 10 \\ \hline 00 \\ 400 \\ \hline 400 \end{array}$$

The estimated cost is \$400, so she has enough money in her budget for the blouses.

87. $3 \cdot 2 = 6$ feet in a fathom

89. $6076 \cdot 3 = 18{,}228$ feet in a league

91. From #90, we know that 20,000 leagues would be deeper than the Mariana Trench. It is impossible to be 20,000 leagues under the sea.

93. $5{,}880{,}000{,}000{,}000 \cdot 8 = 47{,}040{,}000{,}000{,}000$
Sirius is about 47,040,000,000,000 miles from Earth.

95. $17 \cdot 8 = 136$ pages
136 pages are printed in the 17 minutes.

97. $256 \cdot 1024 = 262{,}144$
There are 262,144 bytes in 256 kilobytes.

99. $14 \cdot 9 = 126$; A tablespoon of olive oil has 126 calories from fat.

101. $534{,}650 \cdot 31 = 16{,}574{,}150 \approx 16{,}574{,}000$
To the nearest thousand, the consumption is 16,574,000 gallons in 31 days.

103. Cost $= 464 \cdot 48 = 22{,}272$
Gross income $= 464 \cdot 106 = 49{,}184$
Net income $= 49{,}184 - 22{,}272 = 26{,}912$

105. $380{,}270{,}577 \cdot 2 = 760{,}541{,}154$
The movie would have been the top grossing movie if it had doubled its gross earnings.

107. You could explain that $3(8) = 24$ by counting the total number of circles in three groups of 8 circles each.

109. A product is the result of multiplying numbers. The product of 2 and 3 is 6.

111. Amount harvested $= 1750 \cdot 82 = 143{,}500$ bushels
Amount sold $= 143{,}500 \cdot 31 = \$4{,}448{,}500 \approx \$4{,}449{,}000$
To the nearest 1000, the crop is worth $4,449,000.

113.

$$\begin{array}{ccccc} & & 1 & 1 & \\ & 1 & \mathrm{A} & 5 & 7 \\ \times & & & 4 & 2 \\ \hline & \mathrm{B} & 7 & 1 & \mathrm{C} \end{array}$$

$7 \cdot 2 = 14 \Rightarrow \mathrm{C} = 4$
$2 \cdot \mathrm{A} + 1 = 7 \Rightarrow \mathrm{A} = 3$
($\mathrm{A} = 8$ seems to work as well. However, this solution ends up not working...)
$2 \cdot 1 = \mathrm{B} \Rightarrow \mathrm{B} = 2$

$$\begin{array}{cccccc} & & 1 & 2 & 2 & \\ & & & \not{1} & \not{1} & \\ & & 1 & 3 & 5 & 7 \\ \times & & & & 4 & 2 \\ \hline & & 2 & 7 & 1 & 4 \\ & \mathrm{D} & 4 & 2 & 8 & 0 \\ \hline & 5 & 6 & 9 & \mathrm{E} & 4 \end{array}$$

$1 + 8 = \mathrm{E} \Rightarrow \mathrm{E} = 9$
$4 \cdot 1 + 1 = \mathrm{D} \Rightarrow D = 5$

$\mathrm{A} = 3, \mathrm{B} = 2, \mathrm{C} = 4, \mathrm{D} = 5, \mathrm{E} = 9$

Exercises 1.4 (page 52)

1.
$$\begin{array}{r} 9 \\ 8\overline{)72} \\ \underline{72} \\ 0 \end{array}$$

3.
$$\begin{array}{r} 13 \\ 6\overline{)78} \\ \underline{6} \\ 18 \\ \underline{18} \\ 0 \end{array}$$

5.
$$\begin{array}{r} 87 \\ 5\overline{)435} \\ \underline{40} \\ 35 \\ \underline{35} \\ 0 \end{array}$$

7.
$$\begin{array}{r} 91 \\ 5\overline{)455} \\ \underline{45} \\ 05 \\ \underline{5} \\ 0 \end{array}$$

SECTION 1.4

9.
```
   1 7
8|1 3 6
  8
  ---
  5 6
  5 6
  ---
    0
```

11.
```
      4 0
22|8 8 0
   8 8
   ---
   0 0
     0
   ---
     0
```

13.
```
   8 2
6|4 9 2
  4 8
  ---
    1 2
    1 2
    ---
      0
```

15.
```
   5 R 1
7|3 6
  3 5
  ---
    1
```

17.
```
     4 R 13
17|8 1
   6 8
   ---
   1 3
```

19. zero

21.
```
  3 0 5 1
6|1 8 3 0 6
  1 8
  ---
    0 3
      0
    ---
      3 0
      3 0
      ---
        0 6
          6
        ---
          0
```

23.
```
      3 2
24|7 6 8
   7 2
   ---
     4 8
     4 8
     ---
       0
```

25.
```
       5 4
46|2 4 8 4
   2 3 0
   -----
     1 8 4
     1 8 4
     -----
         0
```

27.
```
       8 7
46|4 0 0 2
   3 6 8
   -----
     3 2 2
     3 2 2
     -----
         0
```

29.
```
          7 6
542|4 1 1 9 2
    3 7 9 4
    -------
      3 2 5 2
      3 2 5 2
      -------
            0
```

31.
```
           3 9 1
355|1 3 8 8 0 5
    1 0 6 5
    -------
      3 2 3 0
      3 1 9 5
      -------
          3 5 5
          3 5 5
          -----
              0
```

33.
```
      1 8 1 R 39
43|7 8 2 2
   4 3
   ---
   3 5 2
   3 4 4
   -----
       8 2
       4 3
       ---
       3 9
```

35.
```
      1 5 R 52
57|9 0 7
   5 7
   ---
   3 3 7
   2 8 5
   -----
     5 2
```

37. $(78)(?) = 1872$
```
       2 4
78|1 8 7 2
   1 5 6
   -----
     3 1 2
     3 1 2
     -----
         0
```
$(78)\boxed{(24)} = 1872$

SECTION 1.4

39.
$$\begin{array}{r} 12802\text{ R}18 \\ 27\overline{)345672} \\ \underline{27} \\ 75 \\ \underline{54} \\ 216 \\ \underline{216} \\ 07 \\ \underline{0} \\ 72 \\ \underline{54} \\ 18 \end{array}$$

41.
$$\begin{array}{r} 873 \\ 64\overline{)55892} \\ \underline{512} \\ 469 \\ \underline{448} \\ 212 \\ \underline{192} \\ 20 \end{array}$$
$873 \approx 870$

43.
$$\begin{array}{r} 54* \\ 415\overline{)225954} \\ \underline{2075} \\ 1845 \\ \underline{1660} \\ 185 \end{array}$$
$54* \approx 500$

45. $24{,}986{,}988 \div 4563$
$$\begin{array}{r} 5476 \\ 4563\overline{)24986988} \\ \underline{22815} \\ 21719 \\ \underline{18252} \\ 34678 \\ \underline{31941} \\ 27378 \\ \underline{27378} \\ 0 \end{array}$$
The actual taxes paid per return in Week 1 are $5476.

47. $48{,}660{,}040 \div 11{,}765$
$$\begin{array}{r} 413* \\ 11765\overline{)48660040} \\ \underline{47060} \\ 16000 \\ \underline{11765} \\ 42354 \\ \underline{35295} \\ 7059 \end{array}$$
The taxes paid per return in Week 3 are about $4100.

49. $1890 \div 14 = 135$
There are 135 trees per acre to harvest.

51. $1260 \div 45 =$ $28 per radio
$72 \times 28 = 2016$; She will pay $2016.

53. $90 \times 365 = 32{,}850$ days; $57{,}000{,}000{,}000 \div 32{,}850 \approx 1{,}735{,}159$; You would need to spend $1,700,000 per day, rounded to the nearest hundred thousand.

55. $1{,}335{,}962{,}000 \div 9{,}596{,}960 \approx 139$
The population density of China was about 139 people per square kilometer.

57. $305{,}967{,}000 \div 9{,}629{,}091 \approx 32$
The population density of the US was about 32 people per square kilometer.

59. $6{,}642{,}000{,}000 \div 82{,}000{,}000 = 81$
There average cost per cat is $81.

61. $33{,}145{,}121 \div 53 \approx 625{,}380$
Each representative from CA represents about 625,380 people.

63. $1{,}850{,}000{,}000{,}000 \div 36{,}756{,}666 \approx 50{,}331$
The GSP in CA is about $50,300 per person.

65. $630 \div 14 = 45$
There are 45 calories per serving.

67. $2700 \div 325 \approx 8$
Juan can take 8 capsules per day.

69. $407{,}000{,}000 \div 8 \approx 400{,}000{,}000 \div 8$
$= 50{,}000{,}000$
About 50,000,000 tickets were sold at $8 each for *Spider-Man*.

71. $119{,}176{,}821 \div 83 \approx 1{,}435{,}865$
The average salary was about $1,436,000.

73. The remainder after division is the amount left over after all groups the size of the divisor have been formed.

75. US
$\#$ packages $= 171{,}000 \div 8 = 21{,}375$
Gross receipts $= 21{,}375 \times 3 = 64{,}125$

France
$\#$ packages $= 171{,}000 \div 12 = 14{,}250$
Gross receipts $= 14{,}250 \times 5 = 71{,}250$

$71{,}250 - 64{,}125 = 7125$
The company will get the greater gross receipts in France. The difference is $7125.

77.
$$\begin{array}{r} 2\ 1\ \text{B} \\ \text{A}\ 3\overline{)4\ \text{C}\ \text{C}\ 1} \end{array}$$
$2 \cdot \text{A3}$ must be less than 4C, so A must equal 1 or 2. Since 1 appears in the problem, A must equal 2.

$$\begin{array}{r} 2\ 1\ \text{B} \\ 2\ 3\overline{)4\ \text{C}\ \text{C}\ 1} \end{array}$$
$23 \cdot 21\text{B} = 4\text{CC}1$. Then $3 \cdot \text{B}$ must end in 1. This only happens when $\text{B} = 7$. Then $23 \cdot 217 = 4991$, so $\text{C} = 9$.
$\boxed{\text{A} = 2, \text{B} = 7, \text{C} = 9}$

Getting Ready for Algebra (page 58)

1.
$$3x = 15$$
$$\frac{3x}{3} = \frac{15}{3}$$
$$x = 5$$
Check:
$$3x = 15$$
$$3(5) \stackrel{?}{=} 15$$
$$15 = 15$$

3.
$$\frac{c}{3} = 6$$
$$3 \cdot \frac{c}{3} = 3 \cdot 6$$
$$c = 18$$
Check:
$$\frac{c}{3} = 6$$
$$\frac{18}{3} \stackrel{?}{=} 6$$
$$6 = 6$$

5.
$$13x = 52$$
$$\frac{13x}{13} = \frac{52}{13}$$
$$x = 4$$
Check:
$$13x = 52$$
$$13(4) \stackrel{?}{=} 52$$
$$52 = 52$$

7.
$$\frac{b}{2} = 23$$
$$2 \cdot \frac{b}{2} = 2 \cdot 23$$
$$b = 46$$
Check:
$$\frac{b}{2} = 23$$
$$\frac{46}{2} \stackrel{?}{=} 23$$
$$23 = 23$$

9.
$$12x = 144$$
$$\frac{12x}{12} = \frac{144}{12}$$
$$x = 12$$
Check:
$$12x = 144$$
$$12(12) \stackrel{?}{=} 144$$
$$144 = 144$$

11.
$$\frac{y}{13} = 24$$
$$13 \cdot \frac{y}{13} = 13 \cdot 24$$
$$y = 312$$
Check:
$$\frac{y}{13} = 24$$
$$\frac{312}{13} \stackrel{?}{=} 24$$
$$24 = 24$$

13.
$$27x = 648$$
$$\frac{27x}{27} = \frac{648}{27}$$
$$x = 24$$
Check:
$$27x = 648$$
$$27(24) \stackrel{?}{=} 648$$
$$648 = 648$$

15.
$$\frac{b}{12} = 2034$$
$$12 \cdot \frac{b}{12} = 12 \cdot 2034$$
$$b = 24{,}408$$
Check:
$$\frac{b}{12} = 2034$$
$$\frac{24{,}408}{12} \stackrel{?}{=} 2034$$
$$2034 = 2034$$

17.
$$1098 = 18x$$
$$\frac{1098}{18} = \frac{18x}{18}$$
$$61 = x$$
Check: $1098 = 18x$
$$1098 \stackrel{?}{=} 18(61)$$
$$1098 = 1098$$

19.
$$34 = \frac{w}{23}$$
$$23 \cdot 34 = 23 \cdot \frac{w}{23}$$
$$782 = w$$
Check: $34 = \frac{w}{23}$
$$34 \stackrel{?}{=} \frac{782}{23}$$
$$34 = 34$$

21.
$$A = lw$$
$$595 = 35w$$
$$\frac{595}{35} = \frac{35w}{35}$$
$17 = w$; The width is 17 ft.

23. Let $w =$ # of lb sold.
$$2w = 4680$$
$$\frac{2w}{2} = \frac{4680}{2}$$
$w = 2340$; He sells 2340 lb of crab.

25.
$$C = np$$
$$5580 = 18p$$
$$\frac{5580}{18} = \frac{18p}{18}$$
$$310 = p$$
The wholesale cost is \$310.

27. Let $L =$ the low temperature in July.
Let $H =$ the high temperature in January.
$$2H = L$$
$$2H = 60$$
$$\frac{2H}{2} = \frac{60}{2}$$
$$H = 30$$
The average high temperature in Jan. is 30° F.

Exercises 1.5 (page 63)

1. $16(16)(16)(16)(16)(16) = 16^6$

3. $7^2 = 7 \cdot 7 = 49$

5. $2^3 = 2 \cdot 2 \cdot 2 = 8$

7. $15^1 = 15$

9. base; exponent; power or value

11. $8^3 = 8 \cdot 8 \cdot 8 = 512$

SECTION 1.5

13. $32^2 = 32 \cdot 32 = 1024$

15. $10^3 = 10 \cdot 10 \cdot 100 = 1000$

17. $9^3 = 9 \cdot 9 \cdot 9 = 729$

19. $3^8 = 3 \cdot 3 \cdot 3 \cdot 3 \cdot 3 \cdot 3 \cdot 3 \cdot 3 = 6561$

21. $88 \times 10^2 = 8800$

23. $21 \times 10^5 = 2{,}100{,}000$

25. $2300 \div 10^2 = 23$

27. $780{,}000 \div 10^3 = 780$

29. exponent

31. $202 \times 10^5 = 20{,}200{,}000$

33. $7{,}270{,}000 \div 10^3 = 7270$

35. $6734 \times 10^4 = 67{,}340{,}000$

37. $\dfrac{528{,}000}{10^3} = 528$

39. $9610 \times 10^7 = 96{,}100{,}000{,}000$

41. $450{,}000{,}000 \div 10^5 = 4500$

43. $10(10)(10)(10)(10)(10)(10)(10)(10)(10)(10) = 10^{11}$

45. $11^5 = 11 \cdot 11 \cdot 11 \cdot 11 \cdot 11 = 161{,}051$

47. $14^5 = 537{,}824$

49. $47{,}160 \times 10^9 = 47{,}160{,}000{,}000{,}000$

51. $\dfrac{680{,}000{,}000}{10^7} = 68$

53. The size of Salvador's lot is 10,800 ft^2.

55. $30 \times 10^4 = 300{,}000$; The skate parks' operating budget is \$300,000.

57. $255 \times 10^{11} = 25{,}500{,}000{,}000{,}000$
The distance from Earth to Alpha Centauri is about 25,500,000,000,000 miles.

59. 6 trillion $= 6{,}000{,}000{,}000{,}000 = 6 \times 10^{12}$

61. $4^4 = 256$; There are 4^6, or 256, bacteria after 3 hours.

63. $4^5 = 1024$; The number of bacteria will exceed 1000 during the 4th hour.

65. $642 \times 10^5 = 64{,}200{,}000$ square miles

67. 1 gigabyte $= 1{,}000{,}000{,}000 = 10^9$ bytes

69. $\$437{,}212{,}000 \approx \$437{,}000{,}000 = \$437 \times 10^6$

71. $60^4 = 12{,}960{,}000$; $60^5 = 777{,}600{,}000$
60^5 is larger than the earnings from *Titanic*.

73. $4^{10} = 4 \cdot 4 \cdot 4 \cdot 4 \cdot 4 \cdot 4 \cdot 4 \cdot 4 \cdot 4 \cdot 4$

75.

Birthday	Amount deposited	Total Amount
1	4	4
2	$4^2 = 16$	$4 + 16 = 20$
3	$4^3 = 64$	$20 + 64 = 84$
4	$4^4 = 256$	$84 + 256 = 340$
5	$4^5 = 1024$	$340 + 1024 = 1364$
6	$4^6 = 4096$	$1364 + 4096 = 5460$

continued on next page...

75. continued

Birthday	Amount deposited	Total Amount
7	$4^7 = 16{,}384$	$5460 + 16{,}384 = 21{,}844$
8	$4^8 = 65{,}536$	$21{,}844 + 65{,}536 = 87{,}380$
9	$4^9 = 262{,}144$	$87{,}380 + 262{,}144 = 349{,}524$
10	$4^{10} = 1{,}048{,}576$	$349{,}524 + 1{,}048{,}576 = 1{,}398{,}100$

They will deposit $1,048,576 on his 10th birthday. The total amount deposited will be $1,398,100.

77. $5^3 + 10^3 + 12^3 + 24^3 = 16{,}677;\ 2^5 + 4^5 + 6^5 + 7^5 = 25{,}639$
$25{,}639 - 16{,}677 = 8962$

Exercises 1.6 (page 69)

1. $3 \cdot 9 + 12 = 27 + 12 = 39$

3. $46 - 5 \cdot 9 = 46 - 45 = 1$

5. $42 + 48 \div 6 = 42 + 8 = 50$

7. $53 - (23 + 2) = 53 - 25 = 28$

9. $84 \div 7 \times 4 = 12 \times 4 = 48$

11. $44 + 5 \cdot 4 - 12 = 44 + 20 - 12$
$= 64 - 12 = 52$

13. $2^3 - 5 + 3^3 = 8 - 5 + 3^3$
$= 8 - 5 + 27 = 3 + 27 = 30$

15. $5 \cdot 7 + 3 \cdot 6 = 35 + 3 \cdot 6$
$= 35 + 18 = 53$

17. $4^2 - 5 \cdot 2 + 7 \cdot 6 = 16 - 5 \cdot 2 + 7 \cdot 6$
$= 16 - 10 + 7 \cdot 6$
$= 16 - 10 + 42$
$= 6 + 42 = 48$

19. $72 \div 8 + 12 - 5 = 9 + 12 - 5 = 21 - 5$
$= 16$

21. $(32 + 25) - (63 - 14) = 57 - (63 - 14)$
$= 57 - 49 = 8$

23. $64 \div 8 \cdot 2^3 + 6 \cdot 7 = 64 \div 8 \cdot 8 + 6 \cdot 7$
$= 8 \cdot 8 + 6 \cdot 7$
$= 64 + 6 \cdot 7$
$= 64 + 42 = 106$

25. $75 \div 15 \cdot 7 = 5 \cdot 7 = 35$

27. $88 - 3(45 - 37) + 42 - 35 = 88 - 3(8) + 42 - 35 = 88 - 24 + 42 - 35 = 64 + 42 - 35$
$= 106 - 35 = 71$

29. $3^5 + 3^4 = 243 + 81 = 324$

31. $102 \cdot 3^3 - 72 \div 6 + 15 = 102 \cdot 27 - 72 \div 6 + 15$
$= 2754 - 72 \div 6 + 15 = 2754 - 12 + 15 = 2742 + 15 = 2757$

33. $50 - 12 \div 6 - 36 \div 6 + 3 = 50 - 2 - 36 \div 6 + 3$
$= 50 - 2 - 6 + 3 = 48 - 6 + 3 = 42 + 3 = 45$

SECTION 1.6

35. $(990 + 740) - (925 + 210) = 1730 - 1135 = 595$
There were 595 more mallards and canvasbacks than teals and woodducks.

37. $(4 \cdot 210 + 990) - (925 + 740) = (840 + 990) - 1665 = 1830 - 1665 = 165$
There would be 165 more woodducks and mallards than teals and canvasbacks.

39. $$\begin{aligned} 11(2^3 \cdot 3 - 20) \div 4 + 33 &= 11(8 \cdot 3 - 20) \div 4 + 33 \\ &= 11(24 - 20) \div 4 + 33 \\ &= 11(4) \div 4 + 33 = 44 \div 4 + 33 = 11 + 33 = 44 \end{aligned}$$

41. $$\begin{aligned} 9(4^2 \cdot 3 - 22) \div 9 + 28 &= 9(16 \cdot 3 - 22) \div 9 + 28 \\ &= 9(48 - 22) \div 9 + 28 \\ &= 9(26) \div 9 + 28 = 234 \div 9 + 28 = 26 + 28 = 54 \end{aligned}$$

43. $12(95 - 22) + 4(47 - 23) + 14(58 - 17) = 12(73) + 4(24) + 14(41) = 876 + 96 + 574 = 1546$
The value of the supplies used was $1546.

45. Average weekly income $= 29(2400 \div 100) + 17 \cdot 35 = 29(24) + 17 \cdot 35 = 696 + 595 = 1291$
His weekly average income is $1291.
Average yearly income $= 50 \cdot 1291 =$ $64,550; His average yearly income is $64,550.

47. $$\begin{aligned} 3(110) + 2(240) + 2(50) + 2(120) = 330 + 480 + 100 + 240 &= 810 + 100 + 240 \\ &= 910 + 240 = 1150 \end{aligned}$$
The breakfast includes a total of 1150 calories.

49. $2 \cdot 30 + 3 \cdot 3 + 20 + 21 + 4 \cdot 18 + 8 = 60 + 9 + 20 + 21 + 72 + 8 = 190$
Sally's total charge is $190.

51. $100 - (6 + 17 + 13 + 2 \cdot 5 + 2 \cdot 6 + 2 \cdot 2) = 100 - (6 + 17 + 13 + 10 + 12 + 4) = 100 - 62 = 38$
They have $38 left on the gift certificate.

53. $(2 \cdot 1800 - 1650) \div 150 = (3600 - 1650) \div 150 = 1950 \div 150 = 13$
She will need to save for 13 months.

55. $(48 \cdot 5) \cdot 4 - (120 + 12 \cdot 22) = 960 - 384 = 576$; He saves $576 per year.

57. $437{,}212{,}000 + 407{,}681{,}000 \div 2 = 437{,}212{,}000 + 203{,}840{,}500 = 641{,}052{,}500$

59. The answer of 15 is correct. Division must occur before subtraction according to order of operations.

61. $$\begin{aligned} (6 \cdot 3 - 8)^2 - 50 + 2 \cdot 3^2 + 2(9 - 5)^3 &= (18 - 8)^2 - 50 + 2 \cdot 9 + 2(4)^3 \\ &= 10^2 - 50 + 18 + 2(64) = 100 - 50 + 18 + 128 = 196 \end{aligned}$$

63. **Answers may vary.**

Getting Ready for Algebra (page 77)

1.
$$\begin{aligned} 4x - 16 &= 12 \\ 4x - 16 + 16 &= 12 + 16 \\ 4x &= 28 \\ \frac{4x}{4} &= \frac{28}{4} \\ x &= 7 \end{aligned}$$
Check:
$$\begin{aligned} 4x - 16 &= 12 \\ 4(7) - 16 &\stackrel{?}{=} 12 \\ 28 - 16 &\stackrel{?}{=} 12 \\ 12 &= 12 \end{aligned}$$

3.
$$\begin{aligned} \frac{y}{3} - 7 &= 4 \\ \frac{y}{3} - 7 + 7 &= 4 + 7 \\ \frac{y}{3} &= 11 \\ 3 \cdot \frac{y}{3} &= 3(11) \\ y &= 33 \end{aligned}$$
Check:
$$\begin{aligned} \frac{y}{3} - 7 &= 4 \\ \frac{33}{3} - 7 &\stackrel{?}{=} 4 \\ 11 - 7 &\stackrel{?}{=} 4 \\ 4 &= 4 \end{aligned}$$

5.
$$\begin{aligned} 45 &= 6x + 9 \\ 45 - 9 &= 6x + 9 - 9 \\ 36 &= 6x \\ \frac{36}{6} &= \frac{6x}{6} \\ 6 &= x \end{aligned}$$
Check:
$$\begin{aligned} 45 &= 6x + 9 \\ 45 &\stackrel{?}{=} 6(6) + 9 \\ 45 &\stackrel{?}{=} 36 + 9 \\ 45 &= 45 \end{aligned}$$

7.
$$\begin{aligned} \frac{c}{8} + 23 &= 27 \\ \frac{c}{8} + 23 - 23 &= 27 - 23 \\ \frac{c}{8} &= 4 \\ 8 \cdot \frac{c}{8} &= 8(4) \\ c &= 32 \end{aligned}$$
Check:
$$\begin{aligned} \frac{c}{8} + 23 &= 27 \\ \frac{32}{8} + 23 &\stackrel{?}{=} 27 \\ 4 + 23 &\stackrel{?}{=} 27 \\ 27 &= 27 \end{aligned}$$

9.
$$\begin{aligned} 11x + 32 &= 54 \\ 11x + 32 - 32 &= 54 - 32 \\ 11x &= 22 \\ \frac{11x}{11} &= \frac{22}{11} \\ x &= 2 \end{aligned}$$
Check:
$$\begin{aligned} 11x + 32 &= 54 \\ 11(2) + 32 &\stackrel{?}{=} 54 \\ 22 + 32 &\stackrel{?}{=} 54 \\ 54 &= 54 \end{aligned}$$

11.
$$\begin{aligned} 15c - 63 &= 117 \\ 15c - 63 + 63 &= 117 + 63 \\ 15c &= 180 \\ \frac{15c}{15} &= \frac{180}{15} \\ c &= 12 \end{aligned}$$
Check:
$$\begin{aligned} 15c - 63 &= 117 \\ 15(12) - 63 &\stackrel{?}{=} 117 \\ 180 - 63 &\stackrel{?}{=} 117 \\ 117 &= 117 \end{aligned}$$

13.
$$\begin{aligned} 81 &= \frac{a}{14} + 67 \\ 81 - 67 &= \frac{a}{14} + 67 - 67 \\ 14 &= \frac{a}{14} \\ 14 \cdot 14 &= 14 \cdot \frac{a}{14} \\ 196 &= a \end{aligned}$$
Check:
$$\begin{aligned} 81 &= \frac{a}{14} + 67 \\ 81 &\stackrel{?}{=} \frac{196}{14} + 67 \\ 81 &\stackrel{?}{=} 14 + 67 \\ 81 &= 81 \end{aligned}$$

15.
$$\begin{aligned} 673 &= 45b - 272 \\ 673 + 272 &= 45b - 272 + 272 \\ 945 &= 45b \\ \frac{945}{45} &= \frac{45b}{45} \\ 21 &= b \end{aligned}$$
Check:
$$\begin{aligned} 673 &= 45b - 272 \\ 673 &\stackrel{?}{=} 45(21) - 272 \\ 673 &\stackrel{?}{=} 945 - 272 \\ 673 &= 673 \end{aligned}$$

17.
$$\begin{aligned} C &= PN + S \\ 309 &= 43N + 8 \\ 309 - 8 &= 43N + 8 - 8 \\ 301 &= 43N \\ \frac{301}{43} &= \frac{43N}{43} \\ 7 &= N \end{aligned}$$
He bought 7 tickets.

19.
$$\begin{aligned} S &= B + PN \\ 88 &= 40 + 8N \\ 88 - 40 &= 40 - 40 + 8N \\ 48 &= 8N \\ \frac{48}{8} &= \frac{8N}{8} \\ 6 &= N \end{aligned}$$
She completed 6 arrangements.

21.
$$\begin{aligned} 8v &= C \\ 8v &= 72 \\ \frac{8v}{8} &= \frac{72}{8} \\ v &= 9 \end{aligned}$$
She can purchase 9 visits at B-Fit.

23.
$$\begin{aligned} 4v + 32 &= C \\ 4v + 32 &= 72 \\ 4v + 32 - 32 &= 72 - 32 \\ 4v &= 40 \\ \frac{4v}{4} &= \frac{40}{4} \\ v &= 10 \end{aligned}$$
Jessica can purchase 10 visits at Gym-Rats.

Exercises 1.7 (page 82)

1. $(5 + 11) \div 2 = 16 \div 2 = 8$

3. $(10 + 22) \div 2 = 32 \div 2 = 16$

5. $(7 + 13 + 14 + 18) \div 4 = 52 \div 4 = 13$

7. $(11 + 13 + 13 + 15) \div 4 = 52 \div 4 = 13$

9. $(10 + 8 + 5 + 5) \div 4 = 28 \div 4 = 7$

11. $(9 + 11 + 6 + 8 + 11) \div 5 = 45 \div 5 = 9$

13. Total of all scores $= 4(14) = 56$; Sum of known scores $= 11 + 14 + 15 = 40$
Missing score $= 56 - 40 = 16$

15. $(22 + 26 + 40 + 48) \div 4 = 136 \div 4 = 34$

17. $(23 + 33 + 43 + 53) \div 4 = 152 \div 4 = 38$

19. $(14 + 17 + 25 + 34 + 50 + 82) \div 6 = 222 \div 6 = 37$

21.
$$\begin{aligned} (111 + 131 + 113 + 333) \div 4 &= 688 \div 4 \\ &= 172 \end{aligned}$$

23.
$$\begin{aligned} (100 + 151 + 228 + 145) \div 4 &= 624 \div 4 \\ &= 156 \end{aligned}$$

25. $(82 + 95 + 101 + 153 + 281 + 110) \div 6 = 822 \div 6 = 137$

27. Total of all scores $= 5(64) = 320$; Sum of known scores $= 39 + 86 + 57 + 79 = 261$
Missing score $= 320 - 261 = 59$

29. Arrange in order: 10, 58, 61
median $= 58$

31. Arrange in order: 33, 34, 77
median $= 34$

SECTION 1.7

33. Arrange in order: 17, 64, 82, 104
median $= (64 + 82) \div 2 = 146 \div 2 = 73$

35. Arrange in order: 7, 49, 55, 108, 131
median $= 55$

37. Arrange in order: 97, 101, 123, 129, 133, 145; median $= (123 + 129) \div 2 = 252 \div 2 = 126$

39. Arrange in order: 39, 41, 77, 95, 103, 123; median $= (77 + 95) \div 2 = 172 \div 2 = 86$

41. Arrange in order: 17, 17, 38, 65, 72
mode $= 17$

43. Arrange in order: 1, 1, 2, 2, 3, 3, 3, 6, 7, 7
mode $= 3$

45. Arrange in order: 19, 22, 23, 36, 45, 46, 89; mode = none

47. Arrange in order: 26, 27, 33, 33, 43, 43, 43, 54; mode $= 43$

49. Arrange in order: 44, 44, 44, 55, 55, 55, 180; mode $= 44$ and 55

51. Arrange in order: 14, 14, 19, 19, 23, 23; mode = none

53. average $= (105 + 106 + 106 + 99 + 98 + 94 + 107 + 101) \div 8 = 816 \div 8 = 102$
Arrange in order: 94, 98, 99, 101, 105, 106, 106, 107
median $= (101 + 105) \div 2 = 206 \div 2 = 103$
The average number of points scored is 102. The median number of points scored is 103.

55. Arrange in order: 128, 131, 132, 132, 134, 134, 134, 135, 135, 135

$$\begin{aligned}\text{average} &= (128 + 131 + 132 + 132 + 134 + 134 + 134 + 135 + 135 + 135) \div 10 \\ &= 1330 \div 10 = 133\end{aligned}$$

median $= (134 + 134) \div 2 = 268 \div 2 = 134$. The average score per round is 133. The median score per round is 134.

57.

	$2 \cdot 19$	$= 38$
	$4 \cdot 22$	$= 88$
	$2 \cdot 24$	$= 48$
	$4 \cdot 30$	$= 120$
	$5 \cdot 34$	$= 170$
	$4 \cdot 40$	$= 160$
	$3 \cdot 48$	$= 144$
SUM	24	$= 768$

Average $= 768 \div 24 = 32$
Since there are 24 scores, the median is the average of the 12th and 13th scores. Median $= (30 + 34) \div 2 = 32$
The mode is the most common score, 34.
The average and median gas mileage are both 32 mpg, while the mode gas mileage is 34 mpg.

59.

	$1 \cdot 99$	$= 99$
	$3 \cdot 110$	$= 330$
	$2 \cdot 115$	$= 230$
	$5 \cdot 124$	$= 620$
	$3 \cdot 130$	$= 390$
	$2 \cdot 155$	$= 310$
	$2 \cdot 167$	$= 334$
	$1 \cdot 197$	$= 197$
	$1 \cdot 210$	$= 210$
SUM	20	$= 2720$

Average $= 2720 \div 20 = 136$;
The average weight is 136 pounds.
Since there are 20 weights, the median is the average of the 10th and 11th weights. Median $= (124 + 124) \div 2 = 124$
The median weight is 124 pounds.

61. $(780 + 850) \div (15 + 11) = 1630 \div 26 \approx 63$; The average cost is about 63 million dollars per mile.

63. $(80 + 40 + 90) \div 3 = 210 \div 3 = 70$. The average price of the three models is \$70.
One of the models is priced below this average price.

65. $(1{,}998{,}257 + 2{,}600{,}167) \div 2 = 2{,}299{,}212$. The average population was about 2,299,212.

67. From #66, the average increase per year was 75,239.

Year	2000	2001	2002	2003	2004	2005
population	1,998,257	2,073,496	2,148,735	2,223,974	2,299,213	2,374,452

Year	2006	2007	2008
population	2,449,691	2,524,930	2,600,169

69. $(14{,}500{,}000 + 14{,}400{,}000 + 14{,}900{,}000 + 14{,}700{,}000) \div 4 = 14{,}625{,}000$. The average attendance for the four years was about 14,600,000.

71. $(1746 + 1472 + 1231) \div 3 = 1483$. The average assets for the top three banks were \$1,483 billion.

73. $(1746 + 1472 + 1231 + 635 + 539) \div 5 \approx 1125$. The average assets for the top five banks were about \$1,125 billion.

75. $(6 + 17 + 49 + 74 + 34) \div 5 = 36$. The mean number of Internet users is 36%.

77. In order: 6, 17, 34, 49, 74. Median $= 34$. The median number of Internet users is 34%.

79. Half of the houses cost more than \$4 million and half cost less than \$4 million.

81. In 1950, half the men getting married for the first time were 23 or younger. In 2002, half the men getting married for the first time were 27 or younger. Men are waiting longer to get married.

83. $(696{,}000{,}000 + 600{,}788{,}188 + 533{,}184{,}219 + 460{,}998{,}007 + 437{,}212{,}000) \div 5 \approx 545{,}636{,}483$

85. $(696 + 601 + 533 + 461 + 437 + 435 + 431 + 423 + 408 + 380) \div 10 \approx 481$ million
The average earnings for all 10 movies were about \$481 million.

SECTION 1.7

87. Answers may vary.

89. Answers may vary.

Exercises 1.8 (page 94)

1. The shortest bar corresponds to the least number of passengers. $\boxed{\text{Frontier}}$

3. The bar representing Hawaiian has a height that corresponds to about $\boxed{\text{14,000 passengers}}$.

5. Add the heights of the bars:
$81{,}000 + 26{,}000 + 10{,}000 + 14{,}000 + 100{,}000 + 100{,}000 + 44{,}000 = 375{,}000$
There were about $\boxed{375{,}000}$ passengers on the airlines.

7. There are 2 symbols next to the Van label. $2 \cdot 20 = \boxed{40}$ vans in for repair.

9. The most symbols (8) occur next to the $\boxed{\text{Full-Size}}$ label.

11. The total number of symbols is 20. $20 \cdot 20 = \boxed{400}$ total vehicles during the year.

13. The country with the fewest sets has the lowest point on the graph: $\boxed{\text{Australia}}$.

15. The number corresponding to India is about 63. india had $\boxed{\text{63,000,000 sets}}$.

17. The number corresponding to India is about 63. The number corresponding to Italy is about 30.
$63 - 30 = \boxed{\text{33 million sets}}$.

19. Add the heights of the two bars: $40{,}000 + 5{,}000 = 45{,}000$. The total spent on paint and lumber is $\boxed{\$45{,}000}$.

21. Subtract the heights: $35{,}000 - 20{,}000 = 15{,}000$. Steel casing costs $\boxed{\$15{,}000}$ less than plastic.

23. Multiply the height by 2: $2 \cdot 20{,}000 = 40{,}000$. They will spend $\boxed{\$40{,}000}$ on steel casing.

25.

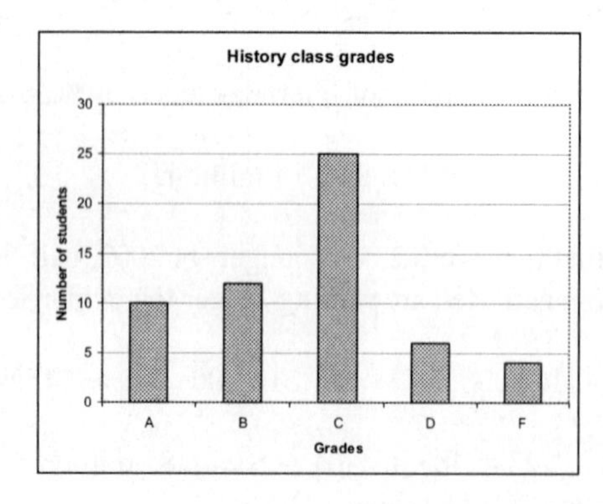

27.

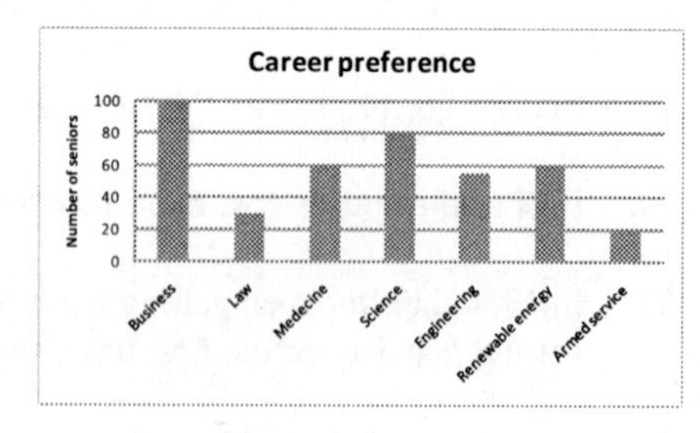

29.

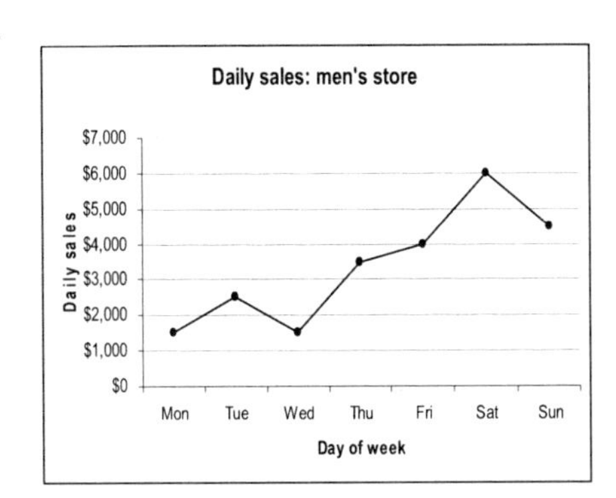

31.

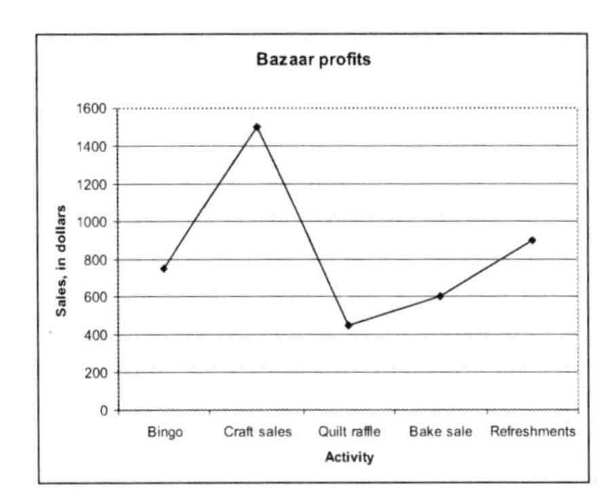

33.

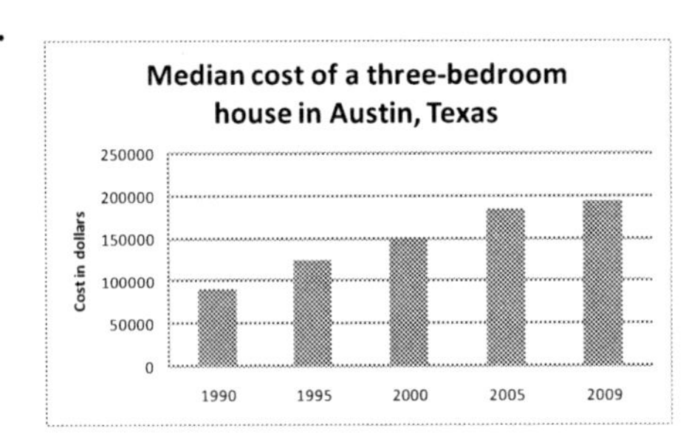

35.

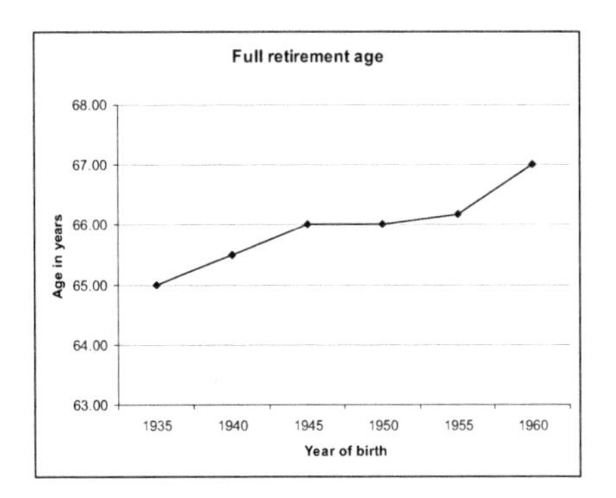

37.

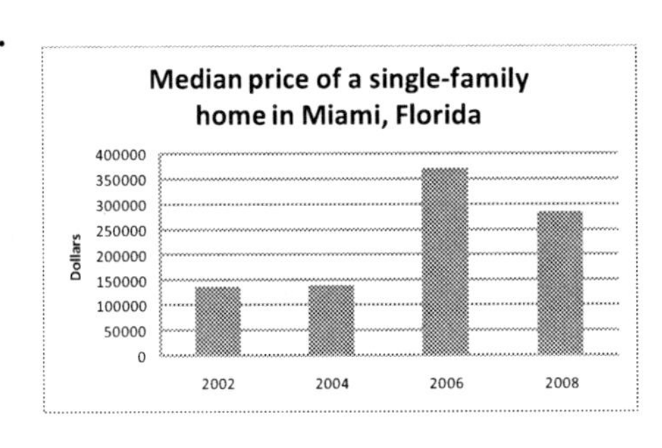

39. Tokyo had the largest population in 2000.

41. Mexico City is expected to grow the most during the 15 year period.

43. The greatest amount of growth was from 1960 to 1970.

45.

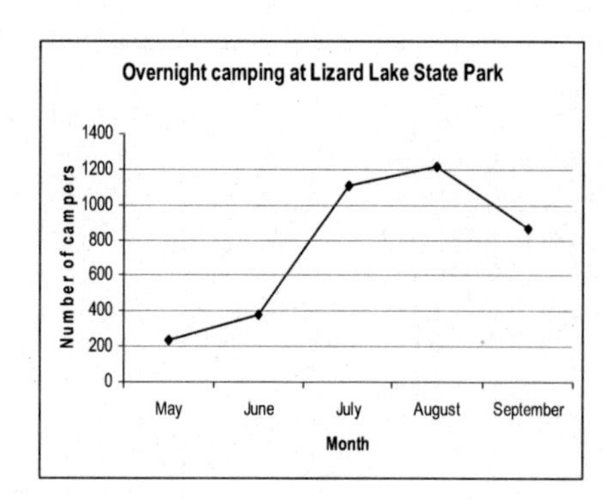

47.

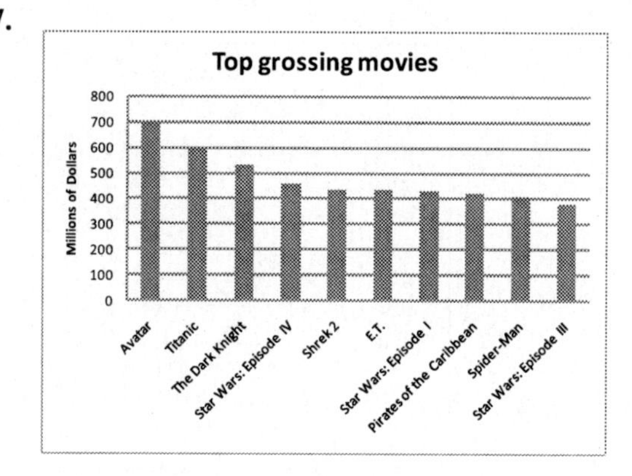

49.

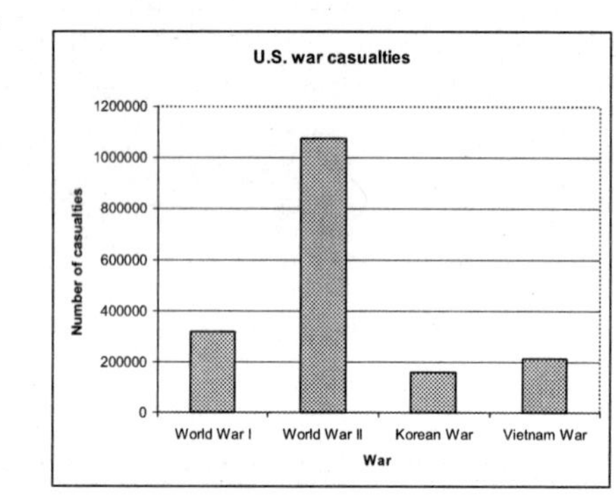

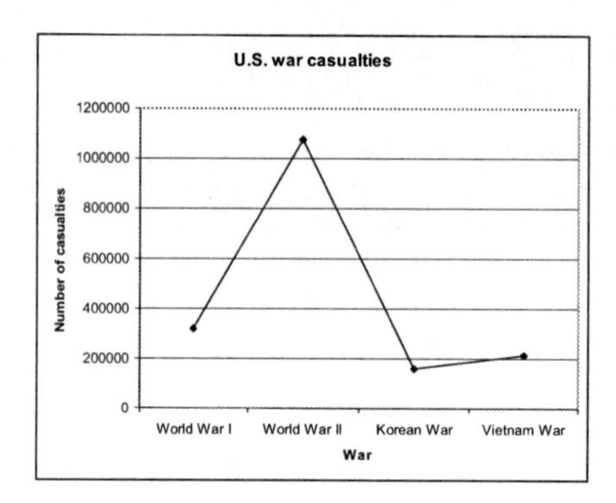

Explanations may vary.

Chapter 1 Review Exercises (page 105)

1. 607,321 = 607 thousands + 321 ones = six hundred seven thousand, three hundred twenty-one

2. 9,070,800 = 9 millions + 70 thousands + 800 ones = nine million, seventy thousand, eight hundred

3. sixty-two thousand, three hundred thirty-seven = 62 thousands + 337 ones = 62,337

4. five million, four hundred forty-four thousand, nineteen
= 5 millions + 444 thousands + 19 ones = 5,444,019

5. 347 is smaller than 351, so $347 < 351$.

6. 76 is larger than 69, so $76 > 69$.

7. 809 is smaller than 811, so $809 < 811$.

CHAPTER 1 REVIEW EXERCISES

8. **TEN**

The digit in the tens position is 3. Since the digit in the ones position is 7, and 7 is greater than 5, round up. Change the 3 in the tens position to 4, and replace the digit in the ones position to 0. The rounded number is 79,440.

HUNDRED

The digit in the hundreds position is 4. Since the digit in the tens position is 3, and 3 is less than 5, round down. Leave the 4 in the hundreds position, and replace the digits in the tens and ones positions with 0. The rounded number is 79,400.

THOUSAND

The digit in the thousands position is 9. Since the digit in the hundreds position is 4, and 4 is less than 5, round down. Leave the 9 in the thousands position, and replace the digits in the hundreds, tens and ones positions with 0. The rounded number is 79,000.

TEN THOUSAND

The digit in the ten thousands position is 7. Since the digit in the thousands position is 9, and 9 is more than 5, round up. Change the 7 in the ten thousands position to 8, and replace the digits in the thousands, hundreds, tens, and ones positions with 0. The rounded number is 80,000.

9. **TEN**

The digit in the tens position is 5. Since the digit in the ones position is 9, and 9 is greater than 5, round up. Change the 5 in the tens position to 6, and replace the digit in the ones position to 0. The rounded number is 183,660.

HUNDRED

The digit in the hundreds position is 6. Since the digit in the tens position is 5, round up. Change the 6 in the hundreds position to 7, and replace the digits in the tens and ones positions with 0. The rounded number is 183,700.

THOUSAND

The digit in the thousands position is 3. Since the digit in the hundreds position is 6, and 6 is greater than 5, round up. Change the 3 in the thousands position to 4, and replace the digits in the hundreds, tens, and ones positions with 0. The rounded number is 184,000.

TEN THOUSAND

The digit in the ten thousands position is 8. Since the digit in the thousands position is 3, and 3 is less than 5, round down. Leave the 8 in the ten thousands position, and replace the digits in the thousands, hundreds, tens, and ones positions with 0. The rounded number is 180,000.

10. Look for the largest entry in the "Number of Residents" columns. The greatest number is 1098. This is in the row labeled "36-50," so the largest age group is 36-50.

11. The number of residents in the "Under 15" row is 472. The number of residents in the "15-25" row is 398. The difference is $472 - 398 = 74$. There are 74 more residents under 15 than there are between 15 and 25.

CHAPTER 1 REVIEW EXERCISES

12. The number of residents in the "26-35" row is 612. The number of residents in the "36-50" row is 1098. The sum is 612 + 1098 = 1710. There are 1710 people between 26 and 50.

13. The number of residents in the "Under 15" row is 472. The number of residents in the "Over 70" row is 89. The difference is 472 – 89 = 383. There are 383 more residents under 15 than there are over 70.

14. Add the number of residents for all classes:
472 + 398 + 612 + 1098 + 602 + 89 = 3271. The population of Hepner is 3271.

15.

$$\begin{array}{cccc} & 1 & 1 & \\ & 3 & 3 & 6 \\ & & 7 & 2 \\ + & 5 & 0 & 9 \\ \hline & 9 & 1 & 7 \end{array}$$

16.

$$\begin{array}{ccccc} & 1 & & 1 & \\ & 3 & 8 & 3 & 4 \\ & & 5 & 1 & 0 \\ + & & 5 & 1 & 9 \\ \hline & 4 & 8 & 6 & 3 \end{array}$$

17.

$$\begin{array}{ccccc} & & 2 & 2 & \\ & & & 3 & 4 \\ & & 4 & 5 & 5 \\ & & & 1 & 7 \\ & & 3 & 7 & 7 \\ + & & 8 & 8 & 1 \\ \hline & 1 & 7 & 6 & 4 \end{array}$$

18.

$$\begin{array}{cccccc} & 1 & 3 & 3 & 1 & \\ & & 6 & 8 & 9 & 1 \\ & 1 & 2 & 0 & 5 & 5 \\ & 5 & 1 & 9 & 3 & 2 \\ & & 4 & 7 & 7 & 2 \\ + & & & 4 & 9 & 2 \\ \hline & 7 & 6 & 1 & 4 & 2 \end{array}$$

19.

$$\begin{array}{cccc} & 9 & 4 & 3 \\ - & 7 & 2 & 2 \\ \hline & 2 & 2 & 1 \end{array}$$

20.

$$\begin{array}{cccc} & 7 & 9 & \\ & & \not{1}\not{0} & 13 \\ & \not{8} & \not{0} & \not{3} \\ - & 7 & 3 & 8 \\ \hline & & 6 & 5 \end{array}$$

21.

$$\begin{array}{ccccc} & 7 & 12 & & \\ & & \not{2} & 11 & \\ & \not{8} & \not{3} & \not{1} & 5 \\ - & 6 & 9 & 8 & 3 \\ \hline & 1 & 3 & 3 & 2 \end{array}$$

22.

$$\begin{array}{ccccccc} & 1 & 13 & & & & \\ & & \not{3} & 16 & & 8 & 12 \\ & \not{2} & \not{4} & \not{6} & 8 & \not{9} & \not{2} \\ - & 1 & 4 & 9 & 5 & 5 & 8 \\ \hline & & 9 & 7 & 3 & 3 & 4 \end{array}$$

23. 18 in. + 18 in. + 25 in. + 16 in. + 25 in. = 102 in.

24.

$$\begin{array}{cccccc} & 3 & 4 & 6 & 8 & 3 \\ & & 5 & 2 & 7 & 8 \\ & 1 & 1 & 4 & 9 & 8 \\ & & & 6 & 7 & 8 \\ + & 5 & 6 & 7 & 2 & 3 \\ \hline \end{array} \Rightarrow \begin{array}{ccccccc} & & 3 & 0 & 0 & 0 & 0 \\ & & 1 & 0 & 0 & 0 & 0 \\ & & 1 & 0 & 0 & 0 & 0 \\ & & & & & & 0 \\ + & & 6 & 0 & 0 & 0 & 0 \\ \hline & 1 & 1 & 0 & 0 & 0 & 0 \end{array} \Rightarrow \begin{array}{ccccccc} & & 1 & & & & \\ & & 3 & 5 & 0 & 0 & 0 \\ & & & 5 & 0 & 0 & 0 \\ & & 1 & 1 & 0 & 0 & 0 \\ & & & 1 & 0 & 0 & 0 \\ + & & 5 & 7 & 0 & 0 & 0 \\ \hline & 1 & 0 & 9 & 0 & 0 & 0 \end{array}$$

25.

$$\begin{array}{ccccc} & 7 & 5 & 3 & 4 \\ - & 4 & 2 & 6 & 7 \\ \hline \end{array} \Rightarrow \begin{array}{ccccc} & 8 & 0 & 0 & 0 \\ - & 4 & 0 & 0 & 0 \\ \hline & 4 & 0 & 0 & 0 \end{array} \Rightarrow \begin{array}{ccccc} & 7 & 5 & 0 & 0 \\ - & 4 & 3 & 0 & 0 \\ \hline & 3 & 2 & 0 & 0 \end{array}$$

CHAPTER 1 REVIEW EXERCISES

26.
$$\begin{array}{rrrr} & & \not{2} & \\ & & 7 & 3 \\ \times & & 2 & 8 \\ \hline & 5 & 8 & 4 \\ 1 & 4 & 6 & 0 \\ \hline 2 & 0 & 4 & 4 \end{array}$$

27.
$$\begin{array}{rrrrr} & & 1 & 1 & \\ & & & \not{5} & \\ & & 4 & 0 & 6 \\ & & & 2 & 9 \\ \hline & 3 & 6 & 5 & 4 \\ & 8 & 1 & 2 & 0 \\ \hline 1 & 1 & 7 & 7 & 4 \end{array}$$

28.
$$\begin{array}{rrrrr} & & & 4 & \\ & & & \not{5} & \\ & & 4 & 0 & 7 \\ \times & & & 6 & 8 \\ \hline & 3 & 2 & 5 & 6 \\ 2 & 4 & 4 & 2 & 0 \\ \hline 2 & 7 & 6 & 7 & 6 \end{array}$$

29.
$$\begin{array}{rrrrr} & & \not{3} & \not{6} & \\ & & 4 & 4 & 9 \\ \times & & 1 & 7 & 1 \\ \hline & & 4 & 4 & 9 \\ 3 & 1 & 4 & 3 & 0 \\ 4 & 4 & 9 & 0 & 0 \\ \hline 7 & 6 & 7 & 7 & 9 \end{array}$$

30. $54 \cdot 24 = 1296$ boxes of cookies; $1296 \cdot 4 = 5184$; The troop grossed \$5184.

31.
$$\begin{array}{rrrr} 5 & 8 & 1 & 0 \\ \times\ 4 & 6 & 2 & \\ \hline \end{array} \Rightarrow \begin{array}{rrrrrrr} & & & 6 & 0 & 0 & 0 \\ \times & & & & 5 & 0 & 0 \\ \hline & & & 0 & 0 & 0 & 0 \\ & & 0 & 0 & 0 & 0 & 0 \\ 3 & 0 & 0 & 0 & 0 & 0 & 0 \\ \hline 3 & 0 & 0 & 0 & 0 & 0 & 0 \end{array}$$

32. Area = length · width
= $66 \cdot 17$
= 1122 square cm

33.
$$\begin{array}{r} 15 \\ 14\overline{)210} \\ \underline{14} \\ 70 \\ \underline{70} \\ 0 \end{array}$$

34.
$$\begin{array}{r} 32 \\ 18\overline{)576} \\ \underline{54} \\ 36 \\ \underline{36} \\ 0 \end{array}$$

35.
$$\begin{array}{r} 86\ \text{R}\,5 \\ 176\overline{)15141} \\ \underline{1408} \\ 1061 \\ \underline{1056} \\ 5 \end{array}$$

36.
$$\begin{array}{r} 3709\ \text{R}\,20 \\ 74\overline{)274486} \\ \underline{222} \\ 524 \\ \underline{518} \\ 68 \\ \underline{0} \\ 686 \\ \underline{666} \\ 20 \end{array}$$

37.
$$\begin{array}{r} 5320 \\ 65\overline{)345892} \\ \underline{325} \\ 208 \\ \underline{195} \\ 139 \\ \underline{130} \end{array}$$
$\boxed{5300}$

38. $4675 \div 32 = 146\ \text{R}\,3 \Rightarrow$ There will be 146 boxes with 3 chocolates left over.

CHAPTER 1 REVIEW EXERCISES

39. $11^3 = 11 \cdot 11 \cdot 11 = 1331$

40. $4^5 = 4 \cdot 4 \cdot 4 \cdot 4 \cdot 4 = 1024$

41. $23 \times 10^3 = 23{,}000$

42. $78{,}000{,}000 \div 10^5 = 780$

43. $712 \times 10^6 = 712{,}000{,}000$

44. $35{,}600{,}000 \div 10^4 = 3560$

45. $\$34 \times 10^8 = \$3{,}400{,}000{,}000$

46. $40 - 24 + 8 = 16 + 8 = 24$

47. $6 \cdot 10 + 5 = 60 + 5 = 65$

48. $$\begin{aligned} 18 \div 2 - 3 \cdot 2 &= 9 - 3 \cdot 2 \\ &= 9 - 6 = 3 \end{aligned}$$

49. $$\begin{aligned} 94 \div 47 + 47 - 16 \cdot 2 + 6 &= 2 + 47 - 16 \cdot 2 + 6 \\ &= 2 + 47 - 32 + 6 = 49 - 32 + 6 = 17 + 6 = 23 \end{aligned}$$

50. $$\begin{aligned} 35 - (25 - 17) + (12 - 10)^2 + 5 \cdot 2 &= 35 - 8 + (12 - 10)^2 + 5 \cdot 2 \\ &= 35 - 8 + 2^2 + 5 \cdot 2 \\ &= 35 - 8 + 4 + 5 \cdot 2 \\ &= 35 - 8 + 4 + 10 = 27 + 4 + 10 = 31 + 10 = 41 \end{aligned}$$

51. $(41 + 64 + 23 + 70 + 87) \div 5 = 285 \div 5 = 57 \Rightarrow$ mean $= 57$
Arrange in order: 23, 41, 64, 70, 87 $\Rightarrow$ median $= 64$; There is no mode.

52. $(93 + 110 + 216 + 317) \div 4 = 736 \div 4 = 184 \Rightarrow$ mean $= 184$
Arrange in order: 93, 110, 216, 317 $\Rightarrow$ median $= (110 + 216) \div 2 = 163$; There is no mode.

53. $(63 + 74 + 53 + 63 + 37 + 82) \div 6 = 372 \div 6 = 62 \Rightarrow$ mean $= 62$
Arrange in order: 37, 53, 63, 63, 74, 82 $\Rightarrow$ median $= (63 + 63) \div 2 = 63$; mode $= 63$

54. $(1086 + 4008 + 3136 + 8312 + 8312 + 1474) \div 6 = 26328 \div 6 = 4388 \Rightarrow$ mean $= 4388$
Arrange in order: 1086, 1474, 3136, 4008, 8312, 8312 $\Rightarrow$ median $= (3136 + 4008) \div 2 = 3572$
mode $= 8312$

55. $(54500 + 45674 + 87420 + 110675 + 63785 + 163782) \div 6 = 525836 \div 6 \approx 87{,}600$
The average salary is about \$87,600.

56. The highest temperature occurs at 3 p.m., while the lowest temperature occurs at 4 a.m.

57. The temperature at 6 a.m. is 60°, while the temperature at 8 p.m. is 75°. The difference is $75° - 60° = 15°$.

58.

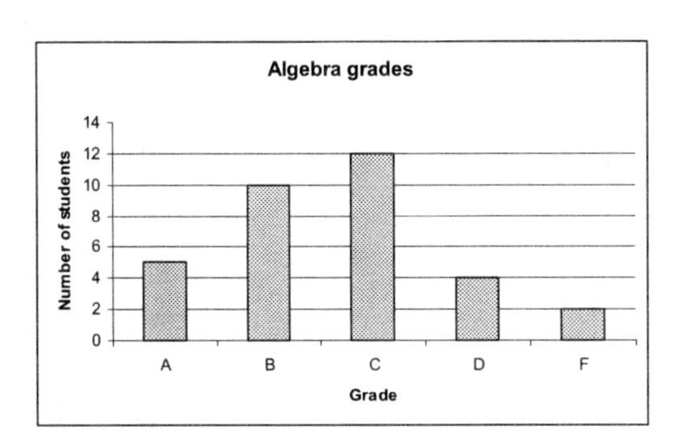

Chapter 1 True/False Concept Review (page 109)

1. false; 1,000,000,000 takes 10 digits.

2. true

3. true

4. false; "Seven is less than twenty-three."

5. false, $2567 > 2566$

6. true

7. true

8. true

9. false; The sum is 87.

10. true

11. true

12. false; The product is 45.

13. true

14. true

15. true

16. false; A number multiplied by 0 is 0.

17. true

18. false; The quotient is 26.

19. true

20. true

21. false; Division by zero is undefined.

22. false; The value of 7^2 is 49.

23. true

24. true

25. false; The value is 450,000.

26. true

27. false; In $(3 + 1)^2$ the addition is done 1st.

28. false; In $(5 - 2) \cdot 3$ the subtraction is done 1st.

29. true

30. true

31. true

32. true

33. false; The median is 54.

34. true

Chapter 1 Test (page 111)

1.

$$\begin{array}{r} 212 \\ 72\overline{)15264} \\ \underline{144} \\ 86 \\ \underline{72} \\ 144 \\ \underline{144} \\ 0 \end{array}$$

2.

$$\begin{array}{rrrrr} & & 5 & 10 & \\ & & & \not{0} & 15 \\ & 9 & \not{6} & \not{1} & \not{5} \\ - & 6 & 3 & 4 & 9 \\ \hline & 3 & 2 & 6 & 6 \end{array}$$

3. $55 \div 5 + 6 \cdot 4 - 7 = 11 + 6 \cdot 4 - 7$
$= 11 + 24 - 7$
$= 35 - 7 = 28$

4.

$$\begin{array}{rrrrrr} & & & & 2 & \\ & & & \not{1} & \not{5} & \\ & & & 4 & 2 & 8 \\ \times & & & & 3 & 7 \\ \hline & & 2 & 9 & 9 & 6 \\ & 1 & 2 & 8 & 4 & 0 \\ \hline & 1 & 5 & 8 & 3 & 6 \end{array}$$

5. 368 is smaller than 371, so $368 < 371$.

6. $55 \times 10^6 = 55{,}000{,}000$

7.

$$\begin{array}{rrrrrrr} & & & & & 2 & \\ & & & & & \not{7} & \\ & & & & & \not{1} & \\ & & & & 6 & 0 & 8 \\ \times & & & & 3 & 9 & 2 \\ \hline & & & 1 & 2 & 1 & 6 \\ & & 5 & 4 & 7 & 2 & 0 \\ & 1 & 8 & 2 & 4 & 0 & 0 \\ \hline & 2 & 3 & 8 & 3 & 3 & 6 \end{array}$$

8. seven hundred thirty thousand, sixty-one = 730 thousands + 61 ones = 730,061

9. $(3456 + 812 + 4002 + 562 + 1123) \div 5 = 9955 \div 5 = 1991$

10.

$$\begin{array}{rrrrrrr} & & & 4 & 1 & 1 & \\ & & & \not{3} & \not{1} & \not{1} & \\ & & & 5 & 7 & 3 & 3 \\ \times & & & & & 6 & 5 \\ \hline & & 2 & 8 & 6 & 6 & 5 \\ & 3 & 4 & 3 & 9 & 8 & 0 \\ \hline & 3 & 7 & 2 & 6 & 4 & 5 \end{array}$$

$372{,}645 \approx 372{,}600$

11. The digit in the thousands position is 8. Since the digit in the hundreds position is 5, round up. Change the 8 in the thousands position to 9, and replace the digits in the hundreds, tens, and ones position with 0. The rounded number is 39,000.

CHAPTER 1 TEST

12. $$\begin{array}{rrrrrr} & 9 & 5 & 9 & 1 & 4 \\ & 3 & 1 & 3 & 4 & 8 \\ & 6 & 8 & 6 & 9 & 9 \\ + & 3 & 0 & 3 & 4 & 1 \\ \hline \end{array} \Rightarrow \begin{array}{rrrrrr} 1 & 0 & 0 & 0 & 0 & 0 \\ & 3 & 0 & 0 & 0 & 0 \\ & 7 & 0 & 0 & 0 & 0 \\ + & 3 & 0 & 0 & 0 & 0 \\ \hline 2 & 3 & 0 & 0 & 0 & 0 \end{array}$$

13. $9^3 = 9 \cdot 9 \cdot 9 = 729$

14. $$\begin{array}{rrrr} {\scriptstyle 1} & {\scriptstyle 1} & {\scriptstyle 1} & \\ & & 8 & 4 \\ & 7 & 4 & 5 \\ & & 5 & 6 \\ +\,7 & 8 & 0 & 2 \\ \hline 8 & 6 & 8 & 7 \end{array}$$

15. $18 \text{ ft} + 42 \text{ ft} + 18 \text{ ft} + 42 \text{ ft} = 120 \text{ ft}$

16. $$\begin{array}{rrrr} & 7 & 5 & 2 \\ \times & & 3 & 8 \\ \hline \end{array} \Rightarrow \begin{array}{rrrrr} & & 8 & 0 & 0 \\ \times & & & 4 & 0 \\ \hline & & 0 & 0 & 0 \\ 3 & 2 & 0 & 0 & 0 \\ \hline 3 & 2 & 0 & 0 & 0 \end{array}$$

17. $$\begin{array}{rrrr} {\scriptstyle 6} & {\scriptstyle 9} & & \\ & \not{1}\not{0} & {\scriptstyle 14} & \\ & & \not{4} & {\scriptstyle 13} \\ \not{7} & \not{0} & \not{5} & \not{3} \\ - & 8 & 9 & 5 \\ \hline 6 & 1 & 5 & 8 \end{array}$$

18. $4{,}005 = 4$ thousands $+ 5$ ones
$=$ four thousand, five

19. $$\begin{aligned} 65 + 3^3 - 66 \div 11 &= 65 + 27 - 66 \div 11 \\ &= 65 + 27 - 6 \\ &= 92 - 6 = 86 \end{aligned}$$

20. $$\begin{array}{rrrrrr} & {\scriptstyle 1} & {\scriptstyle 3} & {\scriptstyle 2} & {\scriptstyle 2} & \\ & 4 & 2 & 8 & 8 & 8 \\ & 6 & 7 & 9 & 1 & 1 \\ & 9 & 3 & 4 & 6 & 7 \\ & 2 & 3 & 5 & 6 & 7 \\ + & 3 & 1 & 8 & 2 & 3 \\ \hline 2 & 5 & 9 & 6 & 5 & 6 \end{array}$$

21. $7{,}730{,}000{,}000 \div 10^6 = 7730$

22. The digit in the millions position is 5. Since the digit in the hundred thousands position is 9, and 9 is greater than 5, round up. Replace the 5 in the millions position with 6, and replace all digits to the right with 0. The rounded number is 676,000,000.

CHAPTER 1 TEST

23.

$$
\begin{array}{r}
1\,1\,6\,0 \text{ R } 32 \\
65\overline{)\,7\,5\,4\,3\,2} \\
\underline{6\,5} \\
1\,0\,4 \\
\underline{6\,5} \\
3\,9\,3 \\
\underline{3\,9\,0} \\
3\,2 \\
\underline{0} \\
3\,2
\end{array}
$$

24. $95 - 8^2 + 48 \div 4 = 95 - 64 + 48 \div 4$
$= 95 - 64 + 12$
$= 31 + 12 = 43$

25. Area $=$ length $\cdot$ width $= 23 \cdot 15 = 345$ square cm

26. $(5 \cdot 3)^2 + \left(4^2\right)^2 + 11 \cdot 3 = 15^2 + \left(4^2\right)^2 + 11 \cdot 3$
$= 15^2 + 16^2 + 11 \cdot 3$
$= 225 + 16^2 + 11 \cdot 3$
$= 225 + 256 + 11 \cdot 3 = 225 + 256 + 33 = 481 + 33 = 514$

27. $(795 + 576 + 691 + 795 + 416 + 909) \div 6 = 4182 \div 6 = 697 \Rightarrow$ mean $= 697$
Arrange in order: 416, 576, 691, 795, 795, 909 $\Rightarrow$ median $= (691 + 795) \div 2 = 743$; mode $= 795$

28. $18 \cdot 700 = 12{,}600$ total words
$12{,}600 \div 75 = 168 \Rightarrow$ It will take 168 minutes, or 2 hours and 48 minutes, to type the pages.

29. $124{,}758{,}000 \div 9 = 13{,}862{,}000$ per person
$13{,}862{,}000 \div 20 = 693{,}100 \Rightarrow$ Each will receive \$693,100 per year.

30. The price range from \$120,001 to \$250,000 had the greatest number of sales.

31. $25 + 15 = 40 \Rightarrow$ The top 2 ranges accounted for a total of 40 sales.

32. $20 - 15 = 5 \Rightarrow$ The lowest range had 5 more sales than the highest range.

33. Add Day and Night for each:
A: $215 + 175 = 390$
B: $365 + 120 = 485$
C: $95 + 50 = 145$
Division B has the most employees.

34. $215 - 95 = 120$. There are 120 more employees in Division A.

35. Add A, B, and C from #33: $390 + 485 + 145 = 1020$ total employees

Exercises 2.1 (page 121)

1. The last digit of 37 is not even, so it is not divisible by 2.

3. The last digit of 80 is even, so it is divisible by 2.

5. The last digit of 48 is even, so it is divisible by 2.

7. The last digit of 56 is not 0 or 5, so it is not divisible by 5.

9. The last digit of 45 is 5, so it is divisible by 5.

11. The last digit of 551 is not 0 or 5, so it is not divisible by 5.

13. The sum of the digits of 81 is $8 + 1 = 9$, which is divisible by 3, so 81 is divisible by 3.

15. The sum of the digits of 54 is $5 + 4 = 9$, which is divisible by 3, so 54 is divisible by 3.

17. The sum of the digits of 78 is $7 + 8 = 15$, which is divisible by 3, so 78 is divisible by 3.

19. Divisible by 2?
The last digit of 2190 is even, so 2190 is divisible by 2.

Divisible by 3?
The sum of the digits of 2190 is $2 + 1 + 9 + 0 = 12$, which is divisible by 3, so 2190 is divisible by 3.

Divisible by 5?
The last digit of 2190 is 0, so 2190 is divisible by 5.

21. Divisible by 2?
The last digit of 3998 is even, so 3998 is divisible by 2.

Divisible by 3?
The sum of the digits of 3998 is $3 + 9 + 9 + 8 = 29$, which is not divisible by 3, so 3998 is not divisible by 3.

Divisible by 5?
The last digit of 3998 is not 0 or 5, so 3998 is not divisible by 5.

23. Divisible by 2?
The last digit of 7535 is not even, so 7535 is not divisible by 2.

Divisible by 3?
The sum of the digits of 7535 is $7 + 5 + 3 + 5 = 20$, which is not divisible by 3, so 7535 is not divisible by 3.

Divisible by 5?
The last digit of 7535 is 5, so 7535 is divisible by 5.

25. Divisible by 2?
The last digit of 4175 is not even, so 4175 is not divisible by 2.

Divisible by 3?
The sum of the digits of 4175 is $4 + 1 + 7 + 5 = 17$, which is not divisible by 3, so 4175 is not divisible by 3.

Divisible by 5?
The last digit of 4175 is 5, so 4175 is divisible by 5.

27. Divisible by 2?
The last digit of 11,205 is not even, so 11,205 is not divisible by 2.

Divisible by 3?
The sum of the digits of 11,205 is $1 + 1 + 2 + 0 + 5 = 9$, which is divisible by 3, so 11,205 is divisible by 3.

Divisible by 5?
The last digit of 11,205 is 5, so 11,205 is divisible by 5.

29. 114 is divisible by both 2 and 3, so it is divisible by 6.

31. 254 is divisible by 2 but not 3, so it is not divisible by 6.

33. 684 is divisible by both 2 and 3, so it is divisible by 6.

35. The sum of the digits is $3 + 5 + 1 = 9$, which is divisible by 9, so 351 is divisible by 9.

37. The sum of the digits is $9 + 3 + 3 = 15$, which is not divisible by 9, so 933 is not divisible by 9.

39. The sum of the digits is $5 + 8 + 5 = 18$, which is divisible by 9, so 585 is divisible by 9.

41. The last digit of 747 is not 0, so 747 is not divisible by 10.

43. The last digit of 920 is 0, so 920 is divisible by 10.

45. The last digit of 1927 is not 0, so 1927 is not divisible by 10.

47. Divisible by 6?
7470 is divisible by both 2 and 3, so it is divisible by 6.

Divisible by 9?
The sum of the digits is $7 + 4 + 7 + 0 = 18$, which is divisible by 9, so 7470 is divisible by 9.

Divisible by 10?
The last digit of 7470 is 0, so 7470 is divisible by 10.

49. Divisible by 6?
8352 is divisible by both 2 and 3, so it is divisible by 6.

Divisible by 9?
The sum of the digits is $8 + 3 + 5 + 2 = 18$, which is divisible by 9, so 8352 is divisible by 9.

Divisible by 10?
The last digit of 8352 is not 0, so 8352 is not divisible by 10.

51. Divisible by 6?
5555 is not divisible by 2 or 3, so it is not divisible by 6.

Divisible by 9?
The sum of the digits is $5 + 5 + 5 + 5 = 20$, which is not divisible by 9, so 5555 is not divisible by 9.

Divisible by 10?
The last digit of 5555 is not 0, so 5555 is not divisible by 10.

53. Divisible by 6?
5700 is divisible by both 2 and 3, so it is divisible by 6.

Divisible by 9?
The sum of the digits is $5 + 7 + 0 + 0 = 12$, which is not divisible by 9, so 5700 is not divisible by 9.

Divisible by 10?
The last digit of 5700 is 0, so 5700 is divisible by 10.

55. Divisible by 6?
7290 is divisible by both 2 and 3, so it is divisible by 6.

Divisible by 9?
The sum of the digits is $7 + 2 + 9 + 0 = 18$, which is divisible by 9, so 7290 is divisible by 9.

Divisible by 10?
The last digit of 7290 is 0, so 7290 is divisible by 10.

57. 439 is not divisible by 3, so it is not possible for each to spend the same whole number of dollars.

59. 987 is divisible by 3, so it is possible for each to spend the same whole number of dollars.

61. 175 is divisible by 5, but not 3 or 10. Thus, the band can march in rows of 5, but not in rows of 3 or 10.

63. There are $5 \cdot 6 = 30$ bottles of beer. 30 is not divisible by 9, so it is not possible for each to have the same whole number of bottles.

65. 310 is not divisible by 3, and neither is 311. Since 312 is divisible by 3, they must raise \$2 more so that each charity can receive the same whole number of dollars.

67. Since each animal has 2 eyes, and since there are 30 eyes, there are $30 \div 2 = 15$ animals in the pen.

69. Since each elephant and rider has 2 eyes, and since there are 48 eyes, there are $48 \div 2 = 24$ total elephants and riders.

71. A number is divisible by 5 when the remainder after dividing by 5 is 0. 20 is divisible by 5, but 21 is not.

73. To test for divisibility by 2, you only need to determine whether the ones digit is even. To test for divisibility by 3, you need to add all the digits and determine whether that sum is divisible by 3.

75. 35,766 is divisible by both 2 and 3, so it is divisible by 6.

77. If a number is divisible by both 3 and 10, then the number is divisible by 30. 11,370 is divisible by both 3 and 10, so it is divisible by 30.

79. **THOUSAND**
The digit in the thousands position is 7. Since the digit in the hundreds position is 4, and 4 is less than 5, round down. Leave the 7 in the thousands position, and replace the digits in the hundreds, tens and ones positions with 0. The rounded number is 67,000.

TEN THOUSAND
The digit in the ten thousands position is 6. Since the digit in the thousands position is 7, and 7 is greater than 5, round up. Replace the 6 in the ten thousands position with 7, and replace the digits in the thousands, hundreds, tens and ones positions with 0. The rounded number is 70,000.

81.
$$\begin{array}{r|l} & 1\,3 \\ 845 & \overline{1\,0\,9\,8\,5} \\ & \underline{8\,4\,5} \\ & 2\,5\,3\,5 \\ & \underline{2\,5\,3\,5} \\ & 0 \end{array}$$

83. 4(18 cm) = 72 cm

85. $12 \cdot 1 = 12$; $12 \cdot 2 = 24$; $12 \cdot 3 = 36$; $12 \cdot 4 = 48$; $12 \cdot 5 = 60$; $12 \cdot 6 = 7$

87. $123 \cdot 1 = 123$; $123 \cdot 2 = 246$; $123 \cdot 3 = 369$; $123 \cdot 4 = 492$; $123 \cdot 5 = 615$; $123 \cdot 6 = 738$

Exercises 2.2 (page 127)

1. $5 \cdot 1 = 5$
$5 \cdot 2 = 10$
$5 \cdot 3 = 15$
$5 \cdot 4 = 20$
$5 \cdot 5 = 25$

3. $23 \cdot 1 = 23$
$23 \cdot 2 = 46$
$23 \cdot 3 = 69$
$23 \cdot 4 = 92$
$23 \cdot 5 = 115$

5. $21 \cdot 1 = 21$
$21 \cdot 2 = 42$
$21 \cdot 3 = 63$
$21 \cdot 4 = 84$
$21 \cdot 5 = 105$

7. $30 \cdot 1 = 30$
$30 \cdot 2 = 60$
$30 \cdot 3 = 90$
$30 \cdot 4 = 120$
$30 \cdot 5 = 150$

9. $50 \cdot 1 = 50$
$50 \cdot 2 = 100$
$50 \cdot 3 = 150$
$50 \cdot 4 = 200$
$50 \cdot 5 = 250$

11. $53 \cdot 1 = 53$
$53 \cdot 2 = 106$
$53 \cdot 3 = 159$
$53 \cdot 4 = 212$
$53 \cdot 5 = 265$

13. $67 \cdot 1 = 67$
$67 \cdot 2 = 134$
$67 \cdot 3 = 201$
$67 \cdot 4 = 268$
$67 \cdot 5 = 335$

15. $85 \cdot 1 = 85$
$85 \cdot 2 = 170$
$85 \cdot 3 = 255$
$85 \cdot 4 = 340$
$85 \cdot 5 = 425$

17. $155 \cdot 1 = 155$; $155 \cdot 2 = 310$; $155 \cdot 3 = 465$
$155 \cdot 4 = 620$; $155 \cdot 5 = 775$

19. $375 \cdot 1 = 375$; $375 \cdot 2 = 750$; $375 \cdot 3 = 1125$;
$375 \cdot 4 = 1500$; $375 \cdot 5 = 1875$

21. 84 is divisible by both 2 and 3, so it is divisible by 6. 84 is a multiple of 6.

23. 90 is divisible by both 2 and 3, so it is divisible by 6. 90 is a multiple of 6.

25. 95 is divisible by neither 2 nor 3, so it is not divisible by 6. 95 is not a multiple of 6.

27. The sum of the digits of 58 is 13. 13 is not divisible by 9, so 58 is not divisible by 9. 58 is not a multiple of 9.

29. The sum of the digits of 89 is 17. 17 is not divisible by 9, so 89 is not divisible by 9. 89 is not a multiple of 9.

31. The sum of the digits of 324 is 9. 9 is divisible by 9, so 324 is divisible by 9. 324 is a multiple of 9.

33.
$$\begin{array}{r|l} & 1\,2 \\ 7 & \overline{8\,4} \\ & \underline{7} \\ & 1\,4 \\ & \underline{1\,4} \\ & 0 \end{array}$$
84 is a multiple of 7.

35.
$$\begin{array}{r|l} & 1\,3 \\ 7 & \overline{9\,1} \\ & \underline{7} \\ & 2\,1 \\ & \underline{2\,1} \\ & 0 \end{array}$$
91 is a multiple of 7.

37.
$$\begin{array}{r|l} & 1\,8 \\ 7 & \overline{1\,2\,6} \\ & \underline{7} \\ & 5\,6 \\ & \underline{5\,6} \\ & 0 \end{array}$$
126 is a multiple of 7.

SECTION 2.2

39. 432 is divisible by 2 and 3, so it is divisible by 6.

The sum of the digits of 432 is 9, which is divisible by 9. 432 is divisible by 9.

$$\begin{array}{r} 28 \\ 15\overline{)432} \\ \underline{30} \\ 132 \\ \underline{120} \\ 12 \end{array}$$

432 is not divisible by 15.

432 is a multiple of 6 and 9, but not a multiple of 15.

41. 660 is divisible by 2 and 3, so it is divisible by 6.

The sum of the digits of 660 is 12, which is not divisible by 9. 660 is not divisible by 9.

$$\begin{array}{r} 44 \\ 15\overline{)660} \\ \underline{60} \\ 60 \\ \underline{60} \\ 0 \end{array}$$

660 is divisible by 15.

660 is a multiple of 6 and 15, but not a multiple of 9.

43. 780 is divisible by 2 and 3, so it is divisible by 6.

The sum of the digits of 780 is 15, which is not divisible by 9. 780 is not divisible by 9.

$$\begin{array}{r} 52 \\ 15\overline{)780} \\ \underline{75} \\ 30 \\ \underline{30} \\ 0 \end{array}$$

780 is divisible by 15.

780 is a multiple of 6 and 15, but not a multiple of 9.

45.

$$\begin{array}{r} 38 \\ 17\overline{)651} \\ \underline{51} \\ 141 \\ \underline{136} \\ 5 \end{array} \qquad \begin{array}{r} 21 \\ 31\overline{)651} \\ \underline{62} \\ 31 \\ \underline{31} \\ 0 \end{array}$$

651 is a multiple of 31, but not 17.

47.

$$\begin{array}{r} 93 \\ 17\overline{)1581} \\ \underline{153} \\ 51 \\ \underline{51} \\ 0 \end{array} \qquad \begin{array}{r} 51 \\ 31\overline{)1581} \\ \underline{155} \\ 31 \\ \underline{31} \\ 0 \end{array}$$

1581 is a multiple of 17 and 31.

49. 14 is not a multiple of 4, so Jean's car will not be checked.

51. The multiples of 5 between 1 and 55 are 5, 10, 15, 20, 25, 30, 35, 40, 45, 50, 55.

53. Each goat has 4 feet, so the total number of feet must be a multiple of 4. 30 is not a multiple of 4, so there cannot be 30 feet.

55. $780 = 6 \cdot 130$; $780 = 10 \cdot 78$; $780 = 12 \cdot 65$; $780 = 13 \cdot 60$; $780 = 15 \cdot 52$; $780 = 20 \cdot 39$; $780 = 26 \cdot 30$; There could be 6 rows of 130 pots, 10 rows of 78 pots, 12 rows of 65 pots, 13 rows of 60 pots, 15 rows of 52 pots, 20 rows of 39 pots, or 26 rows of 30 pots.

57. 240 is a multiple of 20, so the gear is in its original location.

59. $27 = 3 \cdot 9$, so the number of shares of Yahoo! is a multiple of the number of shares of HP. There are three times as many shares of Yahoo! as shares of HP.

61.
$$3^2 + 4^2 \stackrel{?}{=} 5^2$$
$$9 + 16 \stackrel{?}{=} 25$$
$$25 = 25$$

63. $5^2 + 12^2 = 25 + 144 = 169$; Since $13^2 = 169$, the third number must be 13.

65. Number $= 12 \times$ something $= 2 \times 6 \times$ something
Thus, if a number is a multiple of 12, it is also a multiple of 6.

67. Find the first multiple of 7 greater than or equal to 126. Then repeatedly add 7 to that number until you obtain a number greater than or equal to 175.

69. 9743 is not a multiple of 87.

71. There are 1333 multiples of 3 between 1000 and 5000.

73. 14: 1, 2, 7, 14

75. 18: 1, 2, 3, 6, 9, 18

77. 30: 1, 2, 3, 5, 6, 10, 15, 30

79. $10^2 = 100$, $100 > 87$

81. $23^2 = 529$, $529 > 500$

83. No. 24 is divisible by 2 and 8, but it is not divisible by 16.

Exercises 2.3 (page 134)

1.
$$4^2 = 16 \geq 16$$
$$1 \cdot 16 = 16$$
$$2 \cdot 8 = 16$$
$$\not{3}$$
$$4 \cdot 4 = 16$$

3.
$$6^2 = 36 \geq 31$$
$$1 \cdot 31 = 31$$
$$\not{2}, \not{3}, \not{4}, \not{5}, \not{6}$$

5.
$$6^2 = 36 \geq 33$$
$$1 \cdot 33 = 33$$
$$\not{2}$$
$$3 \cdot 11 = 20$$
$$\not{4}, \not{5}, \not{6}$$

7.
$$7^2 = 49 \geq 46$$
$$1 \cdot 46 = 46$$
$$2 \cdot 23 = 46$$
$$\not{3}, \not{4}, \not{5}, \not{6}, \not{7}$$

9.
$$7^2 = 49 \geq 49$$
$$1 \cdot 49 = 49$$
$$\not{2}, \not{3}, \not{4}, \not{5}, \not{6}$$
$$7 \cdot 7 = 49$$

11.
$$9^2 = 81 \geq 72$$
$$1 \cdot 72 = 72$$
$$2 \cdot 36 = 72$$
$$3 \cdot 24 = 72$$
$$4 \cdot 18 = 72$$
$$\not{5}$$
$$6 \cdot 12 = 72$$
$$\not{7}$$
$$8 \cdot 9 = 72$$

SECTION 2.3

13. $10^2 = 100 \geq 88$
$1 \cdot 88 = 88$
$2 \cdot 44 = 88$
$\not{3}$
$4 \cdot 22 = 88$
$\not{5}, \not{6}, \not{7}$
$8 \cdot 11 = 80$
$\not{9}, \not{10}$

15. $10^2 = 100 \geq 95$
$1 \cdot 95 = 95$
$\not{2}, \not{3}, \not{4}$
$5 \cdot 19 = 95$
$\not{6}, \not{7}, \not{8}, \not{9}, \not{10}$

17. $10^2 = 100 \geq 100$
$1 \cdot 100 = 100$
$2 \cdot 50 = 100$
$\not{3}$
$4 \cdot 25 = 100$
$5 \cdot 20 = 100$
$\not{6}, \not{7}, \not{8}, \not{9}$
$10 \cdot 10 = 100$

19. $11^2 = 121 \geq 105$
$1 \cdot 105 = 105$
$\not{2}$
$3 \cdot 35 = 105$
$\not{4}$
$5 \cdot 21 = 105$
$\not{6}$
$7 \cdot 15 = 105$
$\not{8}, \not{9}, \not{10}, \not{11}$

21. $11^2 = 121 \geq 112$
$1 \cdot 112 = 112$
$2 \cdot 56 = 112$
$\not{3}$
$4 \cdot 28 = 112$
$\not{5}, \not{6}$
$7 \cdot 16 = 112$
$8 \cdot 14 = 112$
$\not{9}, \not{10}, \not{11}$

23. $12^2 = 144 \geq 124$
$1 \cdot 124 = 124$
$2 \cdot 62 = 124$
$\not{3}$
$4 \cdot 31 = 124$
$\not{5}, \not{6}, \not{7}, \not{8}, \not{9}, \not{10}, \not{11}, \not{12}$

25. $21^2 = 441 \geq 441$
$1 \cdot 441 = 441$
$\not{2}$
$3 \cdot 147 = 441$
$\not{4}, \not{5}, \not{6}$
$7 \cdot 63 = 441$
$\not{8}$
$9 \cdot 49 = 441$
$\not{10}, \not{11}, \not{12}, \not{13}, \not{14}, \not{15}, \not{16}$
$\not{17}, \not{18}, \not{19}, \not{20}$
$21 \cdot 21 = 441$

27. $22^2 = 484 \geq 459$
$1 \cdot 459 = 459$
$\not{2}$
$3 \cdot 153 = 459$
$\not{4}, \not{5}, \not{6}, \not{7}, \not{8}$
$9 \cdot 51 = 459$
$\not{10}, \not{11}, \not{12}, \not{13}, \not{14}, \not{15}, \not{16}$
$17 \cdot 27 = 459$
$\not{18}, \not{19}, \not{20}, \not{21}, \not{22}$

29. $4^2 = 16 \geq 15$
$1 \cdot 15 = 15 \quad \not{2}, \not{4}$
$3 \cdot 5 = 15$
1, 3, 5, 15

31. $5^2 = 25 \geq 19$
$1 \cdot 19 = 29 \quad \not{2}, \not{3}, \not{4}, \not{5}$
1, 19

33. $7^2 = 49 \geq 39$
$1 \cdot 39 = 39 \quad \not{2}, \not{4}, \not{5}, \not{6}, \not{7}$
$3 \cdot 13 = 39$
1, 3, 13, 39

35. $7^2 = 49 \geq 46$
$1 \cdot 46 = 46 \quad \not{3}, \not{4}, \not{5}, \not{6}, \not{7}$
$2 \cdot 23 = 46$
1, 2, 23, 46

37. $8^2 = 64 \geq 52$
$1 \cdot 52 = 52 \quad \not{3}, \not{5}, \not{6}, \not{7}, \not{8}$
$2 \cdot 26 = 52$
$4 \cdot 13 = 52$
1, 2, 4, 13, 26, 52

39. $8^2 = 64 \geq 64$
$1 \cdot 64 = 64 \quad \not{3}, \not{5}, \not{6}, \not{7}$
$2 \cdot 32 = 64$
$4 \cdot 16 = 64$
$8 \cdot 8 = 64$
1, 2, 4, 8, 16, 32, 64

SECTION 2.3

41. $9^2 = 81 \geq 72$
$1 \cdot 72 = 72$ $\quad \not{5}, \not{7}$
$2 \cdot 36 = 72$
$3 \cdot 24 = 72$
$4 \cdot 18 = 72$
$6 \cdot 12 = 72$
$8 \cdot 9 = 72$
$9 \cdot 8 = 72$
1, 2, 3, 4, 6, 8, 9, 12,
18, 24, 36, 72

43. $9^2 = 81 \geq 78$
$1 \cdot 78 = 78$ $\quad \not{4}, \not{5}, \not{7}$
$2 \cdot 39 = 78$ $\quad \not{8}, \not{9}$
$3 \cdot 26 = 78$
$6 \cdot 13 = 78$
1, 2, 3, 6, 13, 26, 39, 78

45. $11^2 = 121 \geq 112$
$1 \cdot 112 = 112$ $\quad \not{3}, \not{5}, \not{6}, \not{9}$
$2 \cdot 56 = 112$ $\quad \not{10}, \not{11}$
$4 \cdot 28 = 112$
$7 \cdot 16 = 112$
$8 \cdot 14 = 100$
1, 2, 4, 7, 8, 14,
16, 28, 56, 112

47. $12^2 = 144 \geq 134$
$1 \cdot 134 = 134$ $\quad \not{3}, \not{4}, \not{5},$
$2 \cdot 67 = 122$ $\quad \not{6}, \not{7}, \not{8},$
$\not{9}, \not{10}, \not{11},$
$\not{12}$
1, 2, 67, 134

49. $13^2 = 169 \geq 162$
$1 \cdot 162 = 162$ $\quad \not{4}, \not{5}, \not{7},$
$2 \cdot 81 = 162$ $\quad \not{8}, \not{10},$
$3 \cdot 54 = 162$ $\quad \not{11}, \not{12},$
$6 \cdot 27 = 162$ $\quad \not{13}$
$9 \cdot 18 = 162$
1, 2, 3, 6, 9, 18, 27, 54,
81, 162

51. 444: $1 \cdot 444$, $2 \cdot 222$,
$3 \cdot 148$, $4 \cdot 111$,
$6 \cdot 74$, $12 \cdot 37$

53. 652: $1 \cdot 652$, $2 \cdot 326$, $4 \cdot 163$

55. 680: $1 \cdot 680$, $2 \cdot 340$, $4 \cdot 170$, $5 \cdot 136$,
$8 \cdot 85$, $10 \cdot 68$, $17 \cdot 40$, $20 \cdot 34$

57.

# Programs	1	2	3	4	5	6	8
Length	120	60	40	30	24	20	15

59.

# Players	2	4	13	26	52
# Cards (each)	26	13	4	2	1

61. $65 = 5 \cdot 13$, so each caregiver is in charge of 13 preschoolers. The recommendation is not met.

63. The number of members of the band is one more than a multiple of 4 and one more than a multiple of 5. The first common multiple of 4 and 5 is 20. The smallest possible number of band members is 21.

65. 6: 1, 2, 3, 6
$1 + 2 + 3 = 6 \Rightarrow 6$ is a perfect number.

67. 28: 1, 2, 4, 7, 14, 28
$1 + 2 + 4 + 7 + 14 = 28$
28 is a perfect number.

69. The word factor is usually used to refer to multiplication, while the word divisor is usually used to refer to division.

71. The largest factor is 991.

73.

$$\begin{array}{r} 4 \\ \not{2} \\ 48 \\ \times \quad 63 \\ \hline 144 \\ 2880 \\ \hline 3024 \end{array}$$

75.

$$\begin{array}{r} 7 \\ \not{1} \\ 381 \\ \times \quad 92 \\ \hline 762 \\ 34290 \\ \hline 35052 \end{array}$$

77.

$$\begin{array}{r} 33 \\ 78\overline{)2574} \\ 234 \\ \hline 234 \\ 234 \\ \hline 0 \end{array}$$

79.

$$\begin{array}{l} \quad\; 27\text{ R }3 \\ 306\overline{)8265} \\ \quad\; 612 \\ \hline \quad\; 2145 \\ \quad\; 2142 \\ \hline \qquad\quad 3 \end{array}$$

81.

$$\begin{array}{r} 41 \\ 24\overline{)1000} \\ 96 \\ \hline 40 \\ 24 \\ \hline 16 \end{array}$$

There is enough wire to wire 41 speakers, with 16 feet of wire left over.

Exercises 2.4 (page 140)

1. $3^2 = 9 \geq 6$
$2 \cdot 3 = 6$
6 is composite.

3. $3^2 = 9 \geq 8$
$2 \cdot 4 = 8$
8 is composite.

5. $4^2 = 16 \geq 15$
$3 \cdot 5 = 15$
15 is composite.

7. $5^2 = 25 \geq 17$
$\not{2}, \not{3}, \not{5}$
17 is prime.

9. $5^2 = 25 \geq 22$
$2 \cdot 11 = 22$
22 is composite.

11. $6^2 = 36 \geq 29$
$\not{2}, \not{3}, \not{5}$
29 is prime.

13. $7^2 = 49 \geq 38$
$2 \cdot 19 = 38$
38 is composite.

15. $7^2 = 49 \geq 41$
$\not{2}, \not{3}, \not{5}, \not{7}$
41 is prime.

17. $7^2 = 49 \geq 46$
$2 \cdot 23 = 46$
46 is composite.

19. $8^2 = 64 \geq 53$
$\not{2}, \not{3}, \not{5}, \not{7}$
53 is prime.

21. $8^2 = 64 \geq 58$
$2 \cdot 29 = 58$
58 is composite.

23. $8^2 = 64 \geq 52$
$2 \cdot 26 = 52$
52 is composite.

25. $8^2 = 64 \geq 60$
$2 \cdot 30 = 60$
60 is composite.

27. $9^2 = 81 \geq 70$
$2 \cdot 35 = 70$
70 is composite.

29. $9^2 = 81 \geq 79$
$\not{2}, \not{3}, \not{5}, \not{7}$
79 is prime.

SECTION 2.4

31. $10^2 = 100 \geq 83$
$\not{2}, \not{3}, \not{5}, \not{7}$
83 is prime.

33. $10^2 = 100 \geq 91$
$7 \cdot 13 = 91$
91 is composite.

35. $10^2 = 100 \geq 97$
$\not{2}, \not{3}, \not{5}, \not{7}$
97 is prime.

37. $12^2 = 144 \geq 127$
$\not{2}, \not{3}, \not{5}, \not{7}, \not{11}$
127 is prime.

39. $13^2 = 169 \geq 146$
$2 \cdot 73 = 146$
146 is composite.

41. $14^2 = 196 \geq 183$
$3 \cdot 61 = 183$
183 is composite.

43. $15^2 = 225 \geq 213$
$3 \cdot 71 = 213$
213 is composite.

45. $16^2 = 256 \geq 233$
$\not{2}, \not{3}, \not{5}, \not{7}, \not{11}, \not{13}$
233 is prime.

47. $20^2 = 400 \geq 383$
$\not{2}, \not{3}, \not{5}, \not{7}, \not{11}, \not{13}, \not{17}, \not{19}$
383 is prime.

49. $20^2 = 400 \geq 391$
$17 \cdot 23 = 391$
391 is composite.

51. $25^2 = 625 \geq 581$
$7 \cdot 83 = 581$
581 is composite.

53. $37^2 = 1369 \geq 1323$
$3 \cdot 441 = 1323$
1323 is composite.

55. The next prime year is 2017.

57. 12,345,678 is even, so it is divisible by 2. It has more than 2 factors, so is not prime.

59. 2 and 5 are both prime, but neither 25 nor 52 are prime. The statement is not generally true.

61. $1989 = 3 \cdot 663$, so 1989 is not prime.

63. $2^3 - 1 = 8 - 1 = 7$
$2^5 - 1 = 32 - 1 = 31$
$2^7 - 1 = 128 - 1 = 127$

65. $M_3 = 2^3 - 1 = 8 - 1 = 7$
$M_3 \times 2^{3-1} = 7 \times 2^2 = 7 \times 4 = 28$

67. A prime number has only 2 factors. 7 is a prime number since its only factors are 1 and 7. A composite number has more than 2 factors. 9 is a composite number since its factors are 1, 3, and 9.

69. A composite number can have exactly 4 factors. For example, the composite number 8 has 4 factors: 1, 2, 4, and 8.

71. $37{,}789 = 23 \cdot 1643$
37,789 is composite.

73. $120 \div 2 = 60 \Rightarrow$ divisible by 2
$60 \div 2 = 30$, $30 \div 2 = 15$
15 is the final quotient

75. $1040 \div 2 = 520 \Rightarrow$ divisible by 2
$520 \div 2 = 260$, $260 \div 2 = 130$,
$130 \div 2 = 65$; 65 is the final quotient

77. $1029 \div 3 = 343 \Rightarrow$ not divisible by 3
The final quotient is 343.

79. $2880 \div 5 = 576 \Rightarrow$ not divisible by 5
576 is the final quotient.

81. $1859 \div 13 = 143 \Rightarrow$ divisible by 13;
$143 \div 13 = 11$. The final quotient is 11.

Exercises 2.5 (page 148)

1. $\begin{array}{r|l} 2 & 12 \\ 2 & 6 \\ 3 & 3 \\ & 1 \end{array}$

$12 = 2^2 \times 3$

3. $\begin{array}{r|l} 3 & 15 \\ 5 & 5 \\ & 1 \end{array}$

$15 = 3 \times 5$

5. $\begin{array}{r|l} 3 & 21 \\ 7 & 7 \\ & 1 \end{array}$

$21 = 3 \times 7$

7. $\begin{array}{r|l} 2 & 24 \\ 2 & 12 \\ 2 & 6 \\ 3 & 3 \\ & 1 \end{array}$

$24 = 2^3 \times 3$

9. $\begin{array}{r|l} 2 & 28 \\ 2 & 14 \\ 7 & 7 \\ & 1 \end{array}$

$28 = 2^2 \times 7$

11. $\begin{array}{r|l} 2 & 34 \\ 17 & 17 \\ & 1 \end{array}$

$34 = 2 \times 17$

13. $\begin{array}{r|l} 2 & 46 \\ 23 & 23 \\ & 1 \end{array}$

$46 = 2 \times 23$

15. $\begin{array}{r|l} 2 & 42 \\ 3 & 21 \\ 7 & 7 \\ & 1 \end{array}$

$42 = 2 \times 3 \times 7$

17. $\begin{array}{r|l} 2 & 78 \\ 3 & 39 \\ 13 & 13 \\ & 1 \end{array}$

$78 = 2 \times 3 \times 13$

19. $\begin{array}{r|l} 2 & 82 \\ 41 & 41 \\ & 1 \end{array}$

$82 = 2 \times 41$

21. $\begin{array}{r|l} 2 & 90 \\ 3 & 45 \\ 3 & 15 \\ 5 & 5 \\ & 1 \end{array}$

$90 = 2 \times 3^2 \times 5$

23. $\begin{array}{r|l} 5 & 95 \\ 19 & 19 \\ & 1 \end{array}$

$95 = 5 \times 19$

25. $\begin{array}{r|l} 2 & 108 \\ 2 & 54 \\ 3 & 27 \\ 3 & 9 \\ 3 & 3 \\ & 1 \end{array}$

$108 = 2^2 \times 3^3$

27. 143 is prime.

29. $\begin{array}{r|l} 2 & 156 \\ 2 & 78 \\ 3 & 39 \\ 13 & 13 \\ & 1 \end{array}$

$156 = 2^2 \times 3 \times 13$

31. $\begin{array}{r|l} 2 & 174 \\ 3 & 87 \\ 29 & 29 \\ & 1 \end{array}$

$174 = 2 \times 3 \times 29$

33. $\begin{array}{r|l} 2 & 180 \\ 2 & 90 \\ 3 & 45 \\ 3 & 15 \\ 5 & 5 \\ & 1 \end{array}$

$180 = 2^2 \times 3^2 \times 5$

35. $\begin{array}{r|l} 2 & 200 \\ 2 & 100 \\ 2 & 50 \\ 5 & 25 \\ 5 & 5 \\ & 1 \end{array}$

$200 = 2^3 \times 5^2$

37. $\begin{array}{r|l} 2 & 314 \\ 157 & 157 \\ & 1 \end{array}$

$314 = 2 \times 157$

39. $\begin{array}{r|l} 3 & 345 \\ 5 & 115 \\ 23 & 23 \\ & 1 \end{array}$

$345 = 3 \times 5 \times 23$

SECTION 2.5

41.

2 | 460
2 | 230
5 | 115
23 | 23
1

$460 = 2^2 \times 5 \times 23$

43.

2 | 520
2 | 260
2 | 130
5 | 65
13 | 13
1

$520 = 2^3 \times 5 \times 13$

45.

3 | 459
3 | 153
3 | 51
17 | 17
1

$459 = 3^3 \times 17$

47.

2 | 714
3 | 357
7 | 119
17 | 17
1

$714 =$
$2 \times 3 \times 7 \times 17$

49.

5 | 625
5 | 125
5 | 25
5 | 5
1

$625 = 5^4$

51.

2 | 1230
3 | 615
5 | 205
41 | 41
1

$1230 =$
$2 \times 3 \times 5 \times 41$

53. **Answers may vary.**

55. **Answers may vary.**

57. $119 = 7 \times 17$; $143 = 11 \times 13$
119 and 143 are relatively prime.

59. 97 is the largest number less than 100 that is relatively prime to 180.

61. A number is in prime-factored form when it is either prime itself, or when it is written as a product of prime numbers.

63.

2 | 2256
2 | 1128
2 | 564
2 | 282
3 | 141
47 | 47
1

$2256 = 2^4 \times 3 \times 47$

65.

Even Number	Sum of Primes	Even Number	Sum of Primes
6	$3+3$	26	$13+13$
8	$3+5$	28	$5+23$
10	$3+7$	30	$7+23$
12	$5+7$	32	$3+29$
14	$3+11$	34	$3+31$
16	$3+13$	36	$5+31$
18	$5+13$	38	$7+31$
20	$3+17$	40	$3+37$
22	$3+19$	42	$5+37$
24	$5+19$	44	$3+41$

67. $13(5) = 65$
The area is 65 in.^{2}

69. $2262 = 2 \cdot 1131$; $2262 = 13 \cdot 174$
It is a multiple of both 2 and 13.

71. $6170 = 5 \cdot 1234$; 6170 is not divisible by 13. It is not a multiple of both 5 and 13.

73. $4004 = 2 \cdot 2002$; $4004 = 7 \cdot 572$
It is a multiple of both 2 and 7.

75. Answers will vary. Examples: 30, 60, 90

Exercises 2.6 (page 154)

1.
$$\begin{array}{r|ll} 2 & 4 & 18 \\ \hline & 2 & 9 \end{array}$$
$\text{LCM} = 2 \cdot 2 \cdot 9 = 36$

3.
$$\begin{array}{r|ll} 7 & 7 & 21 \\ \hline & 1 & 3 \end{array}$$
$\text{LCM} = 7 \cdot 1 \cdot 3 = 21$

5.
$$\begin{array}{r|ll} 3 & 3 & 30 \\ \hline & 1 & 10 \end{array}$$
$\text{LCM} = 3 \cdot 1 \cdot 10 = 30$

7.
$$\begin{array}{r|ll} 2 & 8 & 20 \\ \hline 2 & 4 & 10 \\ \hline & 2 & 5 \end{array}$$
$\text{LCM} = 2 \cdot 2 \cdot 2 \cdot 5 = 40$

9. No common factors $\Rightarrow$
$\text{LCM} = 5 \cdot 9 = 45$

11.
$$\begin{array}{r|ll} 3 & 9 & 12 \\ \hline & 3 & 4 \end{array}$$
$\text{LCM} = 3 \cdot 3 \cdot 4 = 36$

13.
$$\begin{array}{r|lll} 2 & 4 & 10 & 20 \\ \hline 2 & 2 & 5 & 10 \\ \hline 5 & 1 & 5 & 5 \\ \hline & 1 & 1 & 1 \end{array}$$
$\text{LCM} = 2 \cdot 2 \cdot 5 \cdot 1 \cdot 1 \cdot 1 = 20$

15.
$$\begin{array}{r|lll} 3 & 3 & 12 & 15 \\ \hline & 1 & 4 & 5 \end{array}$$
$\text{LCM} = 3 \cdot 1 \cdot 4 \cdot 5 = 60$

17.
$$\begin{array}{r|lll} 2 & 2 & 6 & 10 \\ \hline & 1 & 3 & 5 \end{array}$$
$\text{LCM} = 2 \cdot 1 \cdot 3 \cdot 5 = 30$

19.
$$\begin{array}{r|lll} 2 & 4 & 15 & 20 \\ \hline 2 & 2 & 15 & 10 \\ \hline 5 & 1 & 15 & 5 \\ \hline & 1 & 3 & 1 \end{array}$$
$\text{LCM} = 2 \cdot 2 \cdot 5 \cdot 1 \cdot 3 \cdot 1 = 60$

21.
$$\begin{array}{r|ll} 2 & 18 & 24 \\ \hline 3 & 9 & 12 \\ \hline & 3 & 4 \end{array}$$
$\text{LCM} = 2 \cdot 3 \cdot 3 \cdot 4 = 72$

23.
$$\begin{array}{r|ll} 2 & 16 & 24 \\ \hline 2 & 8 & 12 \\ \hline 2 & 4 & 6 \\ \hline & 2 & 3 \end{array}$$
$\text{LCM} = 2 \cdot 2 \cdot 2 \cdot 2 \cdot 3 = 48$

25.
$$\begin{array}{r|ll} 2 & 18 & 20 \\ \hline & 9 & 10 \end{array}$$
$\text{LCM} = 2 \cdot 9 \cdot 10 = 180$

SECTION 2.6

27.

$$\begin{array}{r|ll} 2 & 12 & 16 \\ \hline 2 & 6 & 8 \\ \hline & 3 & 4 \end{array}$$

$\text{LCM} = 2 \cdot 2 \cdot 3 \cdot 4 = 48$

29.

$$\begin{array}{r|ll} 2 & 20 & 24 \\ \hline 2 & 10 & 12 \\ \hline & 5 & 6 \end{array}$$

$\text{LCM} = 2 \cdot 2 \cdot 5 \cdot 6 = 120$

31.

$$\begin{array}{r|lll} 2 & 8 & 12 & 16 \\ \hline 2 & 4 & 6 & 8 \\ \hline 2 & 2 & 3 & 4 \\ \hline & 1 & 3 & 2 \end{array}$$

$\text{LCM} = 2 \cdot 2 \cdot 2 \cdot 1 \cdot 3 \cdot 2 = 48$

33.

$$\begin{array}{r|lll} 3 & 9 & 12 & 15 \\ \hline & 3 & 4 & 5 \end{array}$$

$\text{LCM} = 3 \cdot 3 \cdot 4 \cdot 5 = 180$

35.

$$\begin{array}{r|llll} 2 & 4 & 8 & 12 & 24 \\ \hline 2 & 2 & 4 & 6 & 12 \\ \hline 2 & 1 & 2 & 3 & 6 \\ \hline 3 & 1 & 1 & 3 & 3 \\ \hline & 1 & 1 & 1 & 1 \end{array}$$

$\text{LCM} = 2 \cdot 2 \cdot 2 \cdot 3 \cdot 1 \cdot 1 \cdot 1 \cdot 1 = 24$

37.

$$\begin{array}{r|lll} 2 & 12 & 16 & 24 \\ \hline 2 & 6 & 8 & 12 \\ \hline 2 & 3 & 4 & 6 \\ \hline 3 & 3 & 2 & 3 \\ \hline & 1 & 2 & 1 \end{array}$$

$\text{LCM} = 2 \cdot 2 \cdot 2 \cdot 3 \cdot 1 \cdot 2 \cdot 1 = 48$

39.

$$\begin{array}{r|lll} 2 & 21 & 24 & 56 \\ \hline 2 & 21 & 12 & 28 \\ \hline 2 & 21 & 6 & 14 \\ \hline 3 & 21 & 3 & 7 \\ \hline 7 & 7 & 1 & 7 \\ \hline & 1 & 1 & 1 \end{array}$$

$\text{LCM} = 2 \cdot 2 \cdot 2 \cdot 3 \cdot 7 \cdot 1 \cdot 1 \cdot 1 = 168$

41.

$$\begin{array}{r|llll} 2 & 10 & 14 & 21 & 35 \\ \hline 5 & 5 & 7 & 21 & 35 \\ \hline 7 & 1 & 7 & 21 & 7 \\ \hline & 1 & 1 & 3 & 1 \end{array}$$

$\text{LCM} = 2 \cdot 5 \cdot 7 \cdot 1 \cdot 1 \cdot 3 \cdot 1 = 210$

43.

$$\begin{array}{r|llll} 2 & 12 & 17 & 51 & 68 \\ \hline 2 & 6 & 17 & 51 & 34 \\ \hline 3 & 3 & 17 & 51 & 17 \\ \hline 17 & 1 & 17 & 17 & 17 \\ \hline & 1 & 1 & 1 & 1 \end{array}$$

$\text{LCM} = 2 \cdot 2 \cdot 3 \cdot 17 \cdot 1 \cdot 1 \cdot 1 \cdot 1 = 204$

45.

$$\begin{array}{r|lllll} 2 & 35 & 50 & 56 & 70 & 175 \\ \hline 5 & 35 & 25 & 28 & 35 & 175 \\ \hline 5 & 7 & 5 & 28 & 7 & 35 \\ \hline 7 & 7 & 1 & 28 & 7 & 7 \\ \hline & 1 & 1 & 4 & 1 & 1 \end{array}$$

$\text{LCM} = 2 \cdot 5 \cdot 5 \cdot 7 \cdot 1 \cdot 1 \cdot 4 \cdot 1 \cdot 1 = 1400$

SECTION 2.6

47.
$$\begin{array}{r|lll} 2 & 3 & 4 & 8 \\ \hline 2 & 3 & 2 & 4 \\ \hline & 3 & 1 & 2 \end{array}$$
$\text{LCD} = \text{LCM} = 2 \cdot 2 \cdot 3 \cdot 1 \cdot 2 = 24$

49.
$$\begin{array}{r|lll} 2 & 15 & 12 & 8 \\ \hline 2 & 15 & 6 & 4 \\ \hline 3 & 15 & 3 & 2 \\ \hline & 5 & 1 & 2 \end{array}$$
$\text{LCM} = 2 \cdot 2 \cdot 3 \cdot 5 \cdot 1 \cdot 2 = 120$

51. Mary is off every 5th day, while Frank is off every 6th day. 5 and 6 have no common factors, so the LCM is $5 \cdot 6 = 30$. They will be off together again in 30 days (May 31).

53.
$$\begin{array}{r|ll} 2 & 36 & 24 \\ \hline 2 & 18 & 12 \\ \hline 3 & 9 & 6 \\ \hline & 3 & 2 \end{array}$$
$\text{LCM} = 2 \cdot 2 \cdot 3 \cdot 3 \cdot 2 = 72$
$\text{LCM} \div 36 = 72 \div 36 = 2$
They will return after 2 turns of the 1st gear.

55.
$$\begin{array}{r|ll} 2 & 2 & 12 \\ \hline & 1 & 6 \end{array}$$
$\text{LCM} = 2 \cdot 1 \cdot 6 = 12$
They will both be visible in 12 years.

57.
$$\begin{array}{r|ll} 2 & 6 & 10 \\ \hline & 3 & 5 \end{array}$$
$\text{LCM} = 2 \cdot 3 \cdot 5 = 30$; John must run 5 days, and his friend must run 3 days.

59. **Answers will vary.**

61.
$$\begin{array}{r|lll} 2 & 144 & 240 & 360 \\ \hline 2 & 72 & 120 & 180 \\ \hline 2 & 36 & 60 & 90 \\ \hline 2 & 18 & 30 & 45 \\ \hline 3 & 9 & 15 & 45 \\ \hline 3 & 3 & 5 & 15 \\ \hline 5 & 1 & 5 & 5 \\ \hline & 1 & 1 & 1 \end{array}$$
$\text{LCM} = 2 \cdot 2 \cdot 2 \cdot 2 \cdot 3 \cdot 3 \cdot 5 \cdot 1 \cdot 1 \cdot 1 = 720$

63.
$$\begin{array}{r|llll} 2 & 60 & 210 & 315 & 350 \\ \hline 3 & 30 & 105 & 315 & 175 \\ \hline 5 & 10 & 35 & 105 & 175 \\ \hline 7 & 2 & 7 & 21 & 35 \\ \hline & 2 & 1 & 3 & 5 \end{array}$$
$\text{LCM} = 2 \cdot 3 \cdot 5 \cdot 7 \cdot 2 \cdot 1 \cdot 3 \cdot 5 = 6300$

65. 460: 1, 2, 4, 5, 10, 20, 23, 46, 92, 115, 230, 460

67. 510: 1, 2, 3, 5, 6, 10, 15, 17, 30, 34, 51, 85, 102, 170, 255, 510

69. $9240 = 35 \times 264 \Rightarrow 9240$ is divisible by 35.

71. 9240 is not divisible by 75.

73. $112 \div 14 = 8$; $198 \div 14 = 14$ R 2; They must sell at least 15, or at least 7 more.

Chapter 2 Review Exercises (page 159)

1. 44, 50, and 478 end in even digits, so they are divisible by 2.

2. $6 = 6, 3 + 6 = 9, 6 + 3 = 9, 6 + 3 + 6 = 15, 6 + 6 + 3 = 15 \Rightarrow$ All are divisible by 3.

3. 15, 255, and 525 end in 5 (or 0), so they are divisible by 5.

4. $3 + 6 = 9, 6 + 5 = 11, 1 + 4 + 4 = 9, 7 + 1 + 4 = 12$; 36, 144, and 714 are divisible by 3. They are also all even and divisible by 2, so 36, 144, and 714 are divisible by 6.

5. $6 = 6, 3 + 6 = 9, 6 + 3 = 9, 6 + 3 + 6 = 15, 6 + 6 + 3 = 15 \Rightarrow$ 36 and 63 are divisible by 9.

6. 50 and 550 end in 0, so they are divisible by 10.

7. $4 + 4 + 4 = 12, 5 + 5 + 5 = 15, 6 + 6 + 6 = 18, 7 + 7 + 7 = 21, 8 + 8 + 8 = 24, 9 + 9 + 9 = 27 \Rightarrow$ All are divisible by 3. 444, 666, and 888 end in even digits, so they are divisible by 2 as well.

8. $4 + 5 + 0 = 9, 5 + 5 + 0 = 10, 6 + 6 + 0 = 12, 7 + 7 + 0 = 14, 8 + 8 + 0 = 16, 9 + 9 + 0 = 18 \Rightarrow$ 450 and 990 are divisible by 9. They are both even, so they are divisible by 2 as well.

9. All end in 5 (or 0), so all are divisible by 5. $4 + 4 + 5 = 13, 5 + 4 + 5 = 14, 6 + 4 + 5 = 15, 7 + 4 + 5 = 16, 8 + 4 + 5 = 20 \Rightarrow$ Only 645 is divisible by 3 as well.

10. All end in 0, so all are divisible by 10. $4 + 4 + 0 = 8, 5 + 5 + 0 = 10, 6 + 6 + 0 = 12, 7 + 7 + 0 = 14, 8 + 8 + 0 = 16.$ Only 660 is divisible by 3 as well.

11. Divisible by 6?
The last digit of 567 is not even, so it is not divisible by 2. Thus, it is not divisible by 6.

Divisible by 7?

$$\begin{array}{r} 81 \\ 7\overline{)567} \\ \underline{56} \\ 07 \\ \underline{7} \\ 0 \end{array}$$

567 is divisible by 7.

Divisible by 9?
$5 + 6 + 7 = 18$, so 567 is divisible by 9.

12. Divisible by 6?
The last digit of 576 is even, so it is divisible by 2. $5 + 7 + 6 = 18$, so it is divisible by 3. Thus, it is divisible by 6.

Divisible by 7?

$$\begin{array}{r} 82 \\ 7\overline{)576} \\ \underline{56} \\ 16 \\ \underline{14} \\ 2 \end{array}$$

576 is not divisible by 7.

Divisible by 9?
$5 + 7 + 6 = 18$, so 576 is divisible by 9.

CHAPTER 2 REVIEW EXERCISES

13. Divisible by 6?
The last digit of 693 is not even, so it is not divisible by 2. Thus, it is not divisible by 6.

Divisible by 7?

```
    9 9
7 ) 6 9 3
    6 3
    ---
      6 3
      6 3
      ---
        0
```

693 is divisible by 7.

Divisible by 9?
$6 + 9 + 3 = 18$, so 693 is divisible by 9.

14. Divisible by 6?
The last digit of 765 is not even, so it is not divisible by 2. Thus, it is not divisible by 6.

Divisible by 7?

```
    1 0 9
7 ) 7 6 5
    7
    -
    0 6
      0
      -
      6 5
      6 3
      ---
        2
```

765 is not divisible by 7.

Divisible by 9?
$7 + 6 + 5 = 18$, so 765 is divisible by 9.

15. Divisible by 4?
The last two digits form the number 40, which is divisible by 4. Thus, 840 is divisible by 4.

Divisible by 5?
840 ends in 0 (or 5), so it is divisible by 5.

Divisible by 10?
840 ends in 0, so it is divisible by 10.

16. Divisible by 4?
The last two digits form the number 75, which is not divisible by 4. Thus, 575 is not divisible by 4.

Divisible by 5?
575 ends in 5 (or 0), so it is divisible by 5.

Divisible by 10?
575 does not end in 0, so it is not divisible by 10.

17. Divisible by 6?
The last digit of 705 is not even, so it is not divisible by 2. Thus, it is divisible by 6.

Divisible by 15?
The last digit of 705 is 0 (or 5), so it is divisible by 5. $7 + 0 + 5 = 12$, so it is divisible by 3. Thus, it is divisible by 15.

18. Divisible by 6?
The last digit of 975 is not even, so it is not divisible by 2. Thus, it is not divisible by 6.

Divisible by 15?
The last digit of 975 is 5 (or 0), so it is divisible by 5. $9 + 7 + 5 = 21$, so it is divisible by 3. Thus, it is divisible by 15.

19. The last two digits form the number 68. which is divisible by 4, so 268 is divisible by 4. It is possible.

20. 3060 is divisible by 6, so they can divide the expenses evenly.

CHAPTER 2 REVIEW EXERCISES

21. $7 \cdot 1 = 7$
$7 \cdot 2 = 14$
$7 \cdot 3 = 21$
$7 \cdot 4 = 28$
$7 \cdot 5 = 35$

22. $63 \cdot 1 = 63$
$63 \cdot 2 = 126$
$63 \cdot 3 = 189$
$63 \cdot 4 = 252$
$63 \cdot 5 = 315$

23. $73 \cdot 1 = 73$
$73 \cdot 2 = 146$
$73 \cdot 3 = 219$
$73 \cdot 4 = 292$
$73 \cdot 5 = 365$

24. $64 \cdot 1 = 64$
$64 \cdot 2 = 128$
$64 \cdot 3 = 192$
$64 \cdot 4 = 256$
$64 \cdot 5 = 320$

25. $85 \cdot 1 = 85$
$85 \cdot 2 = 170$
$85 \cdot 3 = 255$
$85 \cdot 4 = 340$
$85 \cdot 5 = 425$

26. $97 \cdot 1 = 97$
$97 \cdot 2 = 194$
$97 \cdot 3 = 291$
$97 \cdot 4 = 388$
$97 \cdot 5 = 485$

27. $131 \cdot 1 = 131$
$131 \cdot 2 = 262$
$131 \cdot 3 = 393$
$131 \cdot 4 = 524$
$131 \cdot 5 = 655$

28. $142 \cdot 1 = 142$
$142 \cdot 2 = 284$
$142 \cdot 3 = 426$
$142 \cdot 4 = 568$
$142 \cdot 5 = 710$

29. $211 \cdot 1 = 211$, $211 \cdot 2 = 422$,
$211 \cdot 3 = 633$, $211 \cdot 4 = 844$,
$211 \cdot 5 = 1055$

30. $252 \cdot 1 = 252$, $252 \cdot 2 = 504$,
$252 \cdot 3 = 756$, $252 \cdot 4 = 1008$,
$252 \cdot 5 = 1260$

31. 84 is divisible by 2 and 3, so it is divisible by 6.

The sum of the digits of 84 is 12, which is not divisible by 9. 84 is not divisible by 9.

$$\begin{array}{r} 5 \\ 15\overline{)84} \\ \underline{75} \\ 9 \end{array}$$

84 is not divisible by 15.

84 is a multiple of 6, but not a multiple of 9 or 15.

32. 135 is not divisible by 2, so it is not divisible by 6.

The sum of the digits of 135 is 9, which is divisible by 9. 135 is divisible by 9.

$$\begin{array}{r} 9 \\ 15\overline{)135} \\ \underline{135} \\ 0 \end{array}$$

135 is divisible by 15.

135 is a multiple of 9 and 15, but not a multiple of 6.

33. 210 is divisible by 2 and 3, so it is divisible by 6.

The sum of the digits of 210 is 3, which is not divisible by 9. 210 is not divisible by 9.

$$\begin{array}{r} 14 \\ 15\overline{)210} \\ \underline{15} \\ 60 \\ \underline{60} \\ 0 \end{array}$$

210 is divisible by 15.

210 is a multiple of 6 and 15, but not a multiple of 9.

34. 310 is not divisible by 3, so it is not divisible by 6.

The sum of the digits of 310 is 4, which is not divisible by 9. 310 is not divisible by 9.

$$\begin{array}{r} 20 \\ 15\overline{)310} \\ \underline{30} \\ 10 \\ \underline{0} \\ 10 \end{array}$$

310 is not divisible by 15.

310 is not a multiple of 6, 9, or 15.

CHAPTER 2 REVIEW EXERCISES

35. 315 is not divisible by 2, so it is not divisible by 6.

The sum of the digits of 315 is 9, which is divisible by 9. 315 is divisible by 9.

$$\begin{array}{r} 21 \\ 15\overline{)315} \\ \underline{30} \\ 15 \\ \underline{15} \\ 0 \end{array}$$

315 is divisible by 15.

315 is a multiple of 9 and 15, but not a multiple of 6.

36. 540 is divisible by 2 and 3, so it is divisible by 6.

The sum of the digits of 540 is 9, which is divisible by 9. 540 is divisible by 9.

$$\begin{array}{r} 36 \\ 15\overline{)540} \\ \underline{45} \\ 90 \\ \underline{90} \\ 0 \end{array}$$

540 is divisible by 15.

540 is a multiple of 6, 9, and 15.

37.

$$\begin{array}{r} 24 \\ 12\overline{)288} \\ \underline{24} \\ 48 \\ \underline{48} \\ 0 \end{array}$$

288 is divisible by 12.

$$\begin{array}{r} 18 \\ 16\overline{)288} \\ \underline{16} \\ 128 \\ \underline{128} \\ 0 \end{array}$$

288 is divisible by 16.

288 is a multiple of 12 and 16.

38.

$$\begin{array}{r} 46 \\ 12\overline{)560} \\ \underline{48} \\ 80 \\ \underline{72} \\ 8 \end{array}$$

560 is not divisible by 12.

$$\begin{array}{r} 35 \\ 16\overline{)560} \\ \underline{48} \\ 80 \\ \underline{80} \\ 0 \end{array}$$

560 is divisible by 16.

560 is a multiple of 16, but not 12.

39. $23 \cdot 117 = 2691$. She bought a whole number of shares.

40. 5678 is not a multiple of 23, so she did not buy a whole number of shares.

41. $15: 1 \cdot 15, 3 \cdot 5$

42. $28: 1 \cdot 28, 2 \cdot 14, 4 \cdot 7$

43. $42: 1 \cdot 42, 2 \cdot 21, 3 \cdot 14, 6 \cdot 7$

44. $136: 1 \cdot 136, 2 \cdot 68, 4 \cdot 34, 8 \cdot 17$

45. $236: 1 \cdot 236, 2 \cdot 118, 4 \cdot 59$

46. $325: 1 \cdot 325, 5 \cdot 65, 13 \cdot 25$

47. $354: 1 \cdot 354, 2 \cdot 177, 3 \cdot 118, 6 \cdot 59$

48. $341: 1 \cdot 341, 11 \cdot 31$

49. $343: 1 \cdot 343, 7 \cdot 49$

50. 33: 1, 3, 11, 33

CHAPTER 2 REVIEW EXERCISES

51. 48: 1, 2, 3, 4, 6, 8, 12, 16, 24, 48

52. 57: 1, 3, 19, 57

53. 60: 1, 2, 3, 4, 5, 6, 10, 12, 15, 20, 30, 60

54. 78: 1, 2, 3, 6, 13, 26, 39, 78

55. 97: 1, 97

56. 99: 1, 3, 9, 11, 33, 99

57. 108: 1, 2, 3, 4, 6, 9, 12, 18, 27, 36, 54, 108

58. 110: 1, 2, 5, 10, 11, 22, 55, 110

59. The only factors of 1590 between 1 and 10 are 1, 2, 3, 5, 6, and 10.

# people	1	2	3	5	6	10
Cost per person	1590	795	530	318	265	159

60. Here are the factors that work to make 240 square feet:

Length	80	60	48	40	30	24	20	16
Width	3	4	5	6	8	10	12	15

61. $5^2 = 25 \geq 17$
$\cancel{2}, \cancel{3}, \cancel{5}$; 17 is prime.

62. $30 = 2 \cdot 15 \Rightarrow 30$ is composite.

63. $7^2 = 49 \geq 43$
$\cancel{2}, \cancel{3}, \cancel{5}, \cancel{7}$; 43 is prime.

64. $8^2 = 64 \geq 53$
$\cancel{2}, \cancel{3}, \cancel{5}, \cancel{7}$; 53 is prime.

65. $57 = 3 \cdot 19 \Rightarrow 57$ is composite.

66. $8^2 = 64 \geq 61$
$\cancel{2}, \cancel{3}, \cancel{5}, \cancel{7}$; 61 is prime.

67. $78 = 3 \cdot 26 \Rightarrow 78$ is composite.

68. $10^2 = 100 \geq 89$
$\cancel{2}, \cancel{3}, \cancel{5}, \cancel{7}$; 89 is prime.

69. $93 = 3 \cdot 31 \Rightarrow 93$ is composite.

70. $111 = 3 \cdot 37 \Rightarrow 111$ is composite.

71. $117 = 3 \cdot 39 \Rightarrow 117$ is composite.

72. $91 = 7 \cdot 13 \Rightarrow 91$ is composite.

73. $19^2 = 361 \geq 337$
$\cancel{2}, \cancel{3}, \cancel{5}, \cancel{7}, \cancel{11}, \cancel{13}, \cancel{17}, \cancel{19}$; 337 is prime.

74. $339 = 3 \cdot 113 \Rightarrow 339$ is composite.

75. $391 = 17 \cdot 23 \Rightarrow 391$ is composite.

76. $19^2 = 361 \geq 359$
$\cancel{2}, \cancel{3}, \cancel{5}, \cancel{7}, \cancel{11}, \cancel{13}, \cancel{17}, \cancel{19}$; 359 is prime.

77. $713 = 23 \cdot 31 \Rightarrow 713$ is composite.

78. $30^2 = 900 \geq 853$
$\cancel{2}, \cancel{3}, \cancel{5}, \cancel{7}, \cancel{11}, \cancel{13}, \cancel{17}, \cancel{19}, \cancel{23}, \cancel{29}$
853 is prime.

79. The next prime number after 2017 is 2027.

80. The last year before 1997 that was a prime number was 1993.

CHAPTER 2 REVIEW EXERCISES

81.
$$\begin{array}{r|r} 2 & 32 \\ 2 & 16 \\ 2 & 8 \\ 2 & 4 \\ 2 & 2 \\ & 1 \end{array}$$
$32 = 2^5$

82.
$$\begin{array}{r|r} 2 & 36 \\ 2 & 18 \\ 3 & 9 \\ 3 & 3 \\ & 1 \end{array}$$
$36 = 2^2 \times 3^2$

83.
$$\begin{array}{r|r} 2 & 38 \\ 19 & 19 \\ & 1 \end{array}$$
$38 = 2 \times 19$

84.
$$\begin{array}{r|r} 2 & 42 \\ 3 & 21 \\ 7 & 7 \\ & 1 \end{array}$$
$42 = 2 \times 3 \times 7$

85.
$$\begin{array}{r|r} 3 & 75 \\ 5 & 25 \\ 5 & 5 \\ & 1 \end{array}$$
$75 = 3 \times 5^2$

86.
$$\begin{array}{r|r} 2 & 96 \\ 2 & 48 \\ 2 & 24 \\ 2 & 12 \\ 2 & 6 \\ 3 & 3 \\ & 1 \end{array}$$
$96 = 2^5 \times 3$

87.
$$\begin{array}{r|r} 2 & 222 \\ 3 & 111 \\ 37 & 37 \\ & 1 \end{array}$$
$222 = 2 \times 3 \times 37$

88.
$$\begin{array}{r|r} 2 & 276 \\ 2 & 138 \\ 3 & 69 \\ 23 & 23 \\ & 1 \end{array}$$
$276 = 2^2 \times 3 \times 23$

89.
$$\begin{array}{r|r} 2 & 252 \\ 2 & 126 \\ 3 & 63 \\ 3 & 21 \\ 7 & 7 \\ & 1 \end{array}$$
$252 = 2^2 \times 3^2 \times 7$

90. $256 = 2^8$

91.
$$\begin{array}{r|r} 2 & 290 \\ 5 & 145 \\ 29 & 29 \\ & 1 \end{array}$$
$290 = 2 \times 5 \times 29$

92.
$$\begin{array}{r|r} 3 & 297 \\ 3 & 99 \\ 3 & 33 \\ 11 & 11 \\ & 1 \end{array}$$
$297 = 3^3 \times 11$

93.
$$\begin{array}{r|r} 2 & 340 \\ 2 & 170 \\ 5 & 85 \\ 17 & 17 \\ & 1 \end{array}$$
$340 = 2^2 \times 5 \times 17$

94.
$$\begin{array}{r|r} 2 & 342 \\ 3 & 171 \\ 3 & 57 \\ 19 & 19 \\ & 1 \end{array}$$
$342 = 2 \times 3^2 \times 19$

CHAPTER 2 REVIEW EXERCISES

95. 373 is prime.

96.
$$\begin{array}{r|l} 2 & 264 \\ \hline 2 & 132 \\ \hline 2 & 66 \\ \hline 3 & 33 \\ \hline 11 & 11 \\ \hline & 1 \end{array}$$
$264 = 2^3 \times 3 \times 11$

97.
$$\begin{array}{r|l} 5 & 265 \\ \hline 53 & 53 \\ \hline & 1 \end{array}$$
$265 = 5 \times 53$

98.
$$\begin{array}{r|l} 2 & 266 \\ \hline 7 & 133 \\ \hline 19 & 19 \\ \hline & 1 \end{array}$$
$266 = 2 \times 7 \times 19$

99. $2012 = 2^2 \times 503$
$2013 = 3 \times 11 \times 61$
$2014 = 2 \times 19 \times 53$
$2015 = 5 \times 403$

100. Answers may vary.

101.
$$\begin{array}{r|ll} 2 & 6 & 8 \\ \hline & 3 & 4 \end{array}$$
$\text{LCM} = 2 \cdot 3 \cdot 4 = 24$

102.
$$\begin{array}{r|ll} 2 & 8 & 10 \\ \hline & 4 & 5 \end{array}$$
$\text{LCM} = 2 \cdot 4 \cdot 5 = 40$

103.
$$\begin{array}{r|ll} 2 & 18 & 40 \\ \hline & 9 & 20 \end{array}$$
$\text{LCM} = 2 \cdot 9 \cdot 20 = 360$

104.
$$\begin{array}{r|ll} 2 & 20 & 36 \\ \hline 2 & 10 & 18 \\ \hline & 5 & 9 \end{array}$$
$\text{LCM} = 2 \cdot 2 \cdot 5 \cdot 9 = 180$

105.
$$\begin{array}{r|ll} 2 & 36 & 44 \\ \hline 2 & 18 & 22 \\ \hline & 9 & 11 \end{array}$$
$\text{LCM} = 2 \cdot 2 \cdot 9 \cdot 11 = 396$

106.
$$\begin{array}{r|ll} 2 & 24 & 72 \\ \hline 2 & 12 & 36 \\ \hline 2 & 6 & 18 \\ \hline 3 & 3 & 9 \\ \hline & 1 & 3 \end{array}$$
$\text{LCM} = 2 \cdot 2 \cdot 2 \cdot 3 \cdot 1 \cdot 3 = 72$

107.
$$\begin{array}{r|ll} 5 & 25 & 35 \\ \hline & 5 & 7 \end{array}$$
$\text{LCM} = 5 \cdot 5 \cdot 7 = 175$

108.
$$\begin{array}{r|ll} 7 & 28 & 35 \\ \hline & 4 & 5 \end{array}$$
$\text{LCM} = 7 \cdot 4 \cdot 5 = 140$

109.
$$\begin{array}{r|ll} 5 & 40 & 45 \\ \hline & 8 & 9 \end{array}$$
$\text{LCM} = 5 \cdot 8 \cdot 9 = 360$

110.
$$\begin{array}{r|lll} 3 & 3 & 9 & 18 \\ \hline 3 & 1 & 3 & 6 \\ \hline & 1 & 1 & 2 \end{array}$$
$\text{LCM} = 3 \cdot 3 \cdot 1 \cdot 1 \cdot 2 = 18$

111.
$$\begin{array}{r|rrr} 2 & 6 & 8 & 12 \\ \hline 2 & 3 & 4 & 6 \\ \hline 3 & 3 & 2 & 3 \\ \hline & 1 & 2 & 1 \end{array}$$
$\text{LCM} = 2 \cdot 2 \cdot 3 \cdot 1 \cdot 2 \cdot 1 = 24$

112.
$$\begin{array}{r|rrr} 2 & 8 & 10 & 12 \\ \hline 2 & 4 & 5 & 6 \\ \hline & 2 & 5 & 3 \end{array}$$
$\text{LCM} = 2 \cdot 2 \cdot 2 \cdot 5 \cdot 3 = 120$

113.
$$\begin{array}{r|rrr} 2 & 9 & 24 & 36 \\ \hline 2 & 9 & 12 & 18 \\ \hline 3 & 9 & 6 & 9 \\ \hline 3 & 3 & 2 & 3 \\ \hline & 1 & 2 & 1 \end{array}$$
$\text{LCM} = 2 \cdot 2 \cdot 3 \cdot 3 \cdot 1 \cdot 2 \cdot 1 = 72$

114.
$$\begin{array}{r|rrr} 2 & 18 & 24 & 36 \\ \hline 2 & 9 & 12 & 18 \\ \hline 3 & 9 & 6 & 9 \\ \hline 3 & 3 & 2 & 3 \\ \hline & 1 & 2 & 1 \end{array}$$
$\text{LCM} = 2 \cdot 2 \cdot 3 \cdot 3 \cdot 1 \cdot 2 \cdot 1 = 72$

115.
$$\begin{array}{r|rrr} 2 & 15 & 50 & 60 \\ \hline 3 & 15 & 25 & 30 \\ \hline 5 & 5 & 25 & 10 \\ \hline & 1 & 5 & 2 \end{array}$$
$\text{LCM} = 2 \cdot 3 \cdot 5 \cdot 1 \cdot 5 \cdot 2 = 300$

116.
$$\begin{array}{r|rrr} 2 & 32 & 40 & 60 \\ \hline 2 & 16 & 20 & 30 \\ \hline 2 & 8 & 10 & 15 \\ \hline 5 & 4 & 5 & 15 \\ \hline & 4 & 1 & 3 \end{array}$$
$\text{LCM} = 2 \cdot 2 \cdot 2 \cdot 5 \cdot 4 \cdot 1 \cdot 3 = 480$

117.
$$\begin{array}{r|rrr} 2 & 40 & 60 & 105 \\ \hline 2 & 20 & 30 & 105 \\ \hline 3 & 10 & 15 & 105 \\ \hline 5 & 10 & 5 & 35 \\ \hline & 2 & 1 & 7 \end{array}$$
$\text{LCM} = 2 \cdot 2 \cdot 3 \cdot 5 \cdot 2 \cdot 1 \cdot 7 = 840$

118.
$$\begin{array}{r|rrr} 2 & 55 & 66 & 90 \\ \hline 3 & 55 & 33 & 45 \\ \hline 5 & 55 & 11 & 15 \\ \hline 11 & 11 & 11 & 3 \\ \hline & 1 & 1 & 3 \end{array}$$
$\text{LCM} = 2 \cdot 3 \cdot 5 \cdot 11 \cdot 1 \cdot 1 \cdot 3 = 990$

119. **Answers may vary.**
Examples: 2 and 49, 14 and 49

120. **Answers may vary.**
Examples: 2 and 51, 3 and 34

Chapter 2 True/False Concept Review (page 162)

1. false; 15 is a multiple of 3.

2. true

3. true

4. true

5. true

6. true

7. false; 300 is the only multiple of 300 that is also a factor.

8. false; $25^2 = 625$

9. false; 26 is not divisible by 6.

10. true

11. false; 29 is not divisible by 3.

12. true

13. false; $7 + 7 + 7 + 7 + 3 = 31$

14. true

15. true

16. false; 2 is prime and even.

17. true

18. false; Every prime number has exactly 2 factors.

19. true

20. true

21. false; The LCM of 2, 3, and 4 is 12, while their product is $2 \cdot 3 \cdot 4 = 24$.

22. true

23. false; The LCM of 2, 3, and 4 is 12. The largest divisor of 12 is 12, not 4.

24. true

Chapter 2 Test (page 163)

1. The last digit of 411,234 is even, so it is divisible by 2. The sum of its digits is 15, so it is divisible by 3. It is divisible by 6.

2. 112: 1, 2, 4, 7, 8, 14, 16, 28, 56, 112

3.

$$\begin{array}{r} 1231 \\ 7\overline{)8617} \\ \underline{7} \\ 16 \\ \underline{14} \\ 21 \\ \underline{21} \\ 07 \\ \underline{7} \\ 0 \end{array}$$

8617 is divisible by 7.

4. The sum of the digits of 2,030,000 is 5, so it is not divisible by 3.

5.

$$\begin{array}{r|ll} 2 & 12 & 32 \\ \hline 2 & 6 & 16 \\ \hline & 3 & 8 \end{array}$$

$\text{LCM} = 2 \cdot 2 \cdot 3 \cdot 8 = 96$

6. $75 = 1 \cdot 75 = 3 \cdot 25 = 5 \cdot 15$

CHAPTER 2 TEST

7.

$$\begin{array}{r|r} 2 & 2\,8\,0 \\ \hline 2 & 1\,4\,0 \\ \hline 2 & 7\,0 \\ \hline 5 & 3\,5 \\ \hline 7 & 7 \\ \hline & 1 \end{array}$$

$280 = 2^3 \times 5 \times 7$

8.

$$\begin{array}{r|rrr} 2 & 18 & 42 & 84 \\ \hline 3 & 9 & 21 & 42 \\ \hline 7 & 3 & 7 & 14 \\ \hline & 3 & 1 & 2 \end{array}$$

$\text{LCM} = 2 \cdot 3 \cdot 7 \cdot 3 \cdot 1 \cdot 2 = 252$

9. The last digit of 15,075 is 5, so it is divisible by 5. The sum of its digits is 18, so it is divisible by 3. It is divisible by 15.

10. 208, 221, 234, 247

11. 200 is a factor of 400, but not a multiple.

12. $11^2 = 121 \geq 109$

$\not{2}, \not{3}, \not{5}, \not{7}, \not{11};$ 109 is prime.

13. $111 = 3 \cdot 37 \Rightarrow 111$ is composite.

14.

$$\begin{array}{r|r} 5 & 6\,0\,5 \\ \hline 11 & 1\,2\,1 \\ \hline 11 & 1\,1 \\ \hline & 1 \end{array}$$

$605 = 5 \times 11^2$

15.

$$\begin{array}{r|rrr} 2 & 18 & 21 & 56 \\ \hline 3 & 9 & 21 & 28 \\ \hline 7 & 3 & 7 & 28 \\ \hline & 3 & 1 & 4 \end{array}$$

$\text{LCM} = 2 \cdot 3 \cdot 7 \cdot 3 \cdot 1 \cdot 4 = 504$

16. 2 is the smallest prime number.

17. 299 is the largest composite number less than 300.

18.

$$\begin{array}{r|rrrr} 2 & 6 & 18 & 24 & 30 \\ \hline 3 & 3 & 9 & 12 & 15 \\ \hline & 1 & 3 & 4 & 5 \end{array}$$

$\text{LCM} = 2 \cdot 3 \cdot 1 \cdot 3 \cdot 4 \cdot 5 = 360$

19. No. A set of prime factors can only have one product.

20. **Answers may vary.**
Examples: 2, 3, and 7; 6, 7, and 42

Exercises 3.1 (page 175)

1. The whole is divided into 8 parts.
5 of the parts are shaded. $\Rightarrow \frac{5}{8}$

3. The whole is divided into 5 parts.
2 of the parts are shaded. $\Rightarrow \frac{2}{5}$

5. The unit is divided into 5 parts.
4 of the parts are measured. $\Rightarrow \frac{4}{5}$

7. The whole is divided into 4 parts.
5 of the parts are shaded. $\Rightarrow \frac{5}{4}$

9. The whole is divided into 8 parts.
11 of the parts are shaded. $\Rightarrow \frac{11}{8}$

11. The whole is divided into 11 parts.
14 of the parts are shaded. $\Rightarrow \frac{14}{11}$

13. proper: $\frac{2}{13}, \frac{4}{13}, \frac{5}{13}, \frac{7}{13}, \frac{10}{13}$
improper: $\frac{13}{13}, \frac{15}{13}, \frac{17}{13}$

15. proper: none
improper: $\frac{17}{16}, \frac{18}{17}, \frac{29}{19}, \frac{30}{21}, \frac{23}{23}$

17. proper: $\frac{8}{9}, \frac{13}{15}, \frac{6}{10}$
improper: $\frac{12}{10}, \frac{10}{9}, \frac{19}{19}, \frac{16}{15}$

19. proper: $\frac{11}{12}, \frac{29}{30}$
improper: $\frac{3}{3}, \frac{6}{5}, \frac{8}{8}, \frac{15}{14}, \frac{19}{19}, \frac{21}{20}, \frac{24}{23}, \frac{32}{31}$

21. $$\begin{array}{r} 3 \\ 6\overline{)23} \\ 18 \\ \hline 5 \end{array} \Rightarrow 3\frac{5}{6}$$

23. $$\begin{array}{r} 12 \\ 7\overline{)89} \\ 7 \\ \hline 19 \\ 14 \\ \hline 5 \end{array} \Rightarrow 12\frac{5}{7}$$

25. $$\begin{array}{r} 11 \\ 11\overline{)123} \\ 121 \\ \hline 2 \end{array} \Rightarrow 11\frac{2}{11}$$

27. $$\begin{array}{r} 6 \\ 13\overline{)89} \\ 78 \\ \hline 11 \end{array} \Rightarrow 6\frac{11}{13}$$

29. $$\begin{array}{r} 15 \\ 27\overline{)413} \\ 27 \\ \hline 143 \\ 135 \\ \hline 8 \end{array} \Rightarrow 15\frac{8}{27}$$

31. $$\begin{array}{r} 22 \\ 15\overline{)331} \\ 30 \\ \hline 31 \\ 30 \\ \hline 1 \end{array} \Rightarrow 22\frac{1}{15}$$

33. $5\frac{5}{7} = \frac{7(5)+5}{7} = \frac{35+5}{7} = \frac{40}{7}$

35. $15 = \frac{15}{1}$

37. $8\frac{4}{9} = \frac{9(8)+4}{9} = \frac{72+4}{9} = \frac{76}{9}$

39. $43\frac{2}{5} = \frac{5(43)+2}{5} = \frac{215+2}{5} = \frac{217}{5}$

41. $42\frac{5}{7} = \frac{7(42)+5}{7} = \frac{294+5}{7} = \frac{299}{7}$

43. $61\frac{5}{8} = \frac{8(61)+5}{8} = \frac{488+5}{8} = \frac{493}{8}$

45. Any whole number (but not 0) can be put in the denominator.

47. $4\frac{2}{3} = \frac{3(4)+2}{3} = \frac{12+2}{3} = \frac{14}{3}$
The error was multiplying the numerator and denominator by the 4.

49. The unit is divided into 10 parts.
7 of the parts are measured. $\Rightarrow \frac{7}{10}$

51. The unit is divided into 8 parts.
11 of the parts are measured. $\Rightarrow \frac{11}{8}$

53. **Answers may vary.**

55. Of the homes, $\frac{23}{35}$ have shake roofs.

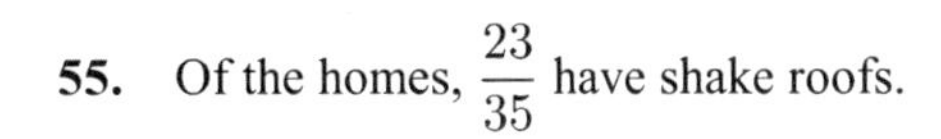

57. $11\frac{1}{2} = \frac{2(11)+1}{2} = \frac{22+1}{2} = \frac{23}{2}$; $6\frac{1}{8} = \frac{8(6)+1}{8} = \frac{48+1}{8} = \frac{49}{8}$; $\frac{1}{4}$ cannot be written as an improper fraction.

59. The unit is divided into 8 parts.
3 of the parts are measured. $\Rightarrow \frac{3}{8}$

61. $12\overline{)85}$ gives quotient 7: $85 - 84 = 1$ $\Rightarrow 7\frac{1}{12}$ The whole number 7 is the closest mark.

63. $34\frac{2}{3} = \frac{3(34)+2}{3} = \frac{102+2}{3} = \frac{104}{3} \Rightarrow$ She can get 104 strips from the wire.

65. $510 - 153 = 357$ bales still in the field. There are $\frac{357}{510}$ of the bales still in the field.

67. The unit is divided into 8 parts.
5 of the parts are measured. $\Rightarrow \frac{5}{8}$

69. This is an improper fraction because the numerator is greater than the denominators.

71. To change $\frac{34}{5}$ to a mixed number, divide 34 by 5. Write the quotient 6 and then the remainder 4 over the divisor 5 $\Rightarrow 6\frac{4}{5}$. To change $7\frac{3}{8}$ to an improper fraction, multiply the denominator 8 by the whole number 7 and then add the numerator 3 to get 59. Write this over 8. $\Rightarrow \frac{59}{8}$

73. A mixed number has a value of 1 or more, while a proper fraction has a value less than 1.

75. $144 \div 16 = 9 \Rightarrow 16 = \frac{144}{9}$
$144 \times 16 = 2304 \Rightarrow 16 = \frac{2304}{144}$

77. $21\frac{3}{4} = \frac{4(21)+3}{4} = \frac{84+3}{4} = \frac{87}{4}$
There are 87 pieces of rope.
$\frac{87}{15} = 5\frac{12}{15} \Rightarrow$ There will be 12 pieces left.

79. **Answers may vary.**

81.
$$\begin{array}{r} 18 \\ 4\overline{)72} \\ \underline{4} \\ 32 \\ \underline{32} \\ 0 \end{array}$$

83.
$$\begin{array}{r} 15 \\ 26\overline{)390} \\ \underline{26} \\ 130 \\ \underline{130} \\ 0 \end{array}$$

85. $21^2 = 441 \geq 419$;
$\cancel{2}, \cancel{3}, \cancel{5}, \cancel{7}, \cancel{11}, \cancel{13}, \cancel{17}, \cancel{19}$
419 is prime.

87. $414 = 2 \times 3^2 \times 23$

89. $(208 + 198 + 214 + 192) \div 4 = 812 \div 4 = 203$; $203 \div 4 = 50$ R 3
She averaged 203 miles per tank and 51 miles per gallon.

Exercises 3.2 (page 184)

1. $\frac{8}{12} = \frac{2 \cdot \cancel{4}}{3 \cdot \cancel{4}} = \frac{2}{3}$

3. $\frac{12}{18} = \frac{2 \cdot \cancel{6}}{3 \cdot \cancel{6}} = \frac{2}{3}$

5. $\frac{18}{20} = \frac{9 \cdot \cancel{2}}{10 \cdot \cancel{2}} = \frac{9}{10}$

7. $\frac{45}{60} = \frac{3 \cdot \cancel{15}}{4 \cdot \cancel{15}} = \frac{3}{4}$

9. $\frac{32}{80} = \frac{2 \cdot \cancel{16}}{5 \cdot \cancel{16}} = \frac{2}{5}$

11. $\frac{9}{30} = \frac{3 \cdot \cancel{3}}{10 \cdot \cancel{3}} = \frac{3}{10}$

13. $\frac{35}{40} = \frac{7 \cdot \cancel{5}}{8 \cdot \cancel{5}} = \frac{7}{8}$

15. $\frac{72}{40} = \frac{9 \cdot \cancel{8}}{5 \cdot \cancel{8}} = \frac{9}{5}$

17. $\frac{42}{77} = \frac{6 \cdot \cancel{7}}{11 \cdot \cancel{7}} = \frac{6}{11}$

19. $\frac{65}{91} = \frac{5 \cdot \cancel{13}}{7 \cdot \cancel{13}} = \frac{5}{7}$

21. $\frac{108}{12} = \frac{9 \cdot \cancel{12}}{1 \cdot \cancel{12}} = \frac{9}{1} = 9$

23. $\frac{93}{62} = \frac{3 \cdot \cancel{31}}{2 \cdot \cancel{31}} = \frac{3}{2}$

25. $\frac{14}{42} = \frac{1 \cdot \cancel{14}}{3 \cdot \cancel{14}} = \frac{1}{3}$

27. $\frac{56}{80} = \frac{7 \cdot \cancel{8}}{10 \cdot \cancel{8}} = \frac{7}{10}$

29. $\frac{27}{36} = \frac{3 \cdot \cancel{9}}{4 \cdot \cancel{9}} = \frac{3}{4}$

31. $\frac{35}{48}$: in simplest form

33. $\frac{50}{75} = \frac{2 \cdot \cancel{25}}{3 \cdot \cancel{25}} = \frac{2}{3}$

35. $\frac{80}{75} = \frac{16 \cdot \cancel{5}}{15 \cdot \cancel{5}} = \frac{16}{15}$

37. $\frac{300}{900} = \frac{1 \cdot \cancel{300}}{3 \cdot \cancel{300}} = \frac{1}{3}$

39. $\frac{75}{80} = \frac{15 \cdot \cancel{5}}{16 \cdot \cancel{5}} = \frac{15}{16}$

41. $\frac{72}{96} = \frac{3 \cdot \cancel{24}}{4 \cdot \cancel{24}} = \frac{3}{4}$

43. $\frac{55}{11} = \frac{5 \cdot \cancel{11}}{1 \cdot \cancel{11}} = \frac{5}{1} = 5$

45. $\frac{85}{105} = \frac{17 \cdot \cancel{5}}{21 \cdot \cancel{5}} = \frac{17}{21}$

47. $\frac{72}{140} = \frac{18 \cdot \cancel{4}}{35 \cdot \cancel{4}} = \frac{18}{35}$

49. $\frac{99}{132} = \frac{3 \cdot \cancel{33}}{4 \cdot \cancel{33}} = \frac{3}{4}$

51. $\frac{16}{81}$ is simplified.

53. $\frac{84}{144} = \frac{7 \cdot \cancel{12}}{12 \cdot \cancel{12}} = \frac{7}{12}$

55. $\frac{196}{210} = \frac{14 \cdot \cancel{14}}{15 \cdot \cancel{14}} = \frac{14}{15}$

57. $\frac{546}{910} = \frac{3 \cdot \cancel{182}}{5 \cdot \cancel{182}} = \frac{3}{5}$

59. The whole foot is divided into 12 inches. 4 inches is then $\frac{4}{12} = \frac{1}{3}$ of a foot. $5'4''$ is $5\frac{1}{3}$ feet.

61. 24 saves out of 28 shots $= \frac{24}{28} = \frac{6}{7}$. The goalie saved $\frac{6}{7}$ of the shots. The other team scored 4 points.

63. 2 out of 6 $= \frac{2}{6} = \frac{1}{3}$ of the plugs are fouled.

65. 140 out of 540 paid $\Rightarrow \frac{140}{540} = \frac{7}{27}$ paid. She still owes $\frac{20}{27}$ of her tuition bill.

67. 64 served, 16 turned away $\Rightarrow$ 80 total clients. $\frac{64}{80} = \frac{4}{5}$ of clients were served.

69. 35 bulls $\Rightarrow 140 - 35 = 105$ cows. $\frac{105}{140} = \frac{3}{4}$ of the elk are cows.

71. $\frac{270}{12}\text{ ft} = \frac{45}{2}\text{ ft} = 22\frac{1}{2}\text{ ft}$

73. Find the prime factorizations of the numerator and the denominator and eliminate the common factors. The product of the remaining factors is the simplified fraction.

$$\frac{525}{1125} = \frac{\cancel{3} \cdot \cancel{5} \cdot \cancel{5} \cdot 7}{\cancel{3} \cdot 3 \cdot \cancel{5} \cdot \cancel{5} \cdot 5} = \frac{7}{15}$$

75. All simplify to $\frac{5}{12}$, so all are equivalent.

77.

$$\begin{array}{r} {}^{1} \\ 42 \\ \times\ \ 6 \\ \hline 252 \end{array}$$

79.

$$\begin{array}{r} 99 \\ \times\ 11 \\ \hline 99 \\ 990 \\ \hline 1089 \end{array}$$

81.

$$\begin{array}{r} {}^{3\,3} \\ 444 \\ \times\ \ \ 8 \\ \hline 3552 \end{array}$$

83.

$$\begin{array}{r} 16 \\ 15\overline{)240} \\ 15 \\ \hline 90 \\ 90 \\ \hline 0 \end{array}$$

85.

$$\begin{array}{r} 16 \\ 21\overline{)336} \\ 21 \\ \hline 126 \\ 126 \\ \hline 0 \end{array}$$

Exercises 3.3 (page 194)

1. $\frac{1}{5} \cdot \frac{2}{5} = \frac{1 \cdot 2}{5 \cdot 5} = \frac{2}{25}$

3. $\frac{3}{2} \cdot \frac{5}{14} = \frac{3 \cdot 5}{2 \cdot 14} = \frac{15}{28}$

5. $\frac{6}{11} \cdot \frac{5}{9} = \frac{\overset{2}{\cancel{6}}}{11} \cdot \frac{5}{\underset{3}{\cancel{9}}} = \frac{2 \cdot 5}{11 \cdot 3} = \frac{10}{33}$

7. $\frac{4}{9} \cdot \frac{9}{16} = \frac{\overset{1}{\cancel{4}}}{\underset{1}{\cancel{9}}} \cdot \frac{\overset{1}{\cancel{9}}}{\underset{4}{\cancel{16}}} = \frac{1 \cdot 1}{1 \cdot 4} = \frac{1}{4}$

SECTION 3.3

9. $14 \cdot \frac{8}{35} = \frac{\overset{2}{\cancel{14}}}{1} \cdot \frac{8}{\underset{5}{\cancel{35}}} = \frac{2 \cdot 8}{1 \cdot 5} = \frac{16}{5}$, or $3\frac{1}{5}$

11. $\frac{3}{16} \cdot \frac{4}{9} \cdot \frac{5}{8} = \frac{\overset{1}{\cancel{3}}}{\underset{4}{\cancel{16}}} \cdot \frac{\overset{1}{\cancel{4}}}{\underset{3}{\cancel{9}}} \cdot \frac{5}{8} = \frac{1}{4} \cdot \frac{1}{3} \cdot \frac{5}{8} = \frac{5}{96}$

13. $8 \cdot \frac{1}{5} \cdot \frac{15}{16} = \frac{\overset{1}{\cancel{8}}}{1} \cdot \frac{1}{\underset{1}{\cancel{5}}} \cdot \frac{\overset{3}{\cancel{15}}}{\underset{2}{\cancel{16}}} = \frac{1 \cdot 1 \cdot 3}{1 \cdot 1 \cdot 2} = \frac{3}{2}$

$= 1\frac{1}{2}$

15. $\frac{11}{3} \cdot \frac{3}{4} \cdot \frac{4}{11} = \frac{\overset{1}{\cancel{11}}}{\underset{1}{\cancel{3}}} \cdot \frac{\overset{1}{\cancel{3}}}{\underset{1}{\cancel{4}}} \cdot \frac{\overset{1}{\cancel{4}}}{\underset{1}{\cancel{11}}} = \frac{1 \cdot 1 \cdot 1}{1 \cdot 1 \cdot 1}$

$= \frac{1}{1} = 1$

17. $\frac{\overset{2}{\cancel{32}}}{\underset{5}{\cancel{55}}} \cdot \frac{\overset{4}{\cancel{44}}}{\underset{3}{\cancel{51}}} \cdot \frac{\overset{2}{\cancel{34}}}{\underset{1}{\cancel{16}}} = \frac{2 \cdot 4 \cdot 2}{5 \cdot 3 \cdot 1} = \frac{16}{15} = 1\frac{1}{15}$

19. $\frac{7}{6} \cdot \frac{8}{23} \cdot \frac{5}{8} \cdot \frac{0}{9} = \frac{7 \cdot 8 \cdot 5 \cdot 0}{6 \cdot 23 \cdot 8 \cdot 9} = \frac{0}{?} = 0$

21. The reciprocal of $\frac{8}{13}$ is $\frac{13}{8}$.

23. The reciprocal of $21 = \frac{21}{1}$ is $\frac{1}{21}$.

25. $\frac{2}{0}$ is undefined, so it has no reciprocal.

27. The reciprocal of $5\frac{1}{12} = \frac{61}{12}$ is $\frac{12}{61}$.

29. The reciprocal of $\frac{1}{17}$ is $\frac{17}{1} = 17$.

31. $\frac{7}{20} \div \frac{14}{15} = \frac{\overset{1}{\cancel{7}}}{\underset{4}{\cancel{20}}} \cdot \frac{\overset{3}{\cancel{15}}}{\underset{2}{\cancel{14}}} = \frac{1 \cdot 3}{4 \cdot 2} = \frac{3}{8}$

33. $\frac{4}{7} \div \frac{3}{8} = \frac{4}{7} \cdot \frac{8}{3} = \frac{32}{21}$ or $1\frac{11}{21}$

35. $\frac{8}{25} \div \frac{16}{25} = \frac{8}{25} \cdot \frac{25}{16} = \frac{\overset{1}{\cancel{8}}}{\underset{1}{\cancel{25}}} \cdot \frac{\overset{1}{\cancel{25}}}{\underset{2}{\cancel{16}}} = \frac{1 \cdot 1}{1 \cdot 2} = \frac{1}{2}$

37. $\frac{7}{44} \div \frac{14}{33} = \frac{7}{44} \cdot \frac{33}{14} = \frac{\overset{1}{\cancel{7}}}{\underset{4}{\cancel{44}}} \cdot \frac{\overset{3}{\cancel{33}}}{\underset{2}{\cancel{14}}} = \frac{1 \cdot 3}{4 \cdot 2}$

$= \frac{3}{8}$

39. $\frac{5}{8} \div \frac{15}{32} = \frac{5}{8} \cdot \frac{32}{15} = \frac{\overset{1}{\cancel{5}}}{\underset{1}{\cancel{8}}} \cdot \frac{\overset{4}{\cancel{32}}}{\underset{3}{\cancel{15}}} = \frac{1 \cdot 4}{1 \cdot 3}$

$= \frac{4}{3}$ or $1\frac{1}{3}$

41. $\frac{21}{40} \div \frac{9}{28} = \frac{21}{40} \cdot \frac{28}{9} = \frac{\overset{7}{\cancel{21}}}{\underset{10}{\cancel{40}}} \cdot \frac{\overset{7}{\cancel{28}}}{\underset{3}{\cancel{9}}} = \frac{7 \cdot 7}{10 \cdot 3} = \frac{49}{30}$, or $1\frac{19}{30}$

43. $\frac{28}{30} \div \frac{14}{15} = \frac{28}{30} \cdot \frac{15}{14} = \frac{\overset{2}{\cancel{28}}}{\underset{2}{\cancel{30}}} \cdot \frac{\overset{1}{\cancel{15}}}{\underset{1}{\cancel{14}}} = \frac{\overset{1}{\cancel{2}} \cdot 1}{\underset{1}{\cancel{2}} \cdot 1}$

$= \frac{1}{1} = 1$

45. $\frac{6}{5} \div \frac{9}{25} = \frac{6}{5} \cdot \frac{25}{9} = \frac{\overset{2}{\cancel{6}}}{\underset{1}{\cancel{5}}} \cdot \frac{\overset{5}{\cancel{25}}}{\underset{3}{\cancel{9}}} = \frac{2 \cdot 5}{1 \cdot 3}$

$= \frac{10}{3}$ or $3\frac{1}{3}$

47. $\frac{90}{55} \div \frac{9}{5} = \frac{90}{55} \cdot \frac{5}{9} = \frac{\overset{10}{\cancel{90}}}{\underset{11}{\cancel{55}}} \cdot \frac{\overset{1}{\cancel{5}}}{\underset{1}{\cancel{9}}} = \frac{10 \cdot 1}{11 \cdot 1} = \frac{10}{11}$

49. $\frac{63}{80} \div \frac{9}{10} = \frac{63}{80} \cdot \frac{10}{9} = \frac{\overset{7}{\cancel{63}}}{\underset{8}{\cancel{80}}} \cdot \frac{\overset{1}{\cancel{10}}}{\underset{1}{\cancel{9}}} = \frac{7 \cdot 1}{8 \cdot 1} = \frac{7}{8}$

SECTION 3.3

51. $\frac{56}{72} \cdot \frac{35}{80} \cdot \frac{6}{14} = \frac{\overset{7}{\cancel{56}}}{\underset{9}{\cancel{72}}} \cdot \frac{\overset{7}{\cancel{35}}}{\underset{16}{\cancel{80}}} \cdot \frac{\overset{3}{\cancel{6}}}{\underset{7}{\cancel{14}}} = \frac{7}{\underset{3}{\cancel{9}}} \cdot \frac{\overset{1}{\cancel{7}}}{16} \cdot \frac{\overset{1}{\cancel{3}}}{\underset{1}{\cancel{7}}} = \frac{7 \cdot 1 \cdot 1}{3 \cdot 16 \cdot 1} = \frac{7}{48}$

53. $\frac{124}{160} \cdot \frac{40}{62} \cdot \frac{44}{11} = \frac{\overset{31}{\cancel{124}}}{\underset{40}{\cancel{160}}} \cdot \frac{\overset{20}{\cancel{40}}}{\underset{31}{\cancel{62}}} \cdot \frac{\overset{4}{\cancel{44}}}{\underset{1}{\cancel{11}}} = \frac{\overset{1}{\cancel{31}}}{\underset{2}{\cancel{40}}} \cdot \frac{\overset{1}{\cancel{20}}}{\underset{1}{\cancel{31}}} \cdot \frac{4}{1} = \frac{1}{\underset{1}{\cancel{2}}} \cdot \frac{1}{1} \cdot \frac{\overset{2}{\cancel{4}}}{1} = \frac{1 \cdot 1 \cdot 2}{1 \cdot 1 \cdot 1} = \frac{2}{1} = 2$

55. $\frac{22}{140} \div \frac{55}{56} = \frac{22}{140} \cdot \frac{56}{55} = \frac{\overset{2}{\cancel{22}}}{\underset{35}{\cancel{140}}} \cdot \frac{\overset{14}{\cancel{56}}}{\underset{5}{\cancel{55}}} = \frac{2 \cdot \overset{2}{\cancel{14}}}{\underset{5}{\cancel{35}} \cdot 5} = \frac{4}{25}$

57. $\frac{16}{81} \div \frac{8}{108} = \frac{16}{81} \cdot \frac{108}{8} = \frac{\overset{2}{\cancel{16}}}{\underset{3}{\cancel{81}}} \cdot \frac{\overset{4}{\cancel{108}}}{\underset{1}{\cancel{8}}} = \frac{2 \cdot 4}{3 \cdot 1} = \frac{8}{3}$, or $2\frac{2}{3}$

59. $\frac{3}{4} \cdot \frac{\boxed{}}{6} = \frac{\overset{1}{\cancel{3}}}{4} \cdot \frac{\boxed{}}{\underset{2}{\cancel{6}}} = \frac{7}{8}$; $\frac{1}{4} \cdot \frac{\boxed{7}}{2} = \frac{7}{8}$

61. $\frac{3}{8} \cdot \frac{4}{5} = \frac{3}{\underset{2}{\cancel{8}}} \cdot \frac{\overset{1}{\cancel{4}}}{5} = \frac{3}{10}$; The error came from inverting before multiplying, which is done only when dividing.

63. $\frac{1}{2}$ of $16 = \frac{1}{2} \cdot 16 = \frac{1}{2} \cdot \frac{16}{1} = \frac{8}{1} = 8$
8 combinations have a boy as firstborn, and 8 have a girl as firstborn.

65. $\frac{1}{4}$ of $16 = \frac{1}{4} \cdot 16 = \frac{1}{4} \cdot \frac{16}{1} = \frac{4}{1} = 4$
$\frac{1}{4}$ of $4 = \frac{1}{4} \cdot 4 = \frac{1}{4} \cdot \frac{4}{1} = \frac{1}{1} = 1$
One combination has 1 boy, then 3 girls.

67. $\frac{3}{5}$ of $100 = \frac{3}{5} \cdot \frac{100}{1} = 60$. When it is three-fifths full, it holds 60 ounces..

69. $\frac{1}{20}$ of $820 = \frac{1}{20} \cdot \frac{820}{1} = 41$
Becky might save 41 gallons per year.

71. $\frac{6}{100}$ of $200{,}000 = \frac{6}{100} \cdot 200{,}000$
$= \frac{6}{100} \cdot \frac{200{,}000}{1}$
$= 12{,}000$
About 12,000 of the patients will survive for 5 years after diagnosis.

73. $\frac{73}{100}$ of $109{,}955{,}000 = \frac{73}{100} \cdot 109{,}955{,}000$
$= \frac{73}{100} \cdot \frac{109{,}955{,}000}{1}$
$= 80{,}267{,}150$
$109{,}955{,}000 - 80{,}267{,}150 = 29{,}687{,}850$
About 29,687,850 Mexicans did not have access to safe drinking water.

75. Multiply $\frac{35}{24}$ by the reciprocal of $\frac{40}{14}$. $\left(\text{Multiply } \frac{35}{24} \text{ by } \frac{14}{40}.\right)$ Then simplify.

77. $\left(\frac{81}{75}\cdot\frac{96}{99}\cdot\frac{55}{125}\right)\div\frac{128}{250}=\left(\frac{81}{75}\cdot\frac{96}{99}\cdot\frac{55}{125}\right)\cdot\frac{250}{128}=\left(\frac{\overset{9}{\cancel{81}}}{\underset{15}{\cancel{75}}}\cdot\frac{96}{\underset{11}{\cancel{99}}}\cdot\frac{\overset{11}{\cancel{55}}}{125}\right)\cdot\frac{250}{128}$

$=\left(\frac{\overset{3}{\cancel{9}}}{\underset{5}{\cancel{15}}}\cdot\frac{96}{\underset{1}{\cancel{11}}}\cdot\frac{\overset{1}{\cancel{11}}}{125}\right)\cdot\frac{250}{128}=\frac{288}{625}\cdot\frac{250}{128}=\frac{\overset{9}{\cancel{288}}}{\underset{5}{\cancel{625}}}\cdot\frac{\overset{2}{\cancel{250}}}{\underset{4}{\cancel{128}}}=\frac{9}{5}\cdot\frac{\overset{1}{\cancel{2}}}{\underset{2}{\cancel{4}}}=\frac{9}{10}$

79. $\begin{array}{r} 4 \\ 5\overline{)\,22} \\ \underline{20} \\ 2 \end{array}\Rightarrow 4\frac{2}{5}$

81. $\begin{array}{r} 3 \\ 13\overline{)\,45} \\ \underline{39} \\ 6 \end{array}\Rightarrow 3\frac{6}{13}$

83. $3\frac{3}{8}=\frac{8(3)+3}{8}=\frac{27}{8}$

85. $7\frac{4}{9}=\frac{9(7)+4}{9}=\frac{67}{9}$

87. $\begin{array}{r} 15 \\ 222\overline{)\,3500} \\ \underline{222} \\ 1280 \\ \underline{1110} \\ 170 \end{array}$ Kevin can haul 15 scoops.

Exercises 3.4 (page 201)

1. $\left(\frac{2}{5}\right)\left(1\frac{2}{5}\right)=\frac{2}{5}\cdot\frac{7}{5}=\frac{14}{25}$

3. $\left(2\frac{3}{8}\right)\left(3\frac{1}{4}\right)=\frac{19}{8}\cdot\frac{13}{4}=\frac{247}{32}=7\frac{23}{32}$

5. $2\left(3\frac{2}{3}\right)=\frac{2}{1}\cdot\frac{11}{3}=\frac{22}{3}=7\frac{1}{3}$

7. $\left(3\frac{3}{5}\right)\left(1\frac{1}{4}\right)(5)=\frac{18}{5}\cdot\frac{5}{4}\cdot\frac{5}{1}=\frac{\overset{9}{\cancel{18}}}{\underset{1}{\cancel{5}}}\cdot\frac{\overset{1}{\cancel{5}}}{\underset{2}{\cancel{4}}}\cdot\frac{5}{1}$

$=\frac{45}{2}=22\frac{1}{2}$

9. $\left(7\frac{1}{2}\right)\left(3\frac{1}{5}\right)=\frac{15}{2}\cdot\frac{16}{5}=\frac{\overset{3}{\cancel{15}}}{\underset{1}{\cancel{2}}}\cdot\frac{\overset{8}{\cancel{16}}}{\underset{1}{\cancel{5}}}$

$=\frac{24}{1}=24$

11. $\left(4\frac{3}{8}\right)\left(\frac{16}{25}\right)=\frac{35}{8}\cdot\frac{16}{25}=\frac{\overset{7}{\cancel{35}}}{\underset{1}{\cancel{8}}}\cdot\frac{\overset{2}{\cancel{16}}}{\underset{5}{\cancel{25}}}$

$=\frac{14}{5}=2\frac{4}{5}$

13. $5\left(3\frac{3}{4}\right)=\frac{5}{1}\cdot\frac{15}{4}=\frac{75}{4}=18\frac{3}{4}$

15. $\left(7\frac{3}{4}\right)\left(\frac{2}{3}\right)(0)=0$

17. $2\left(4\frac{3}{8}\right)\left(3\frac{1}{5}\right)=\frac{2}{1}\cdot\frac{35}{8}\cdot\frac{16}{5}=\frac{2}{1}\cdot\frac{\overset{7}{\cancel{35}}}{\underset{1}{\cancel{8}}}\cdot\frac{\overset{2}{\cancel{16}}}{\underset{1}{\cancel{5}}}=\frac{28}{1}=28$

SECTION 3.4

19. $\left(2\frac{2}{3}\right)\left(\frac{11}{12}\right)\left(6\frac{3}{4}\right) = \frac{8}{3}\cdot\frac{11}{12}\cdot\frac{27}{4} = \frac{\overset{2}{\cancel{8}}}{\underset{1}{\cancel{3}}}\cdot\frac{11}{12}\cdot\frac{\overset{9}{\cancel{27}}}{\underset{1}{\cancel{4}}} = \frac{\overset{1}{\cancel{2}}}{1}\cdot\frac{11}{\underset{6}{\cancel{12}}}\cdot\frac{9}{1} = \frac{1}{1}\cdot\frac{11}{\underset{2}{\cancel{6}}}\cdot\frac{\overset{3}{\cancel{9}}}{1} = \frac{33}{2} = 16\frac{1}{2}$

21. $\left(1\frac{1}{5}\right)\left(4\frac{1}{3}\right)\left(6\frac{1}{6}\right) = \frac{6}{5}\cdot\frac{13}{3}\cdot\frac{37}{6} = \frac{\overset{1}{\cancel{6}}}{5}\cdot\frac{13}{3}\cdot\frac{37}{\underset{1}{\cancel{6}}} = \frac{481}{15} = 32\frac{1}{15}$

23. $(14)\left(6\frac{1}{2}\right)\left(1\frac{2}{13}\right) = \frac{14}{1}\cdot\frac{13}{2}\cdot\frac{15}{13} = \frac{\overset{7}{\cancel{14}}}{1}\cdot\frac{\overset{1}{\cancel{13}}}{\underset{1}{\cancel{2}}}\cdot\frac{15}{\underset{1}{\cancel{13}}} = \frac{105}{1} = 105$

25. $6 \div 2\frac{1}{3} = \frac{6}{1} \div \frac{7}{3} = \frac{6}{1}\cdot\frac{3}{7} = \frac{18}{7} = 2\frac{4}{7}$

27. $3\frac{5}{9} \div 2\frac{1}{18} = \frac{32}{9} \div \frac{37}{18} = \frac{32}{9}\cdot\frac{18}{37} = \frac{32}{\underset{1}{\cancel{9}}}\cdot\frac{\overset{2}{\cancel{18}}}{37} = \frac{64}{37} = 1\frac{27}{37}$

29. $5\frac{2}{3} \div 1\frac{1}{9} = \frac{17}{3} \div \frac{10}{9} = \frac{17}{3}\cdot\frac{9}{10} = \frac{17}{\underset{1}{\cancel{3}}}\cdot\frac{\overset{3}{\cancel{9}}}{10} = \frac{51}{10} = 5\frac{1}{10}$

31. $2\frac{1}{4} \div 1\frac{5}{8} = \frac{9}{4} \div \frac{13}{8} = \frac{9}{4}\cdot\frac{8}{13}$

$= \frac{9}{\underset{1}{\cancel{4}}}\cdot\frac{\overset{2}{\cancel{8}}}{13} = \frac{18}{13} = 1\frac{5}{13}$

33. $3\frac{1}{3} \div 2\frac{1}{2} = \frac{10}{3} \div \frac{5}{2} = \frac{10}{3}\cdot\frac{2}{5}$

$= \frac{\overset{2}{\cancel{10}}}{3}\cdot\frac{2}{\underset{1}{\cancel{5}}} = \frac{4}{3} = 1\frac{1}{3}$

35. $4\frac{4}{15} \div 6\frac{2}{3} = \frac{64}{15} \div \frac{20}{3} = \frac{64}{15}\cdot\frac{3}{20}$

$= \frac{\overset{16}{\cancel{64}}}{\underset{5}{\cancel{15}}}\cdot\frac{\overset{1}{\cancel{3}}}{\underset{5}{\cancel{20}}} = \frac{16}{25}$

37. $\frac{5}{6} \div 4\frac{1}{3} = \frac{5}{6} \div \frac{13}{3} = \frac{5}{6}\cdot\frac{3}{13} = \frac{5}{\underset{2}{\cancel{6}}}\cdot\frac{\overset{1}{\cancel{3}}}{13}$

$= \frac{5}{26}$

39. $8\frac{2}{3} \div 6 = \frac{26}{3} \div \frac{6}{1} = \frac{26}{3}\cdot\frac{1}{6}$

$= \frac{\overset{13}{\cancel{26}}}{3}\cdot\frac{1}{\underset{3}{\cancel{6}}} = \frac{13}{9} = 1\frac{4}{9}$

41. $4\frac{5}{6} \div \frac{1}{3} = \frac{29}{6} \div \frac{1}{3} = \frac{29}{6}\cdot\frac{3}{1} = \frac{29}{\underset{2}{\cancel{6}}}\cdot\frac{\overset{1}{\cancel{3}}}{1}$

$= \frac{29}{2} = 14\frac{1}{2}$

43. $8\frac{1}{10} \div \frac{3}{5} = \frac{81}{10} \div \frac{3}{5} = \frac{81}{10}\cdot\frac{5}{3} = \frac{\overset{27}{\cancel{81}}}{\underset{2}{\cancel{10}}}\cdot\frac{\overset{1}{\cancel{5}}}{\underset{1}{\cancel{3}}} = \frac{27}{2} = 13\frac{1}{2}$

SECTION 3.4

45. $23\frac{1}{4} \div 1\frac{1}{3} = \frac{93}{4} \div \frac{4}{3} = \frac{93}{4} \cdot \frac{3}{4} = \frac{279}{16} = 17\frac{7}{16}$

47. $8\frac{3}{4} \div 3\frac{4}{7} = \frac{35}{4} \div \frac{25}{7} = \frac{35}{4} \cdot \frac{7}{25} = \frac{\overset{7}{\cancel{35}}}{4} \cdot \frac{7}{\underset{5}{\cancel{25}}} = \frac{49}{20} = 2\frac{9}{20}$

49. $12\left(3\frac{3}{8}\right)\left(2\frac{4}{9}\right)\left(3\frac{2}{11}\right) = \frac{12}{1} \cdot \frac{27}{8} \cdot \frac{22}{9} \cdot \frac{35}{11} = \frac{\overset{3}{\cancel{12}}}{1} \cdot \frac{\overset{3}{\cancel{27}}}{\underset{2}{\cancel{8}}} \cdot \frac{\overset{2}{\cancel{22}}}{\underset{1}{\cancel{9}}} \cdot \frac{35}{\underset{1}{\cancel{11}}} = \frac{3}{1} \cdot \frac{3}{\underset{1}{\cancel{2}}} \cdot \frac{\overset{1}{\cancel{2}}}{1} \cdot \frac{35}{1} = \frac{315}{1} = 315$

51. $\left(1\frac{2}{3}\right) \div \left(3\frac{3}{4}\right) \div \left(1\frac{3}{7}\right) = \frac{5}{3} \div \frac{15}{4} \div \frac{10}{7} = \frac{5}{3} \cdot \frac{4}{15} \cdot \frac{7}{10} = \frac{\overset{1}{\cancel{5}}}{3} \cdot \frac{\overset{2}{\cancel{4}}}{\underset{3}{\cancel{15}}} \cdot \frac{7}{\underset{5}{\cancel{10}}} = \frac{14}{45}$

53. In the statement, the whole number parts and the fractional parts are multiplied separately. This is wrong. Change to improper fractions first:

$1\frac{2}{3} \cdot 1\frac{1}{2} = \frac{5}{3} \cdot \frac{3}{2} = \frac{5}{\underset{1}{\cancel{3}}} \cdot \frac{\overset{1}{\cancel{3}}}{2} = \frac{5}{2} = 2\frac{1}{2}$

55. Area = length · width $= 7\frac{2}{3} \cdot 5\frac{1}{2} = \frac{23}{3} \cdot \frac{11}{2} = \frac{253}{6} = 42\frac{1}{6}$ square ft

57. $700 \div \frac{1}{2} = \frac{700}{1} \cdot \frac{2}{1} =$ \$1400 per carat

$3000 \div 1\frac{1}{2} = \frac{3000}{1} \div \frac{3}{2} = \frac{3000}{1} \cdot \frac{2}{3} =$ \$2000 per carat

59. $2\frac{3}{4} \cdot 8 = \frac{11}{4} \cdot \frac{8}{1} = \frac{11}{\underset{1}{\cancel{4}}} \cdot \frac{\overset{2}{\cancel{8}}}{1} = \frac{22}{1} = 22$ parts per million

61. $7 \div 3\frac{1}{2} = \frac{7}{1} \div \frac{7}{2} = \frac{7}{1} \cdot \frac{2}{7} = \frac{\overset{1}{\cancel{7}}}{1} \cdot \frac{2}{\underset{1}{\cancel{7}}} = \frac{2}{1} =$ \$2 per pound

63. $\frac{2}{5} \cdot 2000 = \frac{2}{5} \cdot \frac{2000}{1} = \frac{2}{\underset{1}{\cancel{5}}} \cdot \frac{\overset{400}{\cancel{2000}}}{1} = \frac{800}{1} = 800$ fewer pounds of CO_2

65. $10\frac{1}{2} \div 3\frac{1}{2} = \frac{21}{2} \div \frac{7}{2} = \frac{21}{2} \cdot \frac{2}{7} = 3$; He will need three 2 × 4s.

67. $272 \div 3\frac{1}{2} = \frac{272}{1} \div \frac{7}{2} = \frac{272}{1} \cdot \frac{2}{7} = \frac{544}{7} = 77\frac{5}{7}$; He will need 78 boards.

69. $1\frac{1}{2}$ is larger than 1. Multiplying by a number larger than 1 results in a larger number, while dividing by a number larger than 1 results in a smaller number.

71. $\left(\frac{2}{15}\right)\left(1\frac{6}{7}\right)\left(2\frac{2}{49}\right)\left(16\frac{2}{5}\right)\left(8\frac{3}{4}\right)\left(5\frac{1}{4}\right)\left(3\frac{3}{13}\right) = \frac{2}{15}\cdot\frac{13}{7}\cdot\frac{100}{49}\cdot\frac{82}{5}\cdot\frac{35}{4}\cdot\frac{21}{4}\cdot\frac{42}{13}$

$$= \frac{\overset{1}{\cancel{2}}}{\underset{3}{\cancel{15}}}\cdot\frac{\overset{1}{\cancel{13}}}{\underset{1}{\cancel{7}}}\cdot\frac{\overset{20}{\cancel{100}}}{\underset{7}{\cancel{49}}}\cdot\frac{\overset{41}{\cancel{82}}}{\underset{1}{\cancel{5}}}\cdot\frac{\overset{7}{\cancel{35}}}{\underset{2}{\cancel{4}}}\cdot\frac{\overset{3}{\cancel{21}}}{\underset{2}{\cancel{4}}}\cdot\frac{\overset{6}{\cancel{42}}}{\underset{1}{\cancel{13}}} = \frac{1}{\underset{1}{\cancel{3}}}\cdot\frac{1}{1}\cdot\frac{\overset{10}{\cancel{20}}}{\underset{1}{\cancel{7}}}\cdot\frac{41}{1}\cdot\frac{\overset{1}{\cancel{7}}}{\underset{1}{\cancel{2}}}\cdot\frac{\overset{1}{\cancel{3}}}{\underset{1}{\cancel{2}}}\cdot\frac{\overset{3}{\cancel{6}}}{1} = \frac{1230}{1} = 1230$$

73.

$$\begin{array}{r} 8 \\ 9\overline{)7\,7} \\ \underline{7\,2} \\ 5 \end{array} \Rightarrow 8\frac{5}{9}$$

75. $3\frac{6}{7} = \frac{7(3)+6}{7} = \frac{27}{7}$

77. $14\frac{7}{9} = \frac{9(14)+7}{9} = \frac{133}{9}$

79. $60\left(\frac{1}{40}\right)\left(\frac{3}{20}\right)\left(\frac{4800}{1}\right) = \frac{60}{1}\cdot\frac{1}{40}\cdot\frac{3}{20}\cdot\frac{4800}{1} = \frac{\overset{3}{\cancel{60}}}{1}\cdot\frac{1}{\underset{1}{\cancel{40}}}\cdot\frac{3}{\underset{1}{\cancel{20}}}\cdot\frac{\overset{120}{\cancel{4800}}}{\underset{1}{\cancel{4}}} = 1080$

81. $4+5+6+7+2+6=30$; 30 is a multiple of 3, so 234,572 is divisible by 3. It is even, so it is also divisible by 2, and thus divisible by 6. Jean is eligible for the drawing.

Getting Ready for Algebra (page 207)

1.

$$\begin{aligned} \frac{2x}{3} &= \frac{1}{2} \\ 3\left(\frac{2x}{3}\right) &= 3\left(\frac{1}{2}\right) \\ 2x &= \frac{3}{2} \\ \frac{2x}{2} &= \frac{3}{2(2)} \\ x &= \frac{3}{4} \end{aligned}$$

3.

$$\begin{aligned} \frac{3y}{5} &= \frac{2}{3} \\ 5\left(\frac{3y}{5}\right) &= 5\left(\frac{2}{3}\right) \\ 3y &= \frac{10}{3} \\ \frac{3y}{3} &= \frac{10}{3(3)} \\ y &= \frac{10}{9} = 1\frac{1}{9} \end{aligned}$$

5.

$$\begin{aligned} \frac{4z}{5} &= \frac{1}{4} \\ 5\left(\frac{4z}{5}\right) &= 5\left(\frac{1}{4}\right) \\ 4z &= \frac{5}{4} \\ \frac{4z}{4} &= \frac{5}{4(4)} \\ z &= \frac{5}{16} \end{aligned}$$

7.

$$\begin{aligned} \frac{17}{9} &= \frac{8x}{9} \\ 9\left(\frac{17}{9}\right) &= 9\left(\frac{8x}{9}\right) \\ 17 &= 8x \\ \frac{17}{8} &= x, \text{ or } x = 2\frac{1}{8} \end{aligned}$$

9.
$$\frac{7a}{4} = \frac{5}{2}$$
$$4\left(\frac{7a}{4}\right) = 4\left(\frac{5}{2}\right)$$
$$7a = \frac{20}{2}$$
$$7a = 10$$
$$\frac{7a}{7} = \frac{10}{7}$$
$$a = \frac{10}{7}, \text{ or } a = 1\frac{3}{7}$$

11.
$$\frac{47}{8} = \frac{47b}{12}$$
$$12\left(\frac{47}{8}\right) = 12\left(\frac{47b}{12}\right)$$
$$\frac{\overset{3}{\cancel{12}}}{1} \cdot \frac{47}{\underset{2}{\cancel{4}}} = 47b$$
$$\frac{141}{2} = 47b$$
$$\frac{141}{47(2)} = \frac{47b}{47}$$
$$\frac{141}{94} = b$$
$$\frac{3}{2} = b, \text{ or } b = 1\frac{1}{2}$$

13.
$$\frac{35z}{6} = \frac{35}{12}$$
$$6\left(\frac{35z}{6}\right) = 6\left(\frac{35}{12}\right)$$
$$35z = \frac{\overset{1}{\cancel{6}}}{1} \cdot \frac{35}{\underset{2}{\cancel{12}}}$$
$$35z = \frac{35}{2}$$
$$\frac{35z}{35} = \frac{35}{35(2)}$$
$$z = \frac{1}{2}$$

15.
$$\frac{15a}{2} = \frac{11}{4}$$
$$2\left(\frac{15a}{2}\right) = 2\left(\frac{11}{4}\right)$$
$$15a = \frac{\overset{1}{\cancel{2}}}{1} \cdot \frac{11}{\underset{2}{\cancel{4}}}$$
$$15a = \frac{11}{2}$$
$$\frac{15a}{15} = \frac{11}{15(2)}$$
$$a = \frac{11}{30}$$

17. Let $x =$ the total distance.
two-thirds of the total distance is $\frac{1}{2}$ mile
$$\frac{2}{3}x = \frac{1}{2}$$
$$\frac{2x}{3} = \frac{1}{2}$$
$$3\left(\frac{2x}{3}\right) = 3\left(\frac{1}{2}\right)$$
$$2x = \frac{3}{2}$$
$$\frac{2x}{2} = \frac{3}{2(2)}$$
$$x = \frac{3}{4}$$
The distance is $\frac{3}{4}$ mile.

19. Let $x =$ pounds of tin recycled. $3\frac{1}{2}$ times the lbs of tin was the lbs of glass
$$3\frac{1}{2}x = 630$$
$$\frac{7}{2}x = 630$$
$$\frac{7x}{2} = 630$$
$$2\left(\frac{7x}{2}\right) = 2(630)$$
$$7x = 1260$$
$$\frac{7x}{7} = \frac{1260}{7}$$
$x = 180$; 180 lbs of tin were recycled.

Exercises 3.5 (page 214)

1. $\frac{2}{3}=\frac{2}{3}\cdot\frac{2}{2}=\frac{4}{6}$
$\frac{2}{3}=\frac{2}{3}\cdot\frac{3}{3}=\frac{6}{9}$
$\frac{2}{3}=\frac{2}{3}\cdot\frac{4}{4}=\frac{8}{12}$
$\frac{2}{3}=\frac{2}{3}\cdot\frac{5}{5}=\frac{10}{15}$

3. $\frac{5}{6}=\frac{5}{6}\cdot\frac{2}{2}=\frac{10}{12}$
$\frac{5}{6}=\frac{5}{6}\cdot\frac{3}{3}=\frac{15}{18}$
$\frac{5}{6}=\frac{5}{6}\cdot\frac{4}{4}=\frac{20}{24}$
$\frac{5}{6}=\frac{5}{6}\cdot\frac{5}{5}=\frac{25}{30}$

5. $\frac{4}{9}=\frac{4}{9}\cdot\frac{2}{2}=\frac{8}{18}$
$\frac{4}{9}=\frac{4}{9}\cdot\frac{3}{3}=\frac{12}{27}$
$\frac{4}{9}=\frac{4}{9}\cdot\frac{4}{4}=\frac{16}{36}$
$\frac{4}{9}=\frac{4}{9}\cdot\frac{5}{5}=\frac{20}{45}$

7. $\frac{11}{13}=\frac{11}{13}\cdot\frac{2}{2}=\frac{22}{26}$
$\frac{11}{13}=\frac{11}{13}\cdot\frac{3}{3}=\frac{33}{39}$
$\frac{11}{13}=\frac{11}{13}\cdot\frac{4}{4}=\frac{44}{52}$
$\frac{11}{13}=\frac{11}{13}\cdot\frac{5}{5}=\frac{55}{65}$

9. $\frac{14}{5}=\frac{14}{5}\cdot\frac{2}{2}=\frac{28}{10}$
$\frac{14}{5}=\frac{14}{5}\cdot\frac{3}{3}=\frac{42}{15}$
$\frac{14}{5}=\frac{14}{5}\cdot\frac{4}{4}=\frac{56}{20}$
$\frac{14}{5}=\frac{14}{5}\cdot\frac{5}{5}=\frac{70}{25}$

11. $16\div 2=8$
$\frac{1}{2}=\frac{1}{2}\cdot\frac{8}{8}=\frac{8}{16}$

13. $42\div 7=6$
$\frac{3}{7}=\frac{3}{7}\cdot\frac{6}{6}=\frac{18}{42}$

15. $35\div 5=7$
$\frac{4}{5}=\frac{4}{5}\cdot\frac{7}{7}=\frac{28}{35}$

17. $34\div 17=2$
$\frac{2}{17}=\frac{2}{17}\cdot\frac{2}{2}=\frac{4}{34}$

19. $90\div 15=6$
$\frac{2}{15}=\frac{2}{15}\cdot\frac{6}{6}=\frac{12}{90}$

21. $12\div 3=4$
$\frac{1}{3}=\frac{1}{3}\cdot\frac{4}{4}=\frac{4}{12}$

23. $21\div 7=3$
$\frac{22}{7}=\frac{22}{7}\cdot\frac{3}{3}=\frac{66}{21}$

25. $400\div 8=50$
$\frac{3}{8}=\frac{3}{8}\cdot\frac{50}{50}=\frac{150}{400}$

27. $72\div 18=4$
$\frac{5}{18}=\frac{5}{18}\cdot\frac{4}{4}=\frac{20}{72}$

29. $144\div 18=8$
$\frac{13}{18}=\frac{13}{18}\cdot\frac{8}{8}=\frac{104}{144}$

31. $\frac{2}{23}<\frac{3}{23}<\frac{5}{23}$

33. $\text{LCM}=8$
$\frac{1}{2}=\frac{1}{2}\cdot\frac{4}{4}=\frac{4}{8}$
$\frac{3}{4}=\frac{3}{4}\cdot\frac{2}{2}=\frac{6}{8}$
$\frac{5}{8}=\frac{5}{8}$
$\frac{1}{2}<\frac{5}{8}<\frac{3}{4}$

35. $\text{LCM}=20$
$\frac{1}{4}=\frac{1}{4}\cdot\frac{5}{5}=\frac{5}{20}$
$\frac{2}{5}=\frac{2}{5}\cdot\frac{4}{4}=\frac{8}{20}$
$\frac{3}{10}=\frac{3}{10}\cdot\frac{2}{2}=\frac{6}{20}$
$\frac{1}{4}<\frac{3}{10}<\frac{2}{5}$

SECTION 3.5

37. $\frac{4}{11} < \frac{3}{11} \Rightarrow$ false

39. LCM = 16
$\frac{11}{16} = \frac{11}{16}$
$\frac{7}{8} = \frac{7}{8} \cdot \frac{2}{2} = \frac{14}{16}$
$\frac{11}{16} > \frac{7}{8} \Rightarrow$ false

41. LCM = 30
$\frac{17}{30} = \frac{17}{30}$
$\frac{11}{15} = \frac{11}{15} \cdot \frac{2}{2} = \frac{22}{30}$
$\frac{17}{30} < \frac{11}{15} \Rightarrow$ true

43. LCM = 36
$\frac{5}{9} = \frac{5}{9} \cdot \frac{4}{4} = \frac{20}{36}$
$\frac{11}{18} = \frac{11}{18} \cdot \frac{2}{2} = \frac{22}{36}$
$\frac{7}{12} = \frac{7}{12} \cdot \frac{3}{3} = \frac{21}{36}$
$\frac{5}{9} < \frac{7}{12} < \frac{11}{18}$

45. LCM = 30
$\frac{13}{15} = \frac{13}{15} \cdot \frac{2}{2} = \frac{26}{30}$
$\frac{4}{15} = \frac{4}{15} \cdot \frac{2}{2} = \frac{8}{30}$
$\frac{5}{6} = \frac{5}{6} \cdot \frac{5}{5} = \frac{25}{30}$
$\frac{9}{10} = \frac{9}{10} \cdot \frac{3}{3} = \frac{27}{30}$
$\frac{4}{15} < \frac{5}{6} < \frac{13}{15} < \frac{9}{10}$

47. LCM = 72
$\frac{11}{24} = \frac{11}{24} \cdot \frac{3}{3} = \frac{33}{72}$
$\frac{17}{36} = \frac{17}{36} \cdot \frac{2}{2} = \frac{34}{72}$
$\frac{35}{72} = \frac{35}{72}$
$\frac{11}{24} < \frac{17}{36} < \frac{35}{72}$

49. LCM = 140
$\frac{13}{28} = \frac{13}{28} \cdot \frac{5}{5} = \frac{65}{140}$
$\frac{17}{35} = \frac{17}{35} \cdot \frac{4}{4} = \frac{68}{140}$
$\frac{3}{7} = \frac{3}{7} \cdot \frac{20}{20} = \frac{60}{140}$
$\frac{3}{7} < \frac{13}{28} < \frac{17}{35}$

51. LCM = 80
$1\frac{9}{16} = \frac{25}{16} \cdot \frac{5}{5} = \frac{125}{80}$
$1\frac{13}{20} = \frac{33}{20} \cdot \frac{4}{4} = \frac{132}{80}$
$1\frac{5}{8} = \frac{13}{8} \cdot \frac{10}{10} = \frac{130}{80}$
$1\frac{9}{16} < 1\frac{5}{8} < 1\frac{13}{20}$

53. LCM = 60
$\frac{17}{60} = \frac{17}{60}$
$\frac{4}{15} = \frac{4}{15} \cdot \frac{4}{4} = \frac{16}{60}$
$\frac{17}{60} < \frac{4}{15} \Rightarrow$ false

55. LCM = 24
$\frac{11}{12} = \frac{11}{12} \cdot \frac{2}{2} = \frac{22}{24}$
$\frac{7}{8} = \frac{7}{8} \cdot \frac{3}{3} = \frac{21}{24}$
$\frac{11}{12} > \frac{7}{8} \Rightarrow$ true

57. LCM = 160
$\frac{5}{32} = \frac{5}{32} \cdot \frac{5}{5} = \frac{25}{160}$
$\frac{7}{40} = \frac{7}{40} \cdot \frac{4}{4} = \frac{28}{160}$
$\frac{5}{32} < \frac{7}{40} \Rightarrow$ true

59. LCM = 24
$\frac{1}{2} = \frac{1}{2} \cdot \frac{12}{12} = \frac{12}{24}$
$\frac{2}{3} = \frac{2}{3} \cdot \frac{8}{8} = \frac{16}{24}$
$\frac{1}{6} = \frac{1}{6} \cdot \frac{4}{4} = \frac{4}{24}$
$\frac{5}{8} = \frac{5}{8} \cdot \frac{3}{3} = \frac{15}{24}$

61. LCM = 32; $\frac{1}{8} = \frac{1}{8} \cdot \frac{4}{4} = \boxed{\frac{4}{32}}$; $\frac{3}{32} = \boxed{\frac{3}{32}}$; $\frac{5}{16} = \frac{5}{16} \cdot \frac{2}{2} = \boxed{\frac{10}{32}}$; $\frac{3}{8} = \frac{3}{8} \cdot \frac{4}{4} = \boxed{\frac{12}{32}}$;
$\frac{9}{16} = \frac{9}{16} \cdot \frac{2}{2} = \boxed{\frac{18}{32}}$; $\frac{1}{2} = \frac{1}{2} \cdot \frac{16}{16} = \boxed{\frac{16}{32}}$; $\frac{1}{4} = \frac{1}{4} \cdot \frac{8}{8} = \boxed{\frac{8}{32}}$;
From smallest to largest, the strengths are $\frac{3}{32}, \frac{1}{8}, \frac{1}{4}, \frac{5}{16}, \frac{3}{8}, \frac{1}{2}$, and $\frac{9}{16}$.

63. LCM = 150
$\frac{42}{50} = \frac{42}{50} \cdot \frac{3}{3} = \frac{126}{150}$
$\frac{2}{3} = \frac{2}{3} \cdot \frac{50}{50} = \frac{100}{150}$
$\frac{2}{3} < \frac{42}{50}$
More Americans believe in heaven.

65. LCM = 100
$\frac{9}{20} = \frac{9}{20} \cdot \frac{5}{5} = \frac{45}{100}$
$\frac{12}{50} = \frac{12}{50} \cdot \frac{2}{2} = \frac{24}{100}$
$\frac{12}{50} < \frac{9}{20}$
More Americans believe in ghosts.

67. LCM = 16
$3\frac{1}{8} = \frac{25}{8} \cdot \frac{2}{2} = \frac{50}{16}$
$3\frac{3}{16} = \frac{51}{16}$
$3\frac{1}{4} = \frac{13}{4} \cdot \frac{4}{4} = \frac{52}{16}$
$3\frac{1}{8} < 3\frac{3}{16} < 3\frac{1}{4}$
Chang's measurement is heaviest.

69. LCM = 12; $\frac{1}{4} = \frac{1}{4} \cdot \frac{3}{3} = \frac{3}{12}; \frac{7}{12} = \frac{7}{12}; \frac{1}{6} = \frac{1}{6} \cdot \frac{2}{2} = \frac{2}{12}; \frac{1}{6} < \frac{1}{4} < \frac{7}{12}$
Moe owns the largest part, while Curly owns the smallest part.

71. To simplify a fraction is to find a fraction with a smaller numerator and a smaller denominator that is equivalent to the original fraction. To build a fraction is to find a fraction with a larger numerator and a larger denominator that is equivalent to the original fraction.

73. $70 \div 7 = 10; 91 \div 7 = 13; 161 \div 7 = 23; 784 \div 7 = 112; 4067 \div 7 = 581;$
$\frac{5}{7} = \frac{5}{7} \cdot \frac{10}{10} = \boxed{\frac{50}{70}}; \frac{5}{7} = \frac{5}{7} \cdot \frac{13}{13} = \boxed{\frac{65}{91}}; \frac{5}{7} = \frac{5}{7} \cdot \frac{23}{23} = \boxed{\frac{115}{161}}; \frac{5}{7} = \frac{5}{7} \cdot \frac{112}{112} = \boxed{\frac{560}{784}};$
$\frac{5}{7} = \frac{5}{7} \cdot \frac{581}{581} = \boxed{\frac{2905}{4067}}$

75. LCM = 21

77. LCM = 40

79. LCM = 288

81. $96 = 32 \cdot 3 = 2^5 \cdot 3$

83. $4(625) + 2(365) = 2500 + 730$
$= 3230$; The rental cost \$3230.

Exercises 3.6 (page 221)

1. $\frac{4}{17} + \frac{5}{17} = \frac{9}{17}$

3. $\frac{3}{10} + \frac{4}{10} + \frac{1}{10} = \frac{8}{10} = \frac{4}{5}$

5. $\frac{3}{11} + \frac{8}{11} = \frac{11}{11} = 1$

7. $\frac{4}{15} + \frac{7}{15} + \frac{11}{15} = \frac{22}{15} = 1\frac{7}{15}$

9. $\frac{7}{19} + \frac{6}{19} + \frac{3}{19} = \frac{16}{19}$

11. $\frac{5}{12} + \frac{5}{12} + \frac{5}{12} = \frac{15}{12} = \frac{5}{4} = 1\frac{1}{4}$

13. $\frac{3}{20} + \frac{5}{20} + \frac{7}{20} = \frac{15}{20} = \frac{3}{4}$

15. $\frac{3}{18} + \frac{5}{18} + \frac{4}{18} = \frac{12}{18} = \frac{2}{3}$

17. $\frac{7}{30} + \frac{11}{30} + \frac{3}{30} = \frac{21}{30} = \frac{7}{10}$

19. $\frac{2}{35} + \frac{8}{35} + \frac{4}{35} = \frac{14}{35} = \frac{2}{5}$

SECTION 3.6

21. $\frac{1}{6}+\frac{3}{8}=\frac{1}{6}\cdot\frac{4}{4}+\frac{3}{8}\cdot\frac{3}{3}=\frac{4}{24}+\frac{9}{24}=\frac{13}{24}$

23. $\frac{1}{14}+\frac{3}{7}=\frac{1}{14}+\frac{3}{7}\cdot\frac{2}{2}=\frac{1}{14}+\frac{6}{14}=\frac{7}{14}=\frac{1}{2}$

25. $\frac{7}{16}+\frac{3}{8}=\frac{7}{16}+\frac{3}{8}\cdot\frac{2}{2}=\frac{7}{16}+\frac{6}{16}=\frac{13}{16}$

27. $\frac{1}{3}+\frac{1}{6}+\frac{1}{18}=\frac{1}{3}\cdot\frac{6}{6}+\frac{1}{6}\cdot\frac{3}{3}+\frac{1}{18}=\frac{6}{18}+\frac{3}{18}+\frac{1}{18}=\frac{10}{18}=\frac{5}{9}$

29. $\frac{3}{7}+\frac{1}{14}+\frac{1}{2}=\frac{3}{7}\cdot\frac{2}{2}+\frac{1}{14}+\frac{1}{2}\cdot\frac{7}{7}=\frac{6}{14}+\frac{1}{14}+\frac{7}{14}=\frac{14}{14}=1$

31. $\frac{5}{18}+\frac{11}{24}=\frac{5}{18}\cdot\frac{4}{4}+\frac{11}{24}\cdot\frac{3}{3}=\frac{20}{72}+\frac{33}{72}=\frac{53}{72}$

33. $\frac{3}{5}+\frac{9}{10}+\frac{7}{20}=\frac{3}{5}\cdot\frac{4}{4}+\frac{9}{10}\cdot\frac{2}{2}+\frac{7}{20}=\frac{12}{20}+\frac{18}{20}+\frac{7}{20}=\frac{37}{20}=1\frac{17}{20}$

35. $\frac{2}{3}+\frac{3}{4}+\frac{1}{2}+\frac{5}{12}=\frac{2}{3}\cdot\frac{4}{4}+\frac{3}{4}\cdot\frac{3}{3}+\frac{1}{2}\cdot\frac{6}{6}+\frac{5}{12}=\frac{8}{12}+\frac{9}{12}+\frac{6}{12}+\frac{5}{12}=\frac{28}{12}=\frac{7}{3}=2\frac{1}{3}$

37. $\frac{2}{3}+\frac{5}{12}+\frac{5}{9}+\frac{5}{18}=\frac{2}{3}\cdot\frac{12}{12}+\frac{5}{12}\cdot\frac{3}{3}+\frac{5}{9}\cdot\frac{4}{4}+\frac{5}{18}\cdot\frac{2}{2}=\frac{24}{36}+\frac{15}{36}+\frac{20}{36}+\frac{10}{36}=\frac{69}{36}=\frac{23}{12}=1\frac{11}{12}$

39. $\frac{5}{7}+\frac{2}{3}+\frac{1}{63}+\frac{1}{9}=\frac{5}{7}\cdot\frac{9}{9}+\frac{2}{3}\cdot\frac{21}{21}+\frac{1}{63}+\frac{1}{9}\cdot\frac{7}{7}=\frac{45}{63}+\frac{42}{63}+\frac{1}{63}+\frac{7}{63}=\frac{95}{63}=1\frac{32}{63}$

41. $\frac{3}{5}+\frac{2}{25}+\frac{4}{75}+\frac{2}{3}=\frac{3}{5}\cdot\frac{15}{15}+\frac{2}{25}\cdot\frac{3}{3}+\frac{4}{75}+\frac{2}{3}\cdot\frac{25}{25}=\frac{45}{75}+\frac{6}{75}+\frac{4}{75}+\frac{50}{75}=\frac{105}{75}=\frac{7}{5}=1\frac{2}{5}$

43. $\frac{3}{5}+\frac{7}{8}+\frac{1}{4}+\frac{7}{10}=\frac{3}{5}\cdot\frac{8}{8}+\frac{7}{8}\cdot\frac{5}{5}+\frac{1}{4}\cdot\frac{10}{10}+\frac{7}{10}\cdot\frac{4}{4}=\frac{24}{40}+\frac{35}{40}+\frac{10}{40}+\frac{28}{40}=\frac{97}{40}=2\frac{17}{40}$

45. $\frac{2}{45}+\frac{4}{9}+\frac{1}{15}=\frac{2}{45}+\frac{4}{9}\cdot\frac{5}{5}+\frac{1}{15}\cdot\frac{3}{3}=\frac{2}{45}+\frac{20}{45}+\frac{3}{45}=\frac{25}{45}=\frac{5}{9}$

47. $\frac{11}{40}+\frac{13}{25}+\frac{19}{50}=\frac{11}{40}\cdot\frac{5}{5}+\frac{13}{25}\cdot\frac{8}{8}+\frac{19}{50}\cdot\frac{4}{4}=\frac{55}{200}+\frac{104}{200}+\frac{76}{200}=\frac{235}{200}=1\frac{35}{200}=1\frac{7}{40}$

49. $\frac{25}{36}+\frac{19}{48}=\frac{25}{36}\cdot\frac{4}{4}+\frac{19}{48}\cdot\frac{3}{3}=\frac{100}{144}+\frac{57}{144}=\frac{157}{144}=1\frac{13}{144}$

51. $\frac{1}{20}+\frac{3}{50}+\frac{1}{10}=\frac{1}{20}\cdot\frac{5}{5}+\frac{3}{50}\cdot\frac{2}{2}+\frac{1}{10}\cdot\frac{10}{10}=\frac{5}{100}+\frac{6}{100}+\frac{10}{100}=\frac{21}{100}$. Only $\frac{21}{100}$ of all old tires are not dumped or sent to landfills. This is less than $\frac{1}{4}$ of all old tires.

53. $\frac{1}{200}+\frac{19}{100}=\frac{1}{200}+\frac{19}{100}\cdot\frac{2}{2}=\frac{1}{200}+\frac{38}{200}=\frac{39}{200}$. Home and garden users are more then $\frac{3}{4}$ of all pesticide users.

55. The answer $\frac{3}{10}$ came from adding the numerators and denominators separately. Instead, when a common denominator is present, add the numerators and keep the common denominator in the answer. $\frac{1}{5}+\frac{2}{5}=\frac{3}{5}$.

57. $\frac{1}{4}+\frac{3}{4}+\frac{2}{4}+\frac{1}{4}+\frac{5}{4}+\frac{3}{4}=\frac{15}{4}=3\frac{3}{4}$. The recipe makes $3\frac{3}{4}$ gallons of punch.

59.
$$\frac{7}{8}+\frac{1}{16}+\frac{1}{2}+\frac{1}{8}+\frac{1}{4}=\frac{7}{8}\cdot\frac{2}{2}+\frac{1}{16}+\frac{1}{2}\cdot\frac{8}{8}+\frac{1}{8}\cdot\frac{2}{2}+\frac{1}{4}\cdot\frac{4}{4}=\frac{14}{16}+\frac{1}{16}+\frac{8}{16}+\frac{2}{16}+\frac{4}{16}$$
$$=\frac{29}{16}=1\frac{13}{16}$$

Jonnie needs a bolt that is $1\frac{13}{16}$ inches long.

61. $\frac{1}{8}+\frac{1}{2}+\frac{1}{8}=\frac{1}{8}+\frac{1}{2}\cdot\frac{4}{4}+\frac{1}{8}=\frac{1}{8}+\frac{4}{8}+\frac{1}{8}=\frac{6}{8}=\frac{3}{4}$. The pin is $\frac{3}{4}$ of an inch long.

63.
$$\frac{7}{8}+\frac{3}{16}+\frac{1}{8}+\frac{3}{16}+\frac{1}{8}+\frac{3}{16}+\frac{1}{8}+\frac{3}{16}+\frac{1}{8}+\frac{3}{16}+\frac{1}{8}+\frac{3}{16}+\frac{3}{4}$$
$$=\frac{7}{8}\cdot\frac{2}{2}+\frac{3}{16}+\frac{1}{8}\cdot\frac{2}{2}+\frac{3}{16}+\frac{1}{8}\cdot\frac{2}{2}+\frac{3}{16}+\frac{1}{8}\cdot\frac{2}{2}+\frac{3}{16}+\frac{1}{8}\cdot\frac{2}{2}+\frac{3}{16}+\frac{1}{8}\cdot\frac{2}{2}+\frac{3}{16}+\frac{3}{4}\cdot\frac{4}{4}$$
$$=\frac{14}{16}+\frac{3}{16}+\frac{2}{16}+\frac{3}{16}+\frac{2}{16}+\frac{3}{16}+\frac{2}{16}+\frac{3}{16}+\frac{2}{16}+\frac{3}{16}+\frac{2}{16}+\frac{3}{16}+\frac{12}{16}$$
$$=\frac{54}{16}=3\frac{6}{16}=3\frac{3}{8}$$
The length of the rod is $3\frac{3}{8}$ inches.

65. $\frac{1}{3}+\frac{1}{2}=\frac{1}{3}\cdot\frac{2}{2}+\frac{1}{2}\cdot\frac{3}{3}=\frac{2}{6}+\frac{3}{6}=\frac{5}{6}$; He has $\frac{5}{6}$ of the deck boards installed.

67. The denominator tells you how many parts make up the whole. Thus, if you add numerators from fractions with different denominators, you are adding parts that are not the same size.

69. $124=2^2\cdot 31$; $868=2^2\cdot 7\cdot 31$; LCM $=2^2\cdot 7\cdot 31=868$
$$\frac{67}{124}+\frac{27}{868}=\frac{67}{124}\cdot\frac{7}{7}+\frac{27}{868}=\frac{469}{868}+\frac{27}{868}=\frac{496}{868}=\frac{124\cdot 4}{124\cdot 7}=\frac{4}{7}$$

71.
$$\frac{3}{4}+\frac{13}{16}+\frac{17}{10}+\frac{9}{8}=\frac{3}{4}\cdot\frac{20}{20}+\frac{13}{16}\cdot\frac{5}{5}+\frac{17}{10}\cdot\frac{8}{8}+\frac{9}{8}\cdot\frac{10}{10}$$
$$=\frac{60}{80}+\frac{65}{80}+\frac{136}{80}+\frac{90}{80}=\frac{351}{80}=4\frac{31}{80}\approx 4$$

$4\cdot 9=36$; $\frac{36}{330}=\frac{6}{55}$; He consumes about 4 grams of fat, which is $\frac{6}{55}$ of the number of calories.

73. $5+9+\frac{5}{16}+\frac{5}{24}=14+\frac{5}{16}\cdot\frac{3}{3}+\frac{5}{24}\cdot\frac{2}{2}=14+\frac{15}{48}+\frac{10}{48}=14+\frac{25}{48}=14\frac{25}{48}$

75. $3+11+12+\frac{2}{9}+\frac{2}{3}+\frac{7}{12}=26+\frac{2}{9}\cdot\frac{4}{4}+\frac{2}{3}\cdot\frac{12}{12}+\frac{7}{12}\cdot\frac{3}{3}=26+\frac{8}{36}+\frac{24}{36}+\frac{21}{36}$
$=26+\frac{53}{36}=26+1\frac{17}{36}=27\frac{17}{36}$

77. $4+\frac{1}{12}+1+\frac{3}{4}+3+\frac{1}{8}=8+\frac{1}{12}\cdot\frac{2}{2}+\frac{3}{4}\cdot\frac{6}{6}+\frac{1}{8}\cdot\frac{3}{3}=8+\frac{2}{24}+\frac{18}{24}+\frac{3}{24}$
$=8+\frac{23}{24}=8\frac{23}{24}$

79. $(13-2^3)^3-7(11)=(13-8)^3-7(11)=5^3-7(11)=125-7(11)=125-77=48$

81. $36\left(11\frac{5}{8}\right)=\frac{36}{1}\cdot\frac{93}{8}=\frac{837}{2}=418\frac{1}{2}$; The instructor needs $418\frac{1}{2}$ in. of wire solder.

Exercises 3.7 (page 229)

1. $9\frac{1}{9}+3\frac{7}{9}=9+3+\frac{1}{9}+\frac{7}{9}=12+\frac{8}{9}=12\frac{8}{9}$

3. $3\frac{4}{5}+7\frac{4}{5}=3+7+\frac{4}{5}+\frac{4}{5}=10+\frac{8}{5}=10+1\frac{3}{5}=11\frac{3}{5}$

5. $5\frac{7}{15}+6\frac{2}{3}=5+6+\frac{7}{15}+\frac{2}{3}=11+\frac{7}{15}+\frac{2}{3}\cdot\frac{5}{5}=11+\frac{7}{15}+\frac{10}{15}=11+\frac{17}{15}=11+1\frac{2}{15}$
$=12\frac{2}{15}$

7. $9\frac{9}{14}+3\frac{2}{7}=9+3+\frac{9}{14}+\frac{2}{7}=12+\frac{9}{14}+\frac{2}{7}\cdot\frac{2}{2}=12+\frac{9}{14}+\frac{4}{14}=12+\frac{13}{14}=12\frac{13}{14}$

9. $4\frac{4}{7}+2\frac{11}{14}=4+2+\frac{4}{7}+\frac{11}{14}=6+\frac{4}{7}\cdot\frac{2}{2}+\frac{11}{14}=6+\frac{8}{14}+\frac{11}{14}=6+\frac{19}{14}=6+1\frac{5}{14}=7\frac{5}{14}$

11. $2\frac{8}{15}+7\frac{3}{5}=2+7+\frac{8}{15}+\frac{3}{5}=9+\frac{8}{15}+\frac{3}{5}\cdot\frac{3}{3}=9+\frac{8}{15}+\frac{9}{15}=9+\frac{17}{15}=9+1\frac{2}{15}=10\frac{2}{15}$

13. $10\frac{3}{10}+11\frac{17}{20}=10+11+\frac{3}{10}+\frac{17}{20}=21+\frac{3}{10}\cdot\frac{2}{2}+\frac{17}{20}=21+\frac{6}{20}+\frac{17}{20}=21+\frac{23}{20}$
$=21+1\frac{3}{20}=22\frac{3}{20}$

SECTION 3.7

15. $$2\frac{2}{3}+3\frac{2}{3}+5\frac{1}{6}=2+3+5+\frac{2}{3}+\frac{2}{3}+\frac{1}{6}=10+\frac{2}{3}\cdot\frac{2}{2}+\frac{2}{3}\cdot\frac{2}{2}+\frac{1}{6}=10+\frac{4}{6}+\frac{4}{6}+\frac{1}{6}$$
$$=10+\frac{9}{6}$$
$$=10+1\frac{3}{6}=11\frac{1}{2}$$

17. $$7\frac{1}{5}+6\frac{3}{20}+8+4\frac{1}{4}=7+6+8+4+\frac{1}{5}+\frac{3}{20}+\frac{1}{4}=25+\frac{1}{5}\cdot\frac{4}{4}+\frac{3}{20}+\frac{1}{4}\cdot\frac{5}{5}$$
$$=25+\frac{4}{20}+\frac{3}{20}+\frac{5}{20}$$
$$=25+\frac{12}{20}=25\frac{12}{20}=25\frac{3}{5}$$

19. $$11\frac{2}{3}+6\frac{11}{18}=11+6+\frac{2}{3}+\frac{11}{18}=17+\frac{2}{3}\cdot\frac{6}{6}+\frac{11}{18}=17+\frac{12}{18}+\frac{11}{18}=17+\frac{23}{18}$$
$$=17+1\frac{5}{18}=18\frac{5}{18}$$

21. $$7\frac{3}{7}+5\frac{9}{14}+8\frac{5}{28}=7+5+8+\frac{3}{7}+\frac{9}{14}+\frac{5}{28}=20+\frac{3}{7}\cdot\frac{4}{4}+\frac{9}{14}\cdot\frac{2}{2}+\frac{5}{28}$$
$$=20+\frac{12}{28}+\frac{18}{28}+\frac{5}{28}$$
$$=20+\frac{35}{28}=20+1\frac{7}{28}=21\frac{1}{4}$$

23. $$11\frac{7}{20}+9\frac{7}{30}=11+9+\frac{7}{20}+\frac{7}{30}=20+\frac{7}{20}\cdot\frac{3}{3}+\frac{7}{30}\cdot\frac{2}{2}=20+\frac{21}{60}+\frac{14}{60}=20+\frac{35}{60}$$
$$=20\frac{7}{12}$$

25. $$16\frac{3}{10}+11\frac{7}{15}+7\frac{1}{6}=16+11+7+\frac{3}{10}+\frac{7}{15}+\frac{1}{6}=34+\frac{3}{10}\cdot\frac{3}{3}+\frac{7}{15}\cdot\frac{2}{2}+\frac{1}{6}\cdot\frac{5}{5}$$
$$=34+\frac{9}{30}+\frac{14}{30}+\frac{5}{30}$$
$$=34+\frac{28}{30}=34+\frac{14}{15}=34\frac{14}{15}$$

27. $$119\frac{11}{12}+217\frac{7}{18}=119+217+\frac{11}{12}+\frac{7}{18}=336+\frac{11}{12}\cdot\frac{3}{3}+\frac{7}{18}\cdot\frac{2}{2}=336+\frac{33}{36}+\frac{14}{36}$$
$$=336+\frac{47}{36}$$
$$=336+1\frac{11}{36}=337\frac{11}{36}$$

SECTION 3.7

29. $$59\frac{4}{11}+32\frac{5}{22}+15\frac{5}{6}=59+32+15+\frac{4}{11}+\frac{5}{22}+\frac{5}{6}=106+\frac{4}{11}\cdot\frac{6}{6}+\frac{5}{22}\cdot\frac{3}{3}+\frac{5}{6}\cdot\frac{11}{11}$$
$$=106+\frac{24}{66}+\frac{15}{66}+\frac{55}{66}$$
$$=106+\frac{94}{66}=106+1\frac{28}{66}=107\frac{14}{33}$$

31. $$15\frac{3}{4}+18\frac{2}{3}+21\frac{1}{2}=15+18+21+\frac{3}{4}+\frac{2}{3}+\frac{1}{2}=54+\frac{3}{4}\cdot\frac{3}{3}+\frac{2}{3}\cdot\frac{4}{4}+\frac{1}{2}\cdot\frac{6}{6}$$
$$=54+\frac{9}{12}+\frac{8}{12}+\frac{6}{12}$$
$$=54+\frac{23}{12}=54+1\frac{11}{12}=55\frac{11}{12}$$

33. $$72+15\frac{7}{24}+23\frac{11}{36}=72+15+23+\frac{7}{24}+\frac{11}{36}=110+\frac{7}{24}\cdot\frac{3}{3}+\frac{11}{36}\cdot\frac{2}{2}$$
$$=110+\frac{21}{72}+\frac{22}{72}$$
$$=110+\frac{43}{72}=110\frac{43}{72}$$

35. $$14\frac{9}{10}+15\frac{7}{15}+11=14+15+11+\frac{9}{10}+\frac{7}{15}=40+\frac{9}{10}\cdot\frac{3}{3}+\frac{7}{15}\cdot\frac{2}{2}$$
$$=40+\frac{27}{30}+\frac{14}{30}$$
$$=40+\frac{41}{30}=40+1\frac{11}{30}=41\frac{11}{30}$$

37. $$11\frac{1}{6}+12\frac{3}{10}+13\frac{1}{12}+14\frac{1}{20}=11+12+13+14+\frac{1}{6}+\frac{3}{10}+\frac{1}{12}+\frac{1}{20}$$
$$=50+\frac{1}{6}\cdot\frac{10}{10}+\frac{3}{10}\cdot\frac{6}{6}+\frac{1}{12}\cdot\frac{5}{5}+\frac{1}{20}\cdot\frac{3}{3}$$
$$=50+\frac{10}{60}+\frac{18}{60}+\frac{5}{60}+\frac{3}{60}=50+\frac{36}{60}=50\frac{3}{5}$$

39. $$6\frac{1}{8}+2\frac{3}{4}+\frac{5}{12}+9\frac{1}{3}=6+2+9+\frac{1}{8}+\frac{3}{4}+\frac{5}{12}+\frac{1}{3}$$
$$=17+\frac{1}{8}\cdot\frac{3}{3}+\frac{3}{4}\cdot\frac{6}{6}+\frac{5}{12}\cdot\frac{2}{2}+\frac{1}{3}\cdot\frac{8}{8}$$
$$=17+\frac{3}{24}+\frac{18}{24}+\frac{10}{24}+\frac{8}{24}=17+\frac{39}{24}=17+1\frac{15}{24}=18\frac{5}{8}$$

41. $$12\frac{3}{5}+7\frac{1}{8}+29\frac{3}{4}+14\frac{9}{10}=12+7+29+14+\frac{3}{5}+\frac{1}{8}+\frac{3}{4}+\frac{9}{10}$$
$$=62+\frac{3}{5}\cdot\frac{8}{8}+\frac{1}{8}\cdot\frac{5}{5}+\frac{3}{4}\cdot\frac{10}{10}+\frac{9}{10}\cdot\frac{4}{4}$$
$$=62+\frac{24}{40}+\frac{5}{40}+\frac{30}{40}+\frac{36}{40}=62+\frac{95}{40}=62+2\frac{15}{40}=64\frac{3}{8}$$

43. Perimeter of portrait:

$$18\frac{1}{2} + 24\frac{3}{4} + 18\frac{1}{2} + 24\frac{3}{4} = 18 + 24 + 18 + 24 + \frac{1}{2} + \frac{3}{4} + \frac{1}{2} + \frac{3}{4}$$
$$= 84 + \frac{1}{2} \cdot \frac{2}{2} + \frac{3}{4} + \frac{1}{2} \cdot \frac{2}{2} + \frac{3}{4}$$
$$= 84 + \frac{2}{4} + \frac{3}{4} + \frac{2}{4} + \frac{3}{4} = 84 + \frac{10}{4} = 84 + 2\frac{2}{4} = 86\frac{1}{2}$$

Extra material for 4 corners:

$$8\left(1\frac{1}{2}\right) = \frac{8}{1} \cdot \frac{3}{2} = \frac{24}{2} = 12$$

Total framing material needed:

$$86\frac{1}{2} + 12 = \boxed{98\tfrac{1}{2} \text{ inches of molding are needed.}}$$

45. $$1\frac{3}{8} + 6\frac{1}{2} + 2\frac{3}{4} = 1 + 6 + 2 + \frac{3}{8} + \frac{1}{2} + \frac{3}{4} = 9 + \frac{3}{8} + \frac{1}{2} \cdot \frac{4}{4} + \frac{3}{4} \cdot \frac{2}{2}$$
$$= 9 + \frac{3}{8} + \frac{4}{8} + \frac{6}{8} = 9 + \frac{13}{8} = 9 + 1\frac{5}{8} = 10\frac{5}{8}$$

She needs $10\frac{5}{8}$ yards of fabric.

47. $$3\frac{1}{2} + 2\frac{1}{4} + 4\frac{7}{8} + 3\frac{1}{3} + 5\frac{3}{4} + 2 + 1\frac{3}{8} + 2\frac{2}{3}$$
$$= 3 + 2 + 4 + 3 + 5 + 2 + 1 + 2 + \frac{1}{2} + \frac{1}{4} + \frac{7}{8} + \frac{1}{3} + \frac{3}{4} + \frac{3}{8} + \frac{2}{3}$$
$$= 22 + \frac{1}{2} + \frac{4}{4} + \frac{10}{8} + \frac{3}{3}$$
$$= 22 + \frac{1}{2} + 1 + 1\frac{2}{8} + 1$$
$$= 22 + \frac{1}{2} + 1 + 1 + \frac{1}{4} + 1 = 25 + \frac{1}{2} \cdot \frac{2}{2} + \frac{1}{4} = 25 + \frac{2}{4} + \frac{1}{4} = 25\frac{3}{4}$$

Jeff lost a total of $25\frac{3}{4}$ pounds.

49. $$1\frac{3}{5} + 3\frac{1}{8} + 2\frac{7}{10} = 1 + 3 + 2 + \frac{3}{5} + \frac{1}{8} + \frac{7}{10} = 6 + \frac{3}{5} \cdot \frac{8}{8} + \frac{1}{8} \cdot \frac{5}{5} + \frac{7}{10} \cdot \frac{4}{4}$$
$$= 6 + \frac{24}{40} + \frac{5}{40} + \frac{28}{40}$$
$$= 6 + \frac{57}{40} = 6 + 1\frac{17}{40} = 7\frac{17}{40}$$

The perimeter is $7\frac{17}{40}$ inches.

51. $$3\frac{9}{20} + 5\frac{3}{10} + 3\frac{23}{25} = 3 + 5 + 3 + \frac{9}{20} + \frac{3}{10} + \frac{23}{25} = 11 + \frac{9}{20} \cdot \frac{5}{5} + \frac{3}{10} \cdot \frac{10}{10} + \frac{23}{25} \cdot \frac{4}{4}$$
$$= 11 + \frac{45}{100} + \frac{30}{100} + \frac{92}{100}$$
$$= 11 + \frac{167}{100} = 11 + 1\frac{67}{100} = 12\frac{67}{100}$$

The rainfall for the three months is $12\frac{67}{100}$ inches.

53. $6\frac{11}{16}+8\frac{3}{5}+9\frac{3}{4}+17\frac{1}{2}+5\frac{1}{8}+12\frac{4}{5}+\frac{7}{8}$

$$\begin{aligned}
&=6+8+9+17+5+12+\frac{11}{16}+\frac{3}{5}+\frac{3}{4}+\frac{1}{2}+\frac{1}{8}+\frac{4}{5}+\frac{7}{8}\\
&=57+\frac{11}{16}\cdot\frac{5}{5}+\frac{3}{5}\cdot\frac{16}{16}+\frac{3}{4}\cdot\frac{20}{20}+\frac{1}{2}\cdot\frac{40}{40}+\frac{1}{8}\cdot\frac{10}{10}+\frac{4}{5}\cdot\frac{16}{16}+\frac{7}{8}\cdot\frac{10}{10}\\
&=57+\frac{55}{80}+\frac{48}{80}+\frac{60}{80}+\frac{40}{80}+\frac{10}{80}+\frac{64}{80}+\frac{70}{80}\\
&=57+\frac{347}{80}=57+4\frac{27}{80}=61\frac{27}{80}
\end{aligned}$$

$$15{,}000\left(61\frac{27}{80}\right)=\frac{15{,}000}{1}\cdot\frac{4907}{80}=\frac{73{,}605{,}000}{80}=920{,}062\frac{1}{2}$$

There are $61\frac{27}{80}$ miles of road to be resurfaced, at a cost of about \$920,000.

55. $16+1\frac{1}{2}=17\frac{1}{2}$; It is $17\frac{1}{2}$ inches from A to D.

57. $8\frac{4}{5}+7\frac{3}{8}=8+7+\frac{32}{40}+\frac{15}{40}=15\frac{47}{40}=16\frac{7}{40}$

$8\frac{4}{5}+7\frac{3}{8}=\frac{44}{5}+\frac{59}{8}=\frac{352}{40}+\frac{295}{40}=\frac{647}{40}=16\frac{7}{40}$; Answers may vary.

59.
$$\begin{aligned}
6\left(6\frac{4}{9}\right)+5\left(2\frac{5}{6}\right)+7\left(4\frac{1}{3}\right)&=\frac{6}{1}\cdot\frac{58}{9}+\frac{5}{1}\cdot\frac{17}{6}+\frac{7}{1}\cdot\frac{13}{3}\\
&=\frac{116}{3}+\frac{85}{6}+\frac{91}{3}\\
&=\frac{116}{3}\cdot\frac{2}{2}+\frac{85}{6}+\frac{91}{3}\cdot\frac{2}{2}=\frac{232}{6}+\frac{85}{6}+\frac{182}{6}=\frac{499}{6}=83\frac{1}{6}
\end{aligned}$$

$$\begin{aligned}
6\left(5\frac{1}{3}\right)+8\left(3\frac{2}{3}\right)+7\left(3\frac{5}{42}\right)&=\frac{6}{1}\cdot\frac{16}{3}+\frac{8}{1}\cdot\frac{11}{3}+\frac{7}{1}\cdot\frac{131}{42}\\
&=32+\frac{88}{3}+\frac{131}{6}\\
&=32+\frac{88}{3}\cdot\frac{2}{2}+\frac{131}{6}\\
&=32+\frac{176}{6}+\frac{131}{6}=32+\frac{307}{6}=32+51\frac{1}{6}=83\frac{1}{6}
\end{aligned}$$

The statement is true.

61.
$$\begin{array}{rrrr}
 & & 9 & \\
 & 0 & \not{1}\not{0} & 13\\
 & \not{1} & \not{0} & \not{3}\\
- & & 7 & 7\\
\hline
 & & 2 & 6
\end{array}$$

63.
$$\begin{array}{rrrrr}
 & & 10 & 10 & \\
 & 0 & \not{0} & \not{0} & 11\\
 & \not{1} & \not{1} & \not{1} & \not{1}\\
- & & 8 & 8 & 9\\
\hline
 & & 2 & 2 & 2
\end{array}$$

65. $40\div 8=5$

$\frac{7}{8}=\frac{7}{8}\cdot\frac{5}{5}=\frac{35}{40}$

67. $\frac{1950}{4095} = \frac{10 \cdot 195}{21 \cdot 195} = \frac{10}{21}$

69. $5 + 4 + 6 + 2 + 3 + 3 + 3 + 5 + 4 = 35$
Millie's score is 35, which is under par.

Exercises 3.8 (page 236)

1. $\frac{5}{6} - \frac{1}{6} = \frac{4}{6} = \frac{2}{3}$

3. $\frac{8}{9} - \frac{5}{9} = \frac{3}{9} = \frac{1}{3}$

5. $\frac{11}{16} - \frac{5}{16} = \frac{6}{16} = \frac{3}{8}$

7. $\frac{5}{7} - \frac{3}{14} = \frac{5}{7} \cdot \frac{2}{2} - \frac{3}{14} = \frac{10}{14} - \frac{3}{14} = \frac{7}{14} = \frac{1}{2}$

9. $\frac{3}{8} - \frac{5}{16} = \frac{3}{8} \cdot \frac{2}{2} - \frac{5}{16} = \frac{6}{16} - \frac{5}{16} = \frac{1}{16}$

11. $\frac{3}{15} - \frac{2}{45} = \frac{3}{15} \cdot \frac{3}{3} - \frac{2}{45} = \frac{9}{45} - \frac{2}{45} = \frac{7}{45}$

13. $\frac{11}{21} - \frac{3}{7} = \frac{11}{21} - \frac{3}{7} \cdot \frac{3}{3} = \frac{11}{21} - \frac{9}{21} = \frac{2}{21}$

15. $\frac{13}{18} - \frac{2}{3} = \frac{13}{18} - \frac{2}{3} \cdot \frac{6}{6} = \frac{13}{18} - \frac{12}{18} = \frac{1}{18}$

17. $\frac{17}{24} - \frac{1}{8} = \frac{17}{24} - \frac{1}{8} \cdot \frac{3}{3} = \frac{17}{24} - \frac{3}{24} = \frac{14}{24} = \frac{7}{12}$

19. $\frac{17}{54} - \frac{1}{18} = \frac{17}{54} - \frac{1}{18} \cdot \frac{3}{3} = \frac{17}{54} - \frac{3}{54} = \frac{14}{54} = \frac{7}{27}$

21. $\frac{7}{8} - \frac{5}{6} = \frac{7}{8} \cdot \frac{3}{3} - \frac{5}{6} \cdot \frac{4}{4} = \frac{21}{24} - \frac{20}{24} = \frac{1}{24}$

23. $\frac{11}{15} - \frac{7}{20} = \frac{11}{15} \cdot \frac{4}{4} - \frac{7}{20} \cdot \frac{3}{3} = \frac{44}{60} - \frac{21}{60} = \frac{23}{60}$

25. $\frac{7}{9} - \frac{5}{18} = \frac{7}{9} \cdot \frac{2}{2} - \frac{5}{18} = \frac{14}{18} - \frac{5}{18} = \frac{9}{18} = \frac{1}{2}$

27. $\frac{9}{16} - \frac{1}{6} = \frac{9}{16} \cdot \frac{3}{3} - \frac{1}{6} \cdot \frac{8}{8} = \frac{27}{48} - \frac{8}{48} = \frac{19}{48}$

29. $\frac{13}{20} - \frac{7}{30} = \frac{13}{20} \cdot \frac{3}{3} - \frac{7}{30} \cdot \frac{2}{2} = \frac{39}{60} - \frac{14}{60} = \frac{25}{60} = \frac{5}{12}$

31. $\frac{11}{12} - \frac{11}{18} = \frac{11}{12} \cdot \frac{3}{3} - \frac{11}{18} \cdot \frac{2}{2} = \frac{33}{36} - \frac{22}{36} = \frac{11}{36}$

33. $\frac{7}{10} - \frac{1}{4} = \frac{7}{10} \cdot \frac{2}{2} - \frac{1}{4} \cdot \frac{5}{5} = \frac{14}{20} - \frac{5}{20} = \frac{9}{20}$

35. $\frac{7}{10} - \frac{4}{15} = \frac{7}{10} \cdot \frac{3}{3} - \frac{4}{15} \cdot \frac{2}{2} = \frac{21}{30} - \frac{8}{30} = \frac{13}{30}$

37. $\frac{17}{48} - \frac{1}{16} = \frac{17}{48} - \frac{1}{16} \cdot \frac{3}{3} = \frac{17}{48} - \frac{3}{48} = \frac{14}{48} = \frac{7}{24}$

SECTION 3.8

39. $\frac{13}{16}-\frac{11}{24}=\frac{13}{16}\cdot\frac{3}{3}-\frac{11}{24}\cdot\frac{2}{2}=\frac{39}{48}-\frac{22}{48}$
$=\frac{17}{48}$

41. $\frac{13}{18}-\frac{7}{12}=\frac{13}{18}\cdot\frac{2}{2}-\frac{7}{12}\cdot\frac{3}{3}=\frac{26}{36}-\frac{21}{36}$
$=\frac{5}{36}$

43. $\frac{17}{50}-\frac{13}{40}=\frac{17}{50}\cdot\frac{4}{4}-\frac{13}{40}\cdot\frac{5}{5}=\frac{68}{200}-\frac{65}{200}$
$=\frac{3}{200}$

45. $\frac{22}{35}-\frac{3}{40}=\frac{22}{35}\cdot\frac{8}{8}-\frac{3}{40}\cdot\frac{7}{7}=\frac{176}{280}-\frac{21}{280}$
$=\frac{155}{280}=\frac{31}{56}$

47. $1-\frac{4}{5}=\frac{5}{5}-\frac{4}{5}=\frac{1}{5}$; $\frac{1}{5}$ of all Caucasian Americans have blonde or red hair.

49. $1-\frac{1}{3}-\frac{1}{4}=\frac{1}{1}\cdot\frac{12}{12}-\frac{1}{3}\cdot\frac{4}{4}-\frac{1}{4}\cdot\frac{3}{3}=\frac{12}{12}-\frac{4}{12}-\frac{3}{12}=\frac{5}{12}$; He has $\frac{5}{12}$ of his check left.
$\frac{5}{12}\cdot 480=\frac{5}{12}\cdot\frac{480}{1}=200$; He has \$200 left.

51. $25-16=9$; In that year, $\frac{9}{25}$ of the population did not hold stock.

53. $25-3=22$; In that period, $\frac{22}{25}$ of the population was employed.

55. If $\frac{17}{25}$ of Americans are female, then the rest are male. $\frac{8}{25}$ are male.

57. $\frac{7}{8}-\frac{3}{16}=\frac{7}{8}\cdot\frac{2}{2}-\frac{3}{16}=\frac{14}{16}-\frac{3}{16}=\frac{11}{16}$; The difference in diameters is $\frac{11}{16}$ of an inch.

59. $\frac{14}{25}+\frac{1}{20}=\frac{14}{25}\cdot\frac{4}{4}+\frac{1}{20}\cdot\frac{5}{5}=\frac{56}{100}+\frac{5}{100}=\frac{61}{100}$; $1-\frac{61}{100}=\frac{100}{100}-\frac{61}{100}=\frac{39}{100}$; Discretionary spending accounted for $\frac{39}{100}$ of the budget.

61. You cannot just subtract the numerators and the denominators because the fact that the denominators are not the same means that the parts have different sizes. $\frac{3}{4}-\frac{1}{2}=\frac{3}{4}-\frac{2}{4}=\frac{1}{4}$

63. $\frac{93}{125}-\frac{247}{625}=\frac{93}{125}\cdot\frac{5}{5}-\frac{247}{625}=\frac{465}{625}-\frac{247}{625}=\frac{218}{625}$

65. Brick plus mortar on one end: $3\frac{3}{4}+\frac{3}{8}=4\frac{1}{8}$; Consider the width to be 6 ft (72 in.) and the length be 10 ft (120 in.), plus 16 in. to account for the width (for a total of 136 in.).
width: $144\div 4\frac{1}{8}=\frac{144}{1}\div\frac{33}{8}=\frac{144}{1}\cdot\frac{8}{33}=\frac{384}{11}=34\frac{10}{11}$, or about 35 bricks for both widths.
length: $272\div 4\frac{1}{8}=\frac{272}{1}\div\frac{33}{8}=\frac{272}{1}\cdot\frac{8}{33}=\frac{2176}{33}=65\frac{31}{33}$, or about 66 bricks for both lengths.
The project will use about $35+66$, or 101 bricks.

67.

$$\begin{array}{rrrrr} {\scriptstyle 1} & {\scriptstyle 11} & & & \\ \not{2} & \not{1} & 5 & 9 & 6 \\ - & 9 & 3 & 9 & 6 \\ \hline 1 & 2 & 2 & 0 & 0 \end{array}$$

69. $(22 - 16) + \left(\frac{2}{3} - \frac{4}{15}\right) = 6 + \left(\frac{2}{3} \cdot \frac{5}{5} - \frac{4}{15}\right) = 6 + \left(\frac{10}{15} - \frac{4}{15}\right) = 6 + \frac{6}{15} = 6 + \frac{2}{5} = 6\frac{2}{5}$

71. $(34 - 27) + \left(\frac{7}{10} - \frac{1}{4}\right) = 7 + \left(\frac{7}{10} \cdot \frac{2}{2} - \frac{1}{4} \cdot \frac{5}{5}\right) = 7 + \left(\frac{14}{20} - \frac{5}{20}\right) = 7 + \frac{9}{20} = 7\frac{9}{20}$

73. $(63 - 36) + \left(\frac{3}{4} - \frac{1}{20}\right) = 27 + \left(\frac{3}{4} \cdot \frac{5}{5} - \frac{1}{20}\right) = 27 + \left(\frac{15}{20} - \frac{1}{20}\right) = 27 + \frac{14}{20} = 27 + \frac{7}{10}$
$= 27\frac{7}{10}$

75. $19{,}080 \div 3 = 6360$ bricks each; $6360 \div 795 = 8$; It will take them 8 days.

Exercises 3.9 (page 243)

1.
$$\begin{array}{r} 12\frac{9}{13} \\ -\ 5\frac{7}{13} \\ \hline 7\frac{2}{13} \end{array}$$

3.
$$\begin{array}{r} 147\frac{49}{90} \\ -\ 135\frac{19}{90} \\ \hline 12\frac{30}{90} = 12\frac{1}{3} \end{array}$$

5.
$$\begin{array}{r} 10\frac{7}{8} \\ -\ 5\frac{3}{4} \\ \hline \end{array} \Rightarrow \begin{array}{r} 10\frac{7}{8} \\ -\ 5\frac{6}{8} \\ \hline 5\frac{1}{8} \end{array}$$

7.
$$\begin{array}{r} 12 \\ -\ 9\frac{4}{7} \\ \hline \end{array} \Rightarrow \begin{array}{r} 11\frac{7}{7} \\ -\ 9\frac{4}{7} \\ \hline 2\frac{3}{7} \end{array}$$

9.
$$\begin{array}{r} 87\frac{5}{12} \\ -\ 71\frac{2}{15} \\ \hline \end{array} \Rightarrow \begin{array}{r} 87\frac{25}{60} \\ -\ 71\frac{8}{60} \\ \hline 16\frac{17}{60} \end{array}$$

11.
$$\begin{array}{r} 5\frac{1}{4} \\ -\ 2\frac{3}{4} \\ \hline \end{array} \Rightarrow \begin{array}{r} 4\frac{5}{4} \\ -\ 2\frac{3}{4} \\ \hline 2\frac{2}{4} = 2\frac{1}{2} \end{array}$$

13.
$$\begin{array}{r} 18\frac{5}{12} \\ -\ 9\frac{7}{12} \\ \hline \end{array} \Rightarrow \begin{array}{r} 17\frac{17}{12} \\ -\ 9\frac{7}{12} \\ \hline 8\frac{10}{12} = 8\frac{5}{6} \end{array}$$

15.
$$\begin{array}{r} 234\frac{7}{9} \\ -\ 62 \\ \hline 172\frac{7}{9} \end{array}$$

SECTION 3.9

17. $$\begin{array}{r} 11\frac{2}{3} \\ -\ 5\frac{1}{4} \\ \hline \end{array} \Rightarrow \begin{array}{r} 11\frac{8}{12} \\ -\ 5\frac{3}{12} \\ \hline 6\frac{5}{12} \end{array}$$

19. $$\begin{array}{r} 31\frac{2}{15} \\ -\ 22\frac{11}{15} \\ \hline \end{array} \Rightarrow \begin{array}{r} 30\frac{17}{15} \\ -\ 22\frac{11}{15} \\ \hline 8\frac{6}{15} = 8\frac{2}{5} \end{array}$$

21. $$\begin{array}{r} 310\frac{23}{24} \\ -\ 254\frac{5}{8} \\ \hline \end{array} \Rightarrow \begin{array}{r} 310\frac{23}{24} \\ -\ 254\frac{15}{24} \\ \hline 56\frac{8}{24} = 56\frac{1}{3} \end{array}$$

23. $$\begin{array}{r} 66\frac{7}{15} \\ -\ 51\frac{1}{12} \\ \hline \end{array} \Rightarrow \begin{array}{r} 66\frac{28}{60} \\ -\ 51\frac{5}{60} \\ \hline 15\frac{23}{60} \end{array}$$

25. $$\begin{array}{r} 64\frac{3}{10} \\ -\ 41\frac{7}{15} \\ \hline \end{array} \Rightarrow \begin{array}{r} 64\frac{9}{30} \\ -\ 41\frac{14}{30} \\ \hline \end{array} \Rightarrow \begin{array}{r} 63\frac{39}{30} \\ -\ 41\frac{14}{30} \\ \hline 22\frac{25}{30} \\ = 22\frac{5}{6} \end{array}$$

27. $$\begin{array}{r} 40\frac{1}{6} \\ -\ 24\frac{3}{16} \\ \hline \end{array} \Rightarrow \begin{array}{r} 40\frac{8}{48} \\ -\ 24\frac{9}{48} \\ \hline \end{array} \Rightarrow \begin{array}{r} 39\frac{56}{48} \\ -\ 24\frac{9}{48} \\ \hline 15\frac{47}{48} \end{array}$$

29. $$\begin{array}{r} 78 \\ -\ 14\frac{4}{5} \\ \hline \end{array} \Rightarrow \begin{array}{r} 77\frac{5}{5} \\ -\ 14\frac{4}{5} \\ \hline 63\frac{1}{5} \end{array}$$

31. $$\begin{array}{r} 3\frac{17}{32} \\ -\ 1\frac{5}{16} \\ \hline \end{array} \Rightarrow \begin{array}{r} 3\frac{17}{32} \\ -\ 1\frac{10}{32} \\ \hline 2\frac{7}{32} \end{array}$$

33. $$\begin{array}{r} 10\frac{2}{5} \\ -\ \ \frac{5}{8} \\ \hline \end{array} \Rightarrow \begin{array}{r} 10\frac{16}{40} \\ -\ \ \frac{25}{40} \\ \hline \end{array} \Rightarrow \begin{array}{r} 9\frac{56}{40} \\ -\ \ \frac{25}{40} \\ \hline 9\frac{31}{40} \end{array}$$

35. $$\begin{array}{r} 93\frac{7}{8} \\ -\ 19 \\ \hline 74\frac{7}{8} \end{array}$$

37. $$\begin{array}{r} 46\frac{14}{15} \\ -\ 19\frac{27}{40} \\ \hline \end{array} \Rightarrow \begin{array}{r} 46\frac{112}{120} \\ -\ 19\frac{81}{120} \\ \hline 27\frac{31}{120} \end{array}$$

39. $$\begin{array}{r} 92\frac{3}{16} \\ -\ 17\frac{11}{24} \\ \hline \end{array} \Rightarrow \begin{array}{r} 92\frac{9}{48} \\ -\ 17\frac{22}{48} \\ \hline \end{array} \Rightarrow \begin{array}{r} 91\frac{57}{48} \\ -\ 17\frac{22}{48} \\ \hline 74\frac{35}{48} \end{array}$$

SECTION 3.9

41. $\begin{array}{r} 34\frac{2}{39} \\ -17\frac{21}{26} \\ \hline \end{array} \Rightarrow \begin{array}{r} 34\frac{4}{78} \\ -17\frac{63}{78} \\ \hline \end{array} \Rightarrow \begin{array}{r} 33\frac{82}{78} \\ -17\frac{63}{78} \\ \hline 16\frac{19}{78} \end{array}$

43. $\begin{array}{r} 6\frac{5}{21} \\ -5\frac{7}{15} \\ \hline \end{array} \Rightarrow \begin{array}{r} 6\frac{25}{105} \\ -5\frac{49}{105} \\ \hline \end{array} \Rightarrow \begin{array}{r} 5\frac{130}{105} \\ -5\frac{49}{105} \\ \hline \frac{81}{105} = \frac{27}{35} \end{array}$

45. To subtract the $\frac{1}{4}$, 1 is "borrowed" from 16. However, then the 16 must be renamed as $15 + \frac{4}{4}$. The answer should be $2\frac{3}{4}$.

47. $\begin{array}{r} 8\frac{1}{2} \\ -6\frac{1}{8} \\ \hline \end{array} \Rightarrow \begin{array}{r} 8\frac{4}{8} \\ -6\frac{1}{8} \\ \hline 2\frac{3}{8} \end{array}$

She trims $2\frac{3}{8}$ pounds.

49. $\begin{array}{r} 30\frac{3}{4} \\ -18\frac{7}{10} \\ \hline \end{array} \Rightarrow \begin{array}{r} 30\frac{15}{20} \\ -18\frac{14}{20} \\ \hline 12\frac{1}{20} \end{array}$

He has $12\frac{1}{20}$ tons left.

51. $\begin{array}{r} 8\frac{1}{4} \\ -2\frac{1}{2} \\ \hline \end{array} \Rightarrow \begin{array}{r} 8\frac{1}{4} \\ -2\frac{2}{4} \\ \hline \end{array} \Rightarrow \begin{array}{r} 7\frac{5}{4} \\ -2\frac{2}{4} \\ \hline 5\frac{3}{4} \end{array}$

He won by $5\frac{3}{4}$ in.

53. $\begin{array}{r} 242\frac{1}{3} \\ -209\frac{1}{2} \\ \hline \end{array} \Rightarrow \begin{array}{r} 242\frac{2}{6} \\ -209\frac{3}{6} \\ \hline \end{array} \Rightarrow \begin{array}{r} 241\frac{8}{6} \\ -209\frac{3}{6} \\ \hline 32\frac{5}{6} \end{array}$

It is $32\frac{5}{6}$ ft longer.

55. $\begin{array}{r} 42 \\ -22\frac{7}{10} \\ \hline \end{array} \Rightarrow \begin{array}{r} 41\frac{10}{10} \\ -22\frac{7}{10} \\ \hline 19\frac{3}{10} \end{array}$ They have $19\frac{3}{10}$ miles left.

57. $25\frac{3}{8} + 16\frac{3}{4} = 25 + 16 + \frac{3}{8} + \frac{3}{4} = 41 + \frac{3}{8} + \frac{6}{8} = 41 + \frac{9}{8} = 41 + 1\frac{1}{8} = 42\frac{1}{8}$

$\begin{array}{r} 48 \\ -42\frac{1}{8} \\ \hline \end{array} \Rightarrow \begin{array}{r} 47\frac{8}{8} \\ -42\frac{1}{8} \\ \hline 5\frac{7}{8} \end{array}$ There will be $5\frac{7}{8}$ in. of the board left over.

59. $2\left(\frac{7}{8}\right)=\frac{2}{1}\cdot\frac{7}{8}=\frac{7}{4}=1\frac{3}{4}$

$$\begin{array}{r} 2\frac{1}{8} \\ -1\frac{3}{4} \\ \hline \end{array} \Rightarrow \begin{array}{r} 2\frac{1}{8} \\ -1\frac{6}{8} \\ \hline \end{array} \Rightarrow \begin{array}{r} 1\frac{9}{8} \\ -1\frac{6}{8} \\ \hline \frac{3}{8} \end{array}$$

She can use a bolt that is $\frac{3}{8}$ in. in diameter.

61.

$$\begin{array}{r} 4\frac{1}{3} \\ -2\frac{5}{8} \\ \hline \end{array} \Rightarrow \begin{array}{r} 4\frac{8}{24} \\ -2\frac{15}{24} \\ \hline \end{array} \Rightarrow \begin{array}{r} 3\frac{32}{24} \\ -2\frac{15}{24} \\ \hline 1\frac{17}{24} \end{array}$$

63.

$$\begin{aligned} 4\left(5\frac{5}{6}\right)-3\left(2\frac{7}{12}\right) &= \frac{4}{1}\cdot\frac{35}{6}-\frac{3}{1}\cdot\frac{31}{12} \\ &= \frac{140}{6}-\frac{93}{12} \\ &= \frac{140}{6}\cdot\frac{2}{2}-\frac{93}{12} \\ &= \frac{280}{12}-\frac{93}{12} \\ &= \frac{187}{12} \end{aligned}$$

$$\begin{aligned} 6\left(5\frac{1}{2}\right)-5\frac{3}{4} &= \frac{6}{1}\cdot\frac{11}{2}-\frac{23}{4} \\ &= \frac{66}{2}-\frac{23}{4} \\ &= \frac{66}{2}\cdot\frac{2}{2}-\frac{23}{4} \\ &= \frac{132}{4}-\frac{23}{4} \\ &= \frac{109}{4} \end{aligned}$$

They are not equal.

65.

$$\begin{array}{r} 5\frac{1}{3} \\ -1\frac{2}{3} \\ \hline \end{array} \Rightarrow \begin{array}{r} 4\frac{4}{3} \\ -1\frac{2}{3} \\ \hline 3\frac{2}{3} \end{array}$$

$20\div 3\frac{2}{3}=\frac{20}{1}\div\frac{11}{3}=\frac{20}{1}\cdot\frac{3}{11}=\frac{60}{11}=5\frac{5}{11}$

The snail gains $3\frac{2}{3}$ feet in a day. It will take the snail 6 days to gain over 20 feet.

67.

$$\begin{aligned} 8^2-4\cdot 3+10 &= 64-12+10 \\ &= 52+10=62 \end{aligned}$$

69. $14\cdot 3\div 7\cdot 3=42\div 7\cdot 3=6\cdot 3=18$

71. $5+32\div 4^2=5+32\div 16=5+2=7$

73. $\frac{1}{2}\cdot\frac{5}{8}\cdot\frac{3}{5}\cdot 24=\frac{1}{2}\cdot\frac{\cancel{5}^{1}}{\cancel{8}_{1}}\cdot\frac{3}{\cancel{5}_{1}}\cdot\frac{\cancel{24}^{3}}{1}=\frac{9}{2}=4\frac{1}{2}$

75. $7\frac{3}{4}+1\frac{1}{2}=7+1+\frac{3}{4}+\frac{1}{2}=8+\frac{3}{4}+\frac{2}{4}=8\frac{5}{4}=9\frac{1}{4}$;

$24\left(9\frac{1}{4}\right)=\frac{24}{1}\cdot\frac{37}{4}=6(37)=222$; $222+10=232$ oz per case

$5(232)=1160$; The weight is 1160 oz, or $72\frac{1}{2}$ lb.

Getting Ready for Algebra (page 249)

1.
$$\begin{aligned} a+\frac{1}{8}&=\frac{5}{8}\\ a+\frac{1}{8}-\frac{1}{8}&=\frac{5}{8}-\frac{1}{8}\\ a&=\frac{4}{8}=\frac{1}{2} \end{aligned}$$

3.
$$\begin{aligned} c-\frac{3}{16}&=\frac{7}{16}\\ c-\frac{3}{16}+\frac{3}{16}&=\frac{7}{16}+\frac{3}{16}\\ c&=\frac{10}{16}=\frac{5}{8} \end{aligned}$$

5.
$$\begin{aligned} x+\frac{2}{9}&=\frac{3}{8}\\ x+\frac{2}{9}-\frac{2}{9}&=\frac{3}{8}-\frac{2}{9}\\ x&=\frac{27}{72}-\frac{16}{72}\\ x&=\frac{11}{72} \end{aligned}$$

7.
$$\begin{aligned} y-\frac{5}{7}&=\frac{8}{9}\\ y-\frac{5}{7}+\frac{5}{7}&=\frac{8}{9}+\frac{5}{7}\\ y&=\frac{56}{63}+\frac{45}{63}\\ y&=\frac{101}{63}=1\frac{38}{63} \end{aligned}$$

9.
$$\begin{aligned} a+\frac{9}{8}&=\frac{12}{5}\\ a+\frac{9}{8}-\frac{9}{8}&=\frac{12}{5}-\frac{9}{8}\\ a&=\frac{96}{40}-\frac{45}{40}\\ a&=\frac{51}{40}=1\frac{11}{40} \end{aligned}$$

11.
$$\begin{aligned} c-1\frac{1}{8}&=2\frac{1}{3}\\ c-\frac{9}{8}&=\frac{7}{3}\\ c-\frac{9}{8}+\frac{9}{8}&=\frac{7}{3}+\frac{9}{8}\\ c&=\frac{56}{24}+\frac{27}{24}\\ c&=\frac{83}{24}=3\frac{11}{24} \end{aligned}$$

13.
$$\begin{aligned} x+6\frac{3}{4}&=7\frac{7}{9}\\ x+\frac{27}{4}&=\frac{70}{9}\\ x+\frac{27}{4}-\frac{27}{4}&=\frac{70}{9}-\frac{27}{4}\\ x&=\frac{280}{36}-\frac{243}{36}\\ x&=\frac{37}{36}=1\frac{1}{36} \end{aligned}$$

15.
$$\begin{aligned} 12&=w+8\frac{5}{6}\\ 12-8\frac{5}{6}&=w+8\frac{5}{6}-8\frac{5}{6}\\ \frac{12}{1}-\frac{53}{6}&=w\\ \frac{72}{6}-\frac{53}{6}&=w\\ \frac{19}{6}&=w, \text{ or } w=3\frac{1}{6} \end{aligned}$$

17.
$$a - 13\frac{5}{6} = 22\frac{11}{18}$$
$$a - 13\frac{5}{6} + 13\frac{5}{6} = 22\frac{11}{18} + 13\frac{5}{6}$$
$$a = \frac{407}{18} + \frac{83}{6}$$
$$a = \frac{407}{18} + \frac{249}{18}$$
$$a = \frac{656}{18} = 36\frac{8}{18} = 36\frac{4}{9}$$

19.
$$c + 44\frac{13}{21} = 65\frac{5}{7}$$
$$c + 44\frac{13}{21} - 44\frac{13}{21} = 65\frac{5}{7} - 44\frac{13}{21}$$
$$c = \frac{460}{7} - \frac{937}{21}$$
$$c = \frac{1380}{21} - \frac{937}{21}$$
$$c = \frac{443}{21} = 21\frac{2}{21}$$

21. Let x = the original height.
Orig. ht. + Growth = New ht.
$$x + 1\frac{15}{16} = 45\frac{1}{2}$$
$$x + 1\frac{15}{16} - 1\frac{15}{16} = 45\frac{1}{2} - 1\frac{15}{16}$$
$$x = \frac{91}{2} - \frac{31}{16}$$
$$x = \frac{728}{16} - \frac{31}{16}$$
$$x = \frac{697}{16} = 43\frac{9}{16}$$
The height 10 years ago was $43\frac{9}{16}$ feet.

23. Let x = the original amount.
Orig. Amt. − Used = Amt. left
$$x - 18\frac{2}{3} = 27\frac{1}{3}$$
$$x - 18\frac{2}{3} + 18\frac{2}{3} = 27\frac{1}{3} + 18\frac{2}{3}$$
$$x = 27 + 18 + \frac{1}{3} + \frac{2}{3}$$
$$x = 45 + \frac{3}{3}$$
$$x = 45 + 1 = 46$$
She bought 46 pounds of nails.

Exercises 3.10 (page 255)

1. $\frac{4}{11} + \frac{9}{11} - \frac{6}{11} = \frac{13}{11} - \frac{6}{11} = \frac{7}{11}$

3. $\frac{5}{17} - \left(\frac{1}{17} + \frac{2}{17}\right) = \frac{5}{17} - \frac{3}{17} = \frac{2}{17}$

5. $\frac{7}{8} - \frac{3}{4} \cdot \frac{1}{2} = \frac{7}{8} - \frac{3}{8} = \frac{4}{8} = \frac{1}{2}$

7. $\frac{1}{5} + \frac{3}{10} \div \frac{1}{2} = \frac{1}{5} + \frac{3}{10} \cdot \frac{2}{1} = \frac{1}{5} + \frac{6}{10}$
$= \frac{1}{5} + \frac{3}{5} = \frac{4}{5}$

9. $\frac{3}{5} \cdot \frac{1}{4} - \frac{1}{8} = \frac{3}{20} - \frac{1}{8} = \frac{6}{40} - \frac{5}{40} = \frac{1}{40}$

11. $\frac{7}{12} \div \left(\frac{1}{3} + \frac{1}{3}\right) = \frac{7}{12} \div \frac{2}{3} = \frac{7}{12} \cdot \frac{3}{2} = \frac{7}{8}$

13. $\frac{7}{10} + \left(\frac{1}{5} + \frac{1}{5}\right)^2 = \frac{7}{10} + \left(\frac{2}{5}\right)^2 = \frac{7}{10} + \frac{2}{5} \cdot \frac{2}{5} = \frac{7}{10} + \frac{4}{25} = \frac{35}{50} + \frac{8}{50} = \frac{43}{50}$

15. $\frac{1}{2} \div \frac{2}{3} \cdot \frac{5}{6} = \frac{1}{2} \cdot \frac{3}{2} \cdot \frac{5}{6} = \frac{3}{4} \cdot \frac{5}{6} = \frac{15}{24} = \frac{5}{8}$

17. $\frac{11}{12} + \left(\frac{3}{8} \cdot \frac{1}{2}\right) = \frac{11}{12} + \frac{3}{16} = \frac{44}{48} + \frac{9}{48}$
$= \frac{53}{48} = 1\frac{5}{48}$

SECTION 3.10

19. $\frac{3}{5}-\frac{1}{3}\div\frac{3}{4}+\frac{3}{10}=\frac{3}{5}-\frac{1}{3}\cdot\frac{4}{3}+\frac{3}{10}=\frac{3}{5}-\frac{4}{9}+\frac{3}{10}=\frac{54}{90}-\frac{40}{90}+\frac{27}{90}=\frac{14}{90}+\frac{27}{90}=\frac{41}{90}$

21. $\frac{5}{8}\cdot\frac{1}{3}+\frac{1}{2}\div\frac{1}{3}-\frac{7}{8}=\frac{5}{24}+\frac{1}{2}\cdot\frac{3}{1}-\frac{7}{8}=\frac{5}{24}+\frac{3}{2}-\frac{7}{8}=\frac{5}{24}+\frac{36}{24}-\frac{21}{24}=\frac{41}{24}-\frac{21}{24}=\frac{20}{24}=\frac{5}{6}$

23.
$$\begin{aligned}\frac{11}{14}+\left(\frac{2}{7}\right)^2-\frac{17}{49}\cdot\frac{5}{2}=\frac{11}{14}+\frac{2}{7}\cdot\frac{2}{7}-\frac{17}{49}\cdot\frac{5}{2}&=\frac{11}{14}+\frac{4}{49}-\frac{85}{98}\\&=\frac{77}{98}+\frac{8}{98}-\frac{85}{98}=\frac{85}{98}-\frac{85}{98}=0\end{aligned}$$

25.
$$\begin{aligned}\frac{17}{18}-\frac{7}{9}+\frac{1}{9}\div\left(\frac{4}{3}\right)^2=\frac{17}{18}-\frac{7}{9}+\frac{1}{9}\div\left(\frac{4}{3}\cdot\frac{4}{3}\right)&=\frac{17}{18}-\frac{7}{9}+\frac{1}{9}\div\frac{16}{9}\\&=\frac{17}{18}-\frac{7}{9}+\frac{1}{9}\cdot\frac{9}{16}\\&=\frac{17}{18}-\frac{7}{9}+\frac{1}{16}\\&=\frac{136}{144}-\frac{112}{144}+\frac{9}{144}\\&=\frac{24}{144}+\frac{9}{144}=\frac{33}{144}=\frac{11}{48}\end{aligned}$$

27. $\frac{7}{8}-\left(\frac{1}{2}\div\frac{2}{5}-\frac{3}{8}\right)=\frac{7}{8}-\left(\frac{1}{2}\cdot\frac{5}{2}-\frac{3}{8}\right)=\frac{7}{8}-\left(\frac{5}{4}-\frac{3}{8}\right)=\frac{7}{8}-\left(\frac{10}{8}-\frac{3}{8}\right)=\frac{7}{8}-\frac{7}{8}=0$

29. $\frac{4}{11}+\frac{8}{11}=\frac{12}{11}$
$\frac{12}{11}\div 2=\frac{12}{11}\cdot\frac{1}{2}=\frac{6}{11}$

31. $\frac{2}{7}+\frac{4}{7}+\frac{5}{7}=\frac{11}{7}$
$\frac{11}{7}\div 3=\frac{11}{7}\cdot\frac{1}{3}=\frac{11}{21}$

33. $\frac{4}{15}+\frac{8}{15}+\frac{4}{15}=\frac{16}{15}$
$\frac{16}{15}\div 3=\frac{16}{15}\cdot\frac{1}{3}=\frac{16}{45}$

35. $\frac{1}{5}+\frac{2}{5}+\frac{7}{15}=\frac{3}{15}+\frac{6}{15}+\frac{7}{15}=\frac{16}{15}$
$\frac{16}{15}\div 3=\frac{16}{15}\cdot\frac{1}{3}=\frac{16}{45}$

37. $2\frac{1}{3}+3\frac{1}{2}+1\frac{5}{6}=\frac{7}{3}+\frac{7}{2}+\frac{11}{6}=\frac{14}{6}+\frac{21}{6}+\frac{11}{6}=\frac{46}{6}=\frac{23}{3};\ \frac{23}{3}\div 3=\frac{23}{3}\cdot\frac{1}{3}=\frac{23}{9}=2\frac{5}{9}$

39. $\frac{11}{15}+\frac{2}{5}+\frac{1}{3}+\frac{1}{5}=\frac{11}{15}+\frac{6}{15}+\frac{5}{15}+\frac{3}{15}=\frac{25}{15}=\frac{5}{3};\ \frac{5}{3}\div 4=\frac{5}{3}\cdot\frac{1}{4}=\frac{5}{12}$

41. $\frac{4}{9}+\frac{5}{18}+\frac{5}{6}+\frac{2}{3}+\frac{1}{2}=\frac{8}{18}+\frac{5}{18}+\frac{15}{18}+\frac{12}{18}+\frac{9}{18}=\frac{49}{18};\ \frac{49}{18}\div 5=\frac{49}{18}\cdot\frac{1}{5}=\frac{49}{90}$

43. $3\frac{1}{6}+2\frac{1}{3}+4\frac{1}{6}=\frac{19}{6}+\frac{7}{3}+\frac{25}{6}=\frac{19}{6}+\frac{14}{6}+\frac{25}{6}=\frac{58}{6}=\frac{29}{3};\ \frac{29}{3}\div 3=\frac{29}{3}\cdot\frac{1}{3}=\frac{29}{9}=3\frac{2}{9}$

SECTION 3.10

45. $5\frac{1}{3}+6\frac{2}{5}+9\frac{13}{15}=\frac{16}{3}+\frac{32}{5}+\frac{148}{15}=\frac{80}{15}+\frac{96}{15}+\frac{148}{15}=\frac{324}{15};$
$\frac{324}{15}\div 3=\frac{324}{15}\cdot\frac{1}{3}=\frac{324}{45}=7\frac{9}{45}=7\frac{1}{5}$

47.
$$\begin{aligned}\frac{14}{15}-\left(\frac{4}{5}-\frac{1}{2}\div\frac{2}{3}\right)\cdot\frac{4}{5}=\frac{14}{15}-\left(\frac{4}{5}-\frac{1}{2}\cdot\frac{3}{2}\right)\cdot\frac{4}{5}&=\frac{14}{15}-\left(\frac{4}{5}-\frac{3}{4}\right)\cdot\frac{4}{5}\\&=\frac{14}{15}-\left(\frac{16}{20}-\frac{15}{20}\right)\cdot\frac{4}{5}\\&=\frac{14}{15}-\frac{1}{20}\cdot\frac{4}{5}\\&=\frac{14}{15}-\frac{1}{25}=\frac{70}{75}-\frac{3}{75}=\frac{67}{75}\end{aligned}$$

49. $\left(\frac{3}{5}+\frac{7}{10}\cdot\frac{2}{3}\right)\left(\frac{3}{2}\right)^2=\left(\frac{3}{5}+\frac{7}{15}\right)\left(\frac{9}{4}\right)=\left(\frac{9}{15}+\frac{7}{15}\right)\left(\frac{9}{4}\right)=\frac{16}{15}\cdot\frac{9}{4}=\frac{36}{15}=2\frac{6}{15}=2\frac{2}{5}$

51. $3\frac{2}{3}+4\frac{5}{6}+2\frac{5}{9}=\frac{11}{3}+\frac{29}{6}+\frac{23}{9}=\frac{66}{18}+\frac{87}{18}+\frac{46}{18}=\frac{199}{18};\ \frac{199}{18}\div 3=\frac{199}{18}\cdot\frac{1}{3}=\frac{199}{54}=3\frac{37}{54}$

53. $\frac{7}{15}+4+\frac{4}{5}+7=11+\frac{7}{15}+\frac{12}{15}=11+\frac{19}{15}=11+1\frac{4}{15}=12\frac{4}{15}=\frac{184}{15};$
$\frac{184}{15}\div 4=\frac{184}{15}\cdot\frac{1}{4}=\frac{46}{15}=3\frac{1}{15}$

55.
$$\begin{aligned}23\frac{1}{4}+31\frac{5}{8}+42\frac{3}{4}+28\frac{5}{8}+35\frac{3}{4}+40&=23+31+42+28+35+40+\frac{1}{4}+\frac{5}{8}+\frac{3}{4}+\frac{5}{8}+\frac{3}{4}\\&=199+\frac{2}{8}+\frac{5}{8}+\frac{6}{8}+\frac{5}{8}+\frac{6}{8}\\&=199+\frac{24}{8}=199+3=202\end{aligned}$$

$202\div 6=\frac{202}{6}=33\frac{4}{6}=33\frac{2}{3}$; The average length is $33\frac{2}{3}$ inches.

57. $5(6)+10\left(2\frac{1}{2}\right)=30+\frac{10}{1}\cdot\frac{5}{2}=30+25=55$ cups of cereal

$5\left(\frac{1}{4}\right)+10(0)=\frac{5}{1}\cdot\frac{1}{4}+0=\frac{5}{4}=1\frac{1}{4}$ cups of butter

$5\left(1\frac{3}{4}\right)+10\left(2\frac{1}{2}\right)=\frac{5}{1}\cdot\frac{7}{4}+\frac{10}{1}\cdot\frac{5}{2}=\frac{35}{4}+\frac{50}{2}=\frac{35}{4}+\frac{100}{4}=\frac{135}{4}$
$=33\frac{3}{4}$ cups of chocolate chips

$5\left(3\frac{1}{2}\right)+10(1)=\frac{5}{1}\cdot\frac{7}{2}+\frac{10}{1}=\frac{35}{2}+\frac{20}{2}=\frac{55}{2}=27\frac{1}{2}$ cups of marshmallows

$5(0)+10(4)=0+40=40$ cups of pretzels; $5(0)+10(1)=0+10=10$ cups of raisins

59. $69'\,1\frac{1}{4}'' + 67'\,5\frac{1}{2}'' + 67'\,5'' + 69'\,1\frac{1}{8}'' + 69'\,5\frac{1}{2}''$

feet: $69 + 67 + 67 + 69 + 69 = 341$; $341 \div 5 = \frac{341}{1} \cdot \frac{1}{5} = \frac{341}{5} = 68\frac{1}{5}$ feet

$\frac{1}{5}$ foot: $\frac{1}{5} \cdot 12 = \frac{1}{5} \cdot \frac{12}{1} = \frac{12}{5} = 2\frac{2}{5}$ inches $\Rightarrow$ Average for feet: $68'\,2\frac{2}{5}''$

inches: $1\frac{1}{4} + 5\frac{1}{2} + 5 + 1\frac{1}{8} + 5\frac{1}{2} = \frac{5}{4} + \frac{11}{2} + \frac{5}{1} + \frac{9}{8} + \frac{11}{2} = \frac{10}{8} + \frac{44}{8} + \frac{40}{8} + \frac{9}{8} + \frac{44}{8}$

$= \frac{147}{8}$; $\frac{147}{8} \div 5 = \frac{147}{8} \cdot \frac{1}{5} = \frac{147}{40} = 3\frac{27}{40}''$

Total: $68'\,2\frac{2}{5}'' + 3\frac{27}{40}'' = 68' + \left(2\frac{2}{5} + 3\frac{27}{40}\right)'' = 68' + \left(\frac{12}{5} + \frac{147}{40}\right)'' = 68' + \left(\frac{96}{40} + \frac{147}{40}\right)''$

$= 68' + \frac{243}{40}'' = 68' + 6\frac{3}{40}'' = 68'\,6\frac{3}{40}''$

61. $98\frac{3}{5} - 32 = 66\frac{3}{5}$ $66\frac{3}{5} \cdot \frac{5}{9} = \frac{333}{5} \cdot \frac{5}{9} = 37$

Normal body temperature is 37° C.

63. Perimeter $= 2(4) + 2(2) = 8 + 4 = 12$ feet
Length of wood needed $= 2(12) = 24$ feet
He will need three 2×12s, and he will have 6 ft left over.

65. 1. Do operations in parentheses first. (following order in 2, 3, and 4)
2. Do exponents next. 3. Do multiplications or divisions. 4. Do additions or subtractions.
There is no change for fractions.

67.
$$2\frac{5}{8}\left(4\frac{1}{5} - 3\frac{5}{6}\right) \div 2\frac{1}{2}\left(3\frac{1}{7} + 2\frac{1}{5}\right) = \frac{21}{8}\left(\frac{21}{5} - \frac{23}{6}\right) \div \frac{5}{2}\left(\frac{22}{7} + \frac{11}{5}\right)$$
$$= \frac{21}{8}\left(\frac{126}{30} - \frac{115}{30}\right) \div \frac{5}{2}\left(\frac{110}{35} + \frac{77}{35}\right)$$
$$= \frac{21}{8} \cdot \frac{11}{30} \div \frac{5}{2} \cdot \frac{187}{35}$$
$$= \frac{77}{80} \div \frac{5}{2} \cdot \frac{187}{35}$$
$$= \frac{77}{80} \cdot \frac{2}{5} \cdot \frac{187}{35} = \frac{77}{200} \cdot \frac{187}{35} = \frac{14{,}399}{7000} = 2\frac{399}{7000} = 2\frac{57}{1000}$$

69.
$$3\frac{2}{5} + 1\frac{1}{2}\left(3\frac{2}{5}\right) + \frac{7}{8}\left[1\frac{1}{2}\left(3\frac{2}{5}\right)\right] = \frac{17}{5} + \frac{3}{2} \cdot \frac{17}{5} + \frac{7}{8} \cdot \frac{3}{2} \cdot \frac{17}{5}$$
$$= \frac{17}{5} + \frac{51}{10} + \frac{357}{80} = \frac{272}{80} + \frac{408}{80} + \frac{357}{80} = \frac{1037}{80}$$

$\frac{1037}{80} \cdot \frac{1500}{1} = \frac{1{,}555{,}500}{80} = 19{,}443\frac{60}{80} \approx 19{,}444$; The amount paid is about \$19,444.

71. $\begin{array}{r} 3\frac{4}{7} \\ -\ \frac{9}{14} \\ \hline \end{array} \Rightarrow \begin{array}{r} 3\frac{8}{14} \\ -\ \frac{9}{14} \\ \hline \end{array} \Rightarrow \begin{array}{r} 2\frac{22}{14} \\ -\ \frac{9}{14} \\ \hline 2\frac{13}{14} \end{array}$

73. $3\frac{4}{7} \div \left(\frac{9}{14}\right) = \frac{25}{7} \div \frac{9}{14} = \frac{25}{7} \cdot \frac{14}{9}$
$= \frac{50}{9} = 5\frac{5}{9}$

75. $\frac{9}{\underset{3}{\cancel{15}}} \cdot \frac{\overset{1}{\cancel{3}}}{4} \cdot \frac{\overset{7}{\cancel{35}}}{\underset{2}{\cancel{6}}} = \frac{\overset{3}{\cancel{9}}}{\underset{1}{\cancel{3}}} \cdot \frac{1}{4} \cdot \frac{7}{2} = \frac{21}{8} = 2\frac{5}{8}$

77. $975 = 25 \cdot 39 = 5 \cdot 5 \cdot 3 \cdot 13 = 3 \cdot 5^2 \cdot 13$

79. $4\frac{3}{4} + 5\frac{1}{2} + 6\frac{3}{4} + 3\frac{1}{3} + 4\frac{2}{3} = 4 + 5 + 6 + 3 + 4 + \frac{3}{4} + \frac{1}{2} + \frac{3}{4} + \frac{1}{3} + \frac{2}{3}$
$= 22 + \frac{9}{12} + \frac{6}{12} + \frac{9}{12} + \frac{3}{3} = 22 + \frac{24}{2} + 1 = 22 + 2 + 1 = 25$

$25(10) = 250$; He walked 25 miles and raised \$250.

Chapter 3 Review Exercises (page 263)

1. The whole is divided into 5 parts. 4 of the parts are shaded. $\Rightarrow \frac{4}{5}$

2. The whole is divided into 12 parts. 7 of the parts are shaded. $\Rightarrow \frac{7}{12}$

3. The unit is divided into 10 parts. 9 of the parts are measured. $\Rightarrow \frac{9}{10}$

4. The unit is divided into 12 parts. 11 of the parts are measured. $\Rightarrow \frac{11}{12}$

5. improper: $\frac{8}{3}, \frac{9}{9}, \frac{22}{19}$

6. improper: $\frac{5}{2}$

7. $\begin{array}{r} 7 \\ 11\overline{)82} \\ \underline{77} \\ 5 \end{array} \Rightarrow 7\frac{5}{11}$

8. $\begin{array}{r} 8 \\ 9\overline{)76} \\ \underline{72} \\ 4 \end{array} \Rightarrow 8\frac{4}{9}$

9. $\begin{array}{r} 82 \\ 5\overline{)413} \\ \underline{40} \\ 13 \\ \underline{10} \\ 3 \end{array} \Rightarrow 82\frac{3}{5}$

10. $\begin{array}{r} 49 \\ 7\overline{)344} \\ \underline{28} \\ 64 \\ \underline{63} \\ 1 \end{array} \Rightarrow 49\frac{1}{7}$

11. $6\frac{5}{12} = \frac{12(6) + 5}{12} = \frac{72 + 5}{12} = \frac{77}{12}$

12. $4\frac{3}{7} = \frac{7(4) + 3}{7} = \frac{28 + 3}{7} = \frac{31}{7}$

13. $12\frac{5}{6} = \frac{6(12) + 5}{6} = \frac{72 + 5}{6} = \frac{77}{6}$

14. $9\frac{2}{3} = \frac{3(9) + 2}{3} = \frac{27 + 2}{3} = \frac{29}{3}$

CHAPTER 3 REVIEW EXERCISES

15. $17 = \frac{17}{1}$

16. $35 = \frac{35}{1}$

17. $64{,}435 \div 24 = \frac{64{,}435}{24} = 2684\frac{19}{24}$

The wholesaler can pack $2684\frac{19}{24}$ cases.

18. $46{,}325 \div 15 = \frac{46{,}325}{15} = 3088\frac{5}{15} = 3088\frac{1}{3}$

The wholesaler can make $3088\frac{1}{3}$ packs.

19. $\frac{24}{32} = \frac{3 \cdot \cancel{8}}{4 \cdot \cancel{8}} = \frac{3}{4}$

20. $\frac{10}{25} = \frac{2 \cdot \cancel{5}}{5 \cdot \cancel{5}} = \frac{2}{5}$

21. $\frac{60}{90} = \frac{2 \cdot \cancel{30}}{3 \cdot \cancel{30}} = \frac{2}{3}$

22. $\frac{18}{24} = \frac{3 \cdot \cancel{6}}{4 \cdot \cancel{6}} = \frac{3}{4}$

23. $\frac{21}{35} = \frac{3 \cdot \cancel{7}}{5 \cdot \cancel{7}} = \frac{3}{5}$

24. $\frac{35}{70} = \frac{1 \cdot \cancel{35}}{2 \cdot \cancel{35}} = \frac{1}{2}$

25. $\frac{102}{6} = \frac{17 \cdot \cancel{6}}{1 \cdot \cancel{6}} = 17$

26. $\frac{126}{42} = \frac{3 \cdot \cancel{42}}{1 \cdot \cancel{42}} = 3$

27. $\frac{14}{42} = \frac{1 \cdot \cancel{14}}{3 \cdot \cancel{14}} = \frac{1}{3}$

28. $\frac{30}{45} = \frac{2 \cdot \cancel{15}}{3 \cdot \cancel{15}} = \frac{2}{3}$

29. $\frac{78}{96} = \frac{13 \cdot \cancel{6}}{16 \cdot \cancel{6}} = \frac{13}{16}$

30. $\frac{75}{125} = \frac{3 \cdot \cancel{25}}{5 \cdot \cancel{25}} = \frac{3}{5}$

31. $\frac{26}{130} = \frac{1 \cdot \cancel{26}}{5 \cdot \cancel{26}} = \frac{1}{5}$

32. $\frac{96}{144} = \frac{2 \cdot \cancel{48}}{3 \cdot \cancel{48}} = \frac{2}{3}$

33. $\frac{268}{402} = \frac{2 \cdot \cancel{134}}{3 \cdot \cancel{134}} = \frac{2}{3}$

34. $\frac{630}{1050} = \frac{3 \cdot \cancel{210}}{5 \cdot \cancel{210}} = \frac{3}{5}$

35. $\frac{219}{365} = \frac{3 \cdot \cancel{73}}{5 \cdot \cancel{73}} = \frac{3}{5}$

She is $\frac{3}{5}$ of a year old.

36. $42 + 18 = 60$; $\frac{42}{60} = \frac{7}{10}$

$\frac{7}{10}$ were correctly answered.

37. $\frac{3}{7} \cdot \frac{2}{7} = \frac{3 \cdot 2}{7 \cdot 7} = \frac{6}{49}$

38. $\frac{7}{8} \cdot \frac{1}{6} = \frac{7 \cdot 1}{8 \cdot 6} = \frac{7}{48}$

39. $\frac{3}{5} \cdot \frac{6}{11} = \frac{3 \cdot 6}{5 \cdot 11} = \frac{18}{55}$

40. $6 \cdot \frac{2}{3} \cdot \frac{3}{14} = \frac{6}{1} \cdot \frac{\overset{1}{\cancel{2}}}{\underset{1}{\cancel{3}}} \cdot \frac{\overset{1}{\cancel{3}}}{\underset{7}{\cancel{14}}} = \frac{6}{7}$

41. $\frac{21}{5} \cdot \frac{5}{4} \cdot \frac{4}{21} = \frac{\overset{1}{\cancel{21}}}{\underset{1}{\cancel{5}}} \cdot \frac{\overset{1}{\cancel{5}}}{\underset{1}{\cancel{4}}} \cdot \frac{\overset{1}{\cancel{4}}}{\underset{1}{\cancel{21}}} = \frac{1}{1} = 1$

42. $\frac{28}{35} \cdot \frac{3}{8} \cdot \frac{5}{9} = \frac{\overset{4}{\cancel{28}}}{\underset{5}{\cancel{35}}} \cdot \frac{\overset{1}{\cancel{3}}}{8} \cdot \frac{5}{\underset{3}{\cancel{9}}} = \frac{\overset{1}{\cancel{4}}}{\underset{1}{\cancel{5}}} \cdot \frac{1}{\underset{2}{\cancel{8}}} \cdot \frac{\overset{1}{\cancel{5}}}{3} = \frac{1}{6}$

43. The reciprocal of $\frac{3}{8}$ is $\frac{8}{3}$.

44. The reciprocal of $5 = \frac{5}{1}$ is $\frac{1}{5}$.

45. $\frac{9}{16} \div \frac{7}{8} = \frac{9}{16} \cdot \frac{8}{7} = \frac{9}{\underset{2}{\cancel{16}}} \cdot \frac{\overset{1}{\cancel{8}}}{7} = \frac{9}{14}$

46. $\frac{8}{13} \div \frac{2}{13} = \frac{8}{13} \cdot \frac{13}{2} = \frac{\overset{4}{\cancel{8}}}{\underset{1}{\cancel{13}}} \cdot \frac{\overset{1}{\cancel{13}}}{\underset{1}{\cancel{2}}} = \frac{4}{1} = 4$

47. $\frac{15}{18} \div \frac{30}{27} = \frac{15}{18} \cdot \frac{27}{30} = \frac{\overset{1}{\cancel{15}}}{\underset{2}{\cancel{18}}} \cdot \frac{\overset{3}{\cancel{27}}}{\underset{2}{\cancel{30}}} = \frac{3}{4}$

48. $\frac{12}{15} \div \frac{15}{8} = \frac{12}{15} \cdot \frac{8}{15} = \frac{\overset{4}{\cancel{12}}}{15} \cdot \frac{8}{\underset{5}{\cancel{15}}} = \frac{32}{75}$

CHAPTER 3 REVIEW EXERCISES

49. $\frac{33}{50} \div \frac{27}{35} = \frac{33}{50} \cdot \frac{35}{27} = \frac{\overset{11}{\cancel{33}}}{\underset{10}{\cancel{50}}} \cdot \frac{\overset{7}{\cancel{35}}}{\underset{9}{\cancel{27}}} = \frac{77}{90}$

50. $\frac{32}{45} \div \frac{8}{9} = \frac{32}{45} \cdot \frac{9}{8} = \frac{\overset{4}{\cancel{32}}}{\underset{5}{\cancel{45}}} \cdot \frac{\overset{1}{\cancel{9}}}{\underset{1}{\cancel{8}}} = \frac{4}{5}$

51. $\frac{2}{7} \text{ of } \frac{1}{2} = \frac{2}{7} \cdot \frac{1}{2} = \frac{\overset{1}{\cancel{2}}}{7} \cdot \frac{1}{\underset{1}{\cancel{2}}} = \frac{1}{7}$

$\frac{1}{7}$ of the family income goes for rent.

52. $\frac{5}{4} \div \frac{1}{8} = \frac{5}{4} \cdot \frac{8}{1} = \frac{5}{\underset{1}{\cancel{4}}} \cdot \frac{\overset{2}{\cancel{8}}}{1} = \frac{10}{1} = 10$

One package will feed 10 gerbils.

53. $\left(\frac{4}{5}\right)\left(2\frac{4}{5}\right) = \frac{4}{5} \cdot \frac{14}{5} = \frac{56}{25} = 2\frac{6}{25}$

54. $\left(\frac{3}{7}\right)\left(2\frac{5}{7}\right) = \frac{3}{7} \cdot \frac{19}{7} = \frac{57}{49} = 1\frac{8}{49}$

55. $\left(4\frac{3}{4}\right)\left(3\frac{1}{2}\right) = \frac{19}{4} \cdot \frac{7}{2} = \frac{133}{8} = 16\frac{5}{8}$

56. $\left(3\frac{1}{3}\right)\left(1\frac{4}{5}\right) = \frac{10}{3} \cdot \frac{9}{5} = \frac{\overset{2}{\cancel{10}}}{\underset{1}{\cancel{3}}} \cdot \frac{\overset{3}{\cancel{9}}}{\underset{1}{\cancel{5}}} = \frac{6}{1} = 6$

57. $\left(4\frac{3}{8}\right)\left(\frac{9}{14}\right) = \frac{35}{8} \cdot \frac{9}{14} = \frac{\overset{5}{\cancel{35}}}{8} \cdot \frac{9}{\underset{2}{\cancel{14}}}$

$= \frac{45}{16} = 2\frac{13}{16}$

58. $\left(4\frac{4}{9}\right)\left(\frac{12}{25}\right) = \frac{40}{9} \cdot \frac{12}{25} = \frac{\overset{8}{\cancel{40}}}{\underset{3}{\cancel{9}}} \cdot \frac{\overset{4}{\cancel{12}}}{\underset{5}{\cancel{25}}}$

$= \frac{32}{15} = 2\frac{2}{15}$

59. $\left(3\frac{2}{3}\right)\left(\frac{15}{22}\right)\left(7\frac{1}{2}\right) = \frac{11}{3} \cdot \frac{15}{22} \cdot \frac{15}{2} = \frac{\overset{1}{\cancel{11}}}{\underset{1}{\cancel{3}}} \cdot \frac{\overset{5}{\cancel{15}}}{\underset{2}{\cancel{22}}} \cdot \frac{15}{2} = \frac{75}{4} = 18\frac{3}{4}$

60. $\left(4\frac{3}{4}\right)\left(3\frac{1}{5}\right)\left(5\frac{5}{8}\right) = \frac{19}{4} \cdot \frac{16}{5} \cdot \frac{45}{8} = \frac{19}{\underset{1}{\cancel{4}}} \cdot \frac{\overset{4}{\cancel{16}}}{\underset{1}{\cancel{5}}} \cdot \frac{\overset{9}{\cancel{45}}}{8} = \frac{19}{1} \cdot \frac{\overset{1}{\cancel{4}}}{1} \cdot \frac{9}{\underset{2}{\cancel{8}}} = \frac{171}{2} = 85\frac{1}{2}$

61. $10 \div 1\frac{1}{4} = \frac{10}{1} \div \frac{5}{4} = \frac{10}{1} \cdot \frac{4}{5} = \frac{\overset{2}{\cancel{10}}}{1} \cdot \frac{4}{\underset{1}{\cancel{5}}}$

$= \frac{8}{1} = 8$

62. $4 \div 1\frac{1}{4} = \frac{4}{1} \div \frac{5}{4} = \frac{4}{1} \cdot \frac{4}{5} = \frac{16}{5} = 3\frac{1}{5}$

63. $4\frac{2}{7} \div 5 = \frac{30}{7} \div \frac{5}{1} = \frac{30}{7} \cdot \frac{1}{5} = \frac{\overset{6}{\cancel{30}}}{7} \cdot \frac{1}{\underset{1}{\cancel{5}}} = \frac{6}{7}$

64. $2\frac{1}{6} \div 4 = \frac{13}{6} \div \frac{4}{1} = \frac{13}{6} \cdot \frac{1}{4} = \frac{13}{24}$

65. $8\frac{2}{5} \div 2\frac{1}{3} = \frac{42}{5} \div \frac{7}{3} = \frac{42}{5} \cdot \frac{3}{7}$

$= \frac{\overset{6}{\cancel{42}}}{5} \cdot \frac{3}{\underset{1}{\cancel{7}}} = \frac{18}{5} = 3\frac{3}{5}$

66. $3\frac{1}{6} \div 4\frac{3}{4} = \frac{19}{6} \div \frac{19}{4} = \frac{19}{6} \cdot \frac{4}{19}$

$= \frac{\overset{1}{\cancel{19}}}{\underset{3}{\cancel{6}}} \cdot \frac{\overset{2}{\cancel{4}}}{\underset{1}{\cancel{19}}} = \frac{2}{3}$

67. $31\frac{1}{3} \div 1\frac{1}{9} = \frac{94}{3} \div \frac{10}{9} = \frac{94}{3} \cdot \frac{9}{10} = \frac{\overset{47}{\cancel{94}}}{\underset{1}{\cancel{3}}} \cdot \frac{\overset{3}{\cancel{9}}}{\underset{5}{\cancel{10}}} = \frac{141}{5} = 28\frac{1}{5}$

68. $21\frac{3}{7} \div 8\frac{1}{3} = \frac{150}{7} \div \frac{25}{3} = \frac{150}{7} \cdot \frac{3}{25} = \frac{\overset{6}{\cancel{150}}}{7} \cdot \frac{3}{\underset{1}{\cancel{25}}} = \frac{18}{7} = 2\frac{4}{7}$

69. $205\left(151\frac{3}{5}\right) = \frac{205}{1} \cdot \frac{758}{5} = \frac{\overset{41}{\cancel{205}}}{1} \cdot \frac{758}{\underset{1}{\cancel{5}}} = \frac{31{,}078}{1} = 31{,}078$; She harvests 31,078 bushels.

70. $\left(3\frac{1}{3}\right)(7740) = \frac{10}{3} \cdot \frac{7740}{1} = \frac{10}{\underset{1}{\cancel{3}}} \cdot \frac{\overset{2580}{\cancel{7740}}}{1} = \frac{25{,}800}{1} = 25{,}800$; There are 25,800 brant geese.

71.
$\frac{2}{3} = \frac{2}{3} \cdot \frac{2}{2} = \frac{4}{6}$
$\frac{2}{3} = \frac{2}{3} \cdot \frac{3}{3} = \frac{6}{9}$
$\frac{2}{3} = \frac{2}{3} \cdot \frac{5}{5} = \frac{10}{15}$
$\frac{2}{3} = \frac{2}{3} \cdot \frac{8}{8} = \frac{16}{24}$

72.
$\frac{2}{7} = \frac{2}{7} \cdot \frac{2}{2} = \frac{4}{14}$
$\frac{2}{7} = \frac{2}{7} \cdot \frac{3}{3} = \frac{6}{21}$
$\frac{2}{7} = \frac{2}{7} \cdot \frac{5}{5} = \frac{10}{35}$
$\frac{2}{7} = \frac{2}{7} \cdot \frac{8}{8} = \frac{16}{56}$

73.
$\frac{3}{14} = \frac{3}{14} \cdot \frac{2}{2} = \frac{6}{28}$
$\frac{3}{14} = \frac{3}{14} \cdot \frac{3}{3} = \frac{9}{42}$
$\frac{3}{14} = \frac{3}{14} \cdot \frac{5}{5} = \frac{15}{70}$
$\frac{3}{14} = \frac{3}{14} \cdot \frac{8}{8} = \frac{24}{112}$

74.
$\frac{4}{11} = \frac{4}{11} \cdot \frac{2}{2} = \frac{8}{22}$
$\frac{4}{11} = \frac{4}{11} \cdot \frac{3}{3} = \frac{12}{33}$
$\frac{4}{11} = \frac{4}{11} \cdot \frac{5}{5} = \frac{20}{55}$
$\frac{4}{11} = \frac{4}{11} \cdot \frac{8}{8} = \frac{32}{88}$

75. $45 \div 5 = 9$
$\frac{4}{5} = \frac{4}{5} \cdot \frac{9}{9} = \frac{36}{45}$

76. $56 \div 7 = 8$
$\frac{6}{7} = \frac{6}{7} \cdot \frac{8}{8} = \frac{48}{56}$

77. $144 \div 6 = 24$
$\frac{5}{6} = \frac{5}{6} \cdot \frac{24}{24} = \frac{120}{144}$

78. $132 \div 12 = 11$
$\frac{7}{12} = \frac{7}{12} \cdot \frac{11}{11} = \frac{77}{132}$

79. LCM = 10
$\frac{1}{2} = \frac{1}{2} \cdot \frac{5}{5} = \frac{5}{10}$
$\frac{3}{5} = \frac{3}{5} \cdot \frac{2}{2} = \frac{6}{10}$
$\frac{7}{10} = \frac{7}{10}$
$\frac{1}{2} < \frac{3}{5} < \frac{7}{10}$

80. LCM = 24
$\frac{1}{4} = \frac{1}{4} \cdot \frac{6}{6} = \frac{6}{24}$
$\frac{5}{12} = \frac{5}{12} \cdot \frac{2}{2} = \frac{10}{24}$
$\frac{3}{8} = \frac{3}{8} \cdot \frac{3}{3} = \frac{9}{24}$
$\frac{1}{4} < \frac{3}{8} < \frac{5}{12}$

81. LCM = 495
$\frac{2}{9} = \frac{2}{9} \cdot \frac{55}{55} = \frac{110}{495}$
$\frac{1}{5} = \frac{1}{5} \cdot \frac{99}{99} = \frac{99}{495}$
$\frac{3}{11} = \frac{3}{11} \cdot \frac{45}{45} = \frac{135}{495}$
$\frac{1}{5} < \frac{2}{9} < \frac{3}{11}$

CHAPTER 3 REVIEW EXERCISES

82. LCM = 18

$\frac{10}{9} = \frac{10}{9} \cdot \frac{2}{2} = \frac{20}{18}$

$\frac{4}{3} = \frac{4}{3} \cdot \frac{6}{6} = \frac{24}{18}$

$\frac{7}{6} = \frac{7}{6} \cdot \frac{3}{3} = \frac{21}{18}$

$\frac{19}{18} = \frac{19}{18}$

$\frac{19}{18} < \frac{10}{9} < \frac{7}{6} < \frac{4}{3}$

83. LCM = 56

$\frac{3}{7} = \frac{3}{7} \cdot \frac{8}{8} = \frac{24}{56}$

$\frac{5}{14} = \frac{5}{14} \cdot \frac{4}{4} = \frac{20}{56}$

$\frac{11}{28} = \frac{11}{28} \cdot \frac{2}{2} = \frac{22}{56}$

$\frac{21}{56} = \frac{21}{56}$

$\frac{5}{14} < \frac{21}{56} < \frac{11}{28} < \frac{3}{7}$

84. LCM = 24

$7\frac{3}{4} = 7 + \frac{3}{4} \cdot \frac{6}{6} = 7\frac{18}{24}$

$7\frac{7}{8} = 7 + \frac{7}{8} \cdot \frac{3}{3} = 7\frac{21}{24}$

$7\frac{5}{6} = 7 + \frac{5}{6} \cdot \frac{4}{4} = 7\frac{20}{24}$

$7\frac{3}{4} < 7\frac{5}{6} < 7\frac{7}{8}$

85. $\frac{3}{14} < \frac{5}{14} \Rightarrow$ true

86. $\frac{13}{10} > \frac{11}{10} \Rightarrow$ true

87. LCM = 42

$\frac{11}{14} = \frac{11}{14} \cdot \frac{3}{3} = \frac{33}{42}$

$\frac{17}{21} = \frac{17}{21} \cdot \frac{2}{2} = \frac{34}{42}$

$\frac{11}{14} > \frac{17}{21} \Rightarrow$ false

88. LCM = 323

$\frac{12}{17} = \frac{12}{17} \cdot \frac{19}{19} = \frac{228}{323}$

$\frac{14}{19} = \frac{14}{19} \cdot \frac{17}{17} = \frac{238}{323}$

$\frac{12}{17} < \frac{14}{19} \Rightarrow$ true

89. LCM = 16

$\frac{3}{4} = \frac{3}{4} \cdot \frac{4}{4} = \frac{12}{16}$

$\frac{5}{8} = \frac{5}{8} \cdot \frac{2}{2} = \frac{10}{16}$

$\frac{7}{16} = \frac{7}{16}$

$\frac{1}{2} = \frac{1}{2} \cdot \frac{8}{8} = \frac{8}{16}$

The largest is the $\frac{3}{4}$ ton.

The smallest is the $\frac{7}{16}$ ton.

90. $5\left(\frac{3}{16}\right) = \frac{5}{1} \cdot \frac{3}{16} = \frac{15}{16}$

$4\left(\frac{6}{25}\right) = \frac{4}{1} \cdot \frac{6}{25} = \frac{24}{25}$

LCM = 400

$\frac{15}{16} \cdot \frac{25}{25} = \frac{375}{400}$

$\frac{24}{25} \cdot \frac{16}{16} = \frac{384}{400}$

Her brother ate the most fat.

91. $\frac{4}{15} + \frac{7}{15} = \frac{11}{15}$

92. $\frac{3}{10} + \frac{3}{10} = \frac{6}{10} = \frac{3}{5}$

93. $\frac{2}{9} + \frac{2}{9} + \frac{2}{9} = \frac{6}{9} = \frac{2}{3}$

94. $\frac{3}{16} + \frac{2}{16} + \frac{3}{16} = \frac{8}{16} = \frac{1}{2}$

95. $\frac{7}{32} + \frac{8}{32} + \frac{5}{32} = \frac{20}{32} = \frac{5}{8}$

96. $\frac{7}{30} + \frac{7}{30} + \frac{6}{30} = \frac{20}{30} = \frac{2}{3}$

97. $\frac{4}{15} + \frac{1}{3} = \frac{4}{15} + \frac{1}{3} \cdot \frac{5}{5}$

$= \frac{4}{15} + \frac{5}{15} = \frac{9}{15} = \frac{3}{5}$

98. $\frac{5}{17} + \frac{7}{34} = \frac{5}{17} \cdot \frac{2}{2} + \frac{7}{34}$

$= \frac{10}{34} + \frac{7}{34} = \frac{17}{34} = \frac{1}{2}$

99. $\frac{3}{35} + \frac{8}{21} = \frac{3}{35} \cdot \frac{3}{3} + \frac{8}{21} \cdot \frac{5}{5} = \frac{9}{105} + \frac{40}{105} = \frac{49}{105} = \frac{7}{15}$

CHAPTER 3 REVIEW EXERCISES

100. $\frac{11}{30}+\frac{9}{20}+\frac{3}{10}=\frac{11}{30}\cdot\frac{2}{2}+\frac{9}{20}\cdot\frac{3}{3}+\frac{3}{10}\cdot\frac{6}{6}=\frac{22}{60}+\frac{27}{60}+\frac{18}{60}=\frac{67}{60}=1\frac{7}{60}$

101. $\frac{1}{6}+\frac{7}{8}+\frac{7}{12}=\frac{1}{6}\cdot\frac{4}{4}+\frac{7}{8}\cdot\frac{3}{3}+\frac{7}{12}\cdot\frac{2}{2}=\frac{4}{24}+\frac{21}{24}+\frac{14}{24}=\frac{39}{24}=1\frac{15}{24}=1\frac{5}{8}$

102. $\frac{7}{15}+\frac{11}{30}+\frac{5}{6}=\frac{7}{15}\cdot\frac{2}{2}+\frac{11}{30}+\frac{5}{6}\cdot\frac{5}{5}=\frac{14}{30}+\frac{11}{30}+\frac{25}{30}=\frac{50}{30}=1\frac{20}{30}=1\frac{2}{3}$

103. $\frac{1}{2}+\frac{3}{8}+\frac{1}{4}=\frac{1}{2}\cdot\frac{4}{4}+\frac{3}{8}+\frac{1}{4}\cdot\frac{2}{2}=\frac{4}{8}+\frac{3}{8}+\frac{2}{8}=\frac{9}{8}=1\frac{1}{8}$; It grew $1\frac{1}{8}$ inches.

104. $\frac{1}{10}+\frac{3}{10}+\frac{4}{10}+\frac{7}{10}=\frac{15}{10}=1\frac{5}{10}=1\frac{1}{2}$; It will be $1\frac{1}{2}$ inches thick.

105. $3\frac{2}{5}+5\frac{7}{10}=3+5+\frac{2}{5}+\frac{7}{10}=8+\frac{2}{5}\cdot\frac{2}{2}+\frac{7}{10}=8+\frac{4}{10}+\frac{7}{10}=8+\frac{11}{10}=8+1\frac{1}{10}=9\frac{1}{10}$

106. $17\frac{5}{12}+1\frac{5}{6}=17+1+\frac{5}{12}+\frac{5}{6}=18+\frac{5}{12}+\frac{5}{6}\cdot\frac{2}{2}=18+\frac{5}{12}+\frac{10}{12}=18+\frac{15}{12}$

$=18+1\frac{3}{12}=19\frac{1}{4}$

107. $4\frac{5}{7}+10\frac{9}{14}=4+10+\frac{5}{7}+\frac{9}{14}=14+\frac{5}{7}\cdot\frac{2}{2}+\frac{9}{14}=14+\frac{10}{14}+\frac{9}{14}=14+\frac{19}{14}$

$=14+1\frac{5}{14}=15\frac{5}{14}$

108. $4\frac{7}{15}+6\frac{2}{3}=4+6+\frac{7}{15}+\frac{2}{3}=10+\frac{7}{15}+\frac{2}{3}\cdot\frac{5}{5}=10+\frac{7}{15}+\frac{10}{15}=10+\frac{17}{15}$

$=10+1\frac{2}{15}=11\frac{2}{15}$

109. $7\frac{3}{8}+5\frac{5}{6}=7+5+\frac{3}{8}+\frac{5}{6}=12+\frac{3}{8}\cdot\frac{3}{3}+\frac{5}{6}\cdot\frac{4}{4}=12+\frac{9}{24}+\frac{20}{24}=12+\frac{29}{24}$

$=12+1\frac{5}{24}=13\frac{5}{24}$

110. $20\frac{5}{12}+4\frac{17}{18}=20+4+\frac{5}{12}+\frac{17}{18}=24+\frac{5}{12}\cdot\frac{3}{3}+\frac{17}{18}\cdot\frac{2}{2}=24+\frac{15}{36}+\frac{34}{36}=24+\frac{49}{36}$

$=24+1\frac{13}{36}=25\frac{13}{36}$

111. $14\frac{7}{20}+11\frac{3}{16}=14+11+\frac{7}{20}+\frac{3}{16}=25+\frac{7}{20}\cdot\frac{4}{4}+\frac{3}{16}\cdot\frac{5}{5}=25+\frac{28}{80}+\frac{15}{80}$

$=25+\frac{43}{80}=25\frac{43}{80}$

CHAPTER 3 REVIEW EXERCISES

112. $11\frac{7}{24} + 32\frac{7}{18} = 11 + 32 + \frac{7}{24} + \frac{7}{18} = 43 + \frac{7}{24} \cdot \frac{3}{3} + \frac{7}{18} \cdot \frac{4}{4} = 43 + \frac{21}{72} + \frac{28}{72}$

$= 43 + \frac{49}{72} = 43\frac{49}{72}$

113. $17\frac{1}{5} + 18\frac{1}{4} + 19\frac{3}{10} = 17 + 18 + 19 + \frac{1}{5} + \frac{1}{4} + \frac{3}{10} = 54 + \frac{1}{5} \cdot \frac{4}{4} + \frac{1}{4} \cdot \frac{5}{5} + \frac{3}{10} \cdot \frac{2}{2}$

$= 54 + \frac{4}{20} + \frac{5}{20} + \frac{6}{20}$

$= 54 + \frac{15}{20} = 54 + \frac{3}{4} = 54\frac{3}{4}$

114. $18\frac{3}{4} + 19 + 25\frac{7}{12} = 18 + 19 + 25 + \frac{3}{4} + \frac{7}{12} = 62 + \frac{3}{4} \cdot \frac{3}{3} + \frac{7}{12}$

$= 62 + \frac{9}{12} + \frac{7}{12} = 62 + \frac{16}{12} = 62 + 1\frac{4}{12} = 63\frac{1}{3}$

115. $25\frac{2}{3} + 16\frac{1}{6} + 18\frac{3}{4} = 25 + 16 + 18 + \frac{2}{3} + \frac{1}{6} + \frac{3}{4} = 59 + \frac{2}{3} \cdot \frac{4}{4} + \frac{1}{6} \cdot \frac{2}{2} + \frac{3}{4} \cdot \frac{3}{3}$

$= 59 + \frac{8}{12} + \frac{2}{12} + \frac{9}{12}$

$= 59 + \frac{19}{12} = 59 + 1\frac{7}{12} = 60\frac{7}{12}$

116. $29\frac{7}{8} + 19\frac{5}{12} + 32\frac{3}{4} = 29 + 19 + 32 + \frac{7}{8} + \frac{5}{12} + \frac{3}{4} = 80 + \frac{7}{8} \cdot \frac{3}{3} + \frac{5}{12} \cdot \frac{2}{2} + \frac{3}{4} \cdot \frac{6}{6}$

$= 80 + \frac{21}{24} + \frac{10}{24} + \frac{18}{23}$

$= 80 + \frac{49}{24} = 80 + 2\frac{1}{24} = 82\frac{1}{24}$

117. $1\frac{3}{4} + 2\frac{1}{5} + \frac{2}{3} + 1\frac{1}{2} + 3\frac{1}{4} = 1 + 2 + 1 + 3 + \frac{3}{4} + \frac{1}{5} + \frac{2}{3} + \frac{1}{2} + \frac{1}{4}$

$= 7 + \frac{3}{4} \cdot \frac{15}{15} + \frac{1}{5} \cdot \frac{12}{12} + \frac{2}{3} \cdot \frac{20}{20} + \frac{1}{2} \cdot \frac{10}{10} + \frac{1}{4} \cdot \frac{15}{15}$

$= 7 + \frac{45}{60} + \frac{12}{60} + \frac{40}{60} + \frac{10}{60} + \frac{15}{60} = 7 + \frac{142}{60} = 7 + 2\frac{22}{60} = 9\frac{11}{30}$

Russ' total weight loss was $9\frac{11}{30}$ pounds.

118. $6\frac{3}{4} + 1\frac{3}{5} + 2\frac{2}{3} + 5\frac{1}{2} = 6 + 1 + 2 + 5 + \frac{3}{4} + \frac{3}{5} + \frac{2}{3} + \frac{1}{2}$

$= 14 + \frac{3}{4} \cdot \frac{15}{15} + \frac{3}{5} \cdot \frac{12}{12} + \frac{2}{3} \cdot \frac{20}{20} + \frac{1}{2} \cdot \frac{30}{30}$

$= 14 + \frac{45}{60} + \frac{36}{60} + \frac{40}{60} + \frac{30}{60} = 14 + \frac{151}{60} = 14 + 2\frac{31}{60} = 16\frac{31}{60}$

Roona's fish weighed a total of $16\frac{31}{60}$ pounds.

CHAPTER 3 REVIEW EXERCISES

119. $\frac{7}{8}-\frac{5}{16}=\frac{7}{8}\cdot\frac{2}{2}-\frac{5}{16}=\frac{14}{16}-\frac{5}{16}=\frac{9}{16}$

120. $\frac{5}{18}-\frac{2}{9}=\frac{5}{18}-\frac{2}{9}\cdot\frac{2}{2}=\frac{5}{18}-\frac{4}{18}=\frac{1}{18}$

121. $\frac{7}{8}-\frac{1}{4}=\frac{7}{8}-\frac{1}{4}\cdot\frac{2}{2}=\frac{7}{8}-\frac{2}{8}=\frac{5}{8}$

122. $\frac{17}{20}-\frac{1}{5}=\frac{17}{20}-\frac{1}{5}\cdot\frac{4}{4}=\frac{17}{20}-\frac{4}{20}=\frac{13}{20}$

123. $\frac{17}{24}-\frac{1}{6}=\frac{17}{24}-\frac{1}{6}\cdot\frac{4}{4}=\frac{17}{24}-\frac{4}{24}$
$=\frac{13}{24}$

124. $\frac{7}{15}-\frac{3}{20}=\frac{7}{15}\cdot\frac{4}{4}-\frac{3}{20}\cdot\frac{3}{3}=\frac{28}{60}-\frac{9}{60}$
$=\frac{19}{60}$

125. $\frac{5}{6}-\frac{4}{5}=\frac{5}{6}\cdot\frac{5}{5}-\frac{4}{5}\cdot\frac{6}{6}=\frac{25}{30}-\frac{24}{30}$
$=\frac{1}{30}$

126. $\frac{7}{10}-\frac{1}{4}=\frac{7}{10}\cdot\frac{2}{2}-\frac{1}{4}\cdot\frac{5}{5}=\frac{14}{20}-\frac{5}{20}$
$=\frac{9}{20}$

127. $\frac{19}{25}-\frac{8}{15}=\frac{19}{25}\cdot\frac{3}{3}-\frac{8}{15}\cdot\frac{5}{5}=\frac{57}{75}-\frac{40}{75}$
$=\frac{17}{75}$

128. $\frac{21}{32}-\frac{5}{16}=\frac{21}{32}-\frac{5}{16}\cdot\frac{2}{2}=\frac{21}{32}-\frac{10}{32}$
$=\frac{11}{32}$

129. $\frac{57}{90}-\frac{17}{60}=\frac{57}{90}\cdot\frac{2}{2}-\frac{17}{60}\cdot\frac{3}{3}=\frac{114}{180}-\frac{51}{180}$
$=\frac{63}{180}=\frac{7}{20}$

130. $\frac{13}{15}-\frac{9}{20}=\frac{13}{15}\cdot\frac{4}{4}-\frac{9}{20}\cdot\frac{3}{3}=\frac{52}{60}-\frac{27}{60}$
$=\frac{25}{60}=\frac{5}{12}$

131. $\frac{3}{4}-\frac{1}{3}=\frac{3}{4}\cdot\frac{3}{3}-\frac{1}{3}\cdot\frac{4}{4}=\frac{9}{12}-\frac{4}{12}=\frac{5}{12}$; She has $\frac{5}{12}$ of an ounce left.

132. $\frac{13}{16}-\frac{5}{8}=\frac{13}{16}-\frac{5}{8}\cdot\frac{2}{2}=\frac{13}{16}-\frac{10}{16}=\frac{3}{16}$; The carpenter removes $\frac{3}{16}$ of an inch.

133. $\begin{array}{r} 167\frac{3}{5} \\ -\ 82\frac{3}{10} \\ \hline \end{array} \Rightarrow \begin{array}{r} 167\frac{6}{10} \\ -\ 82\frac{3}{10} \\ \hline 85\frac{3}{10} \end{array}$

134. $\begin{array}{r} 6\frac{5}{6} \\ -\ 3\frac{3}{10} \\ \hline \end{array} \Rightarrow \begin{array}{r} 6\frac{25}{30} \\ -\ 3\frac{9}{30} \\ \hline 3\frac{16}{30}=3\frac{8}{15} \end{array}$

135. $\begin{array}{r} 26\frac{1}{10} \\ -\ 10\frac{9}{10} \\ \hline \end{array} \Rightarrow \begin{array}{r} 25\frac{11}{10} \\ -\ 10\frac{9}{10} \\ \hline 15\frac{2}{10}=15\frac{1}{5} \end{array}$

136. $\begin{array}{r} 19\frac{3}{8} \\ -\ 8\frac{5}{8} \\ \hline \end{array} \Rightarrow \begin{array}{r} 18\frac{11}{8} \\ -\ 8\frac{5}{8} \\ \hline 10\frac{6}{8}=10\frac{3}{4} \end{array}$

CHAPTER 3 REVIEW EXERCISES

137. $$8\frac{2}{3} - 5\frac{5}{6} \Rightarrow 8\frac{4}{6} - 5\frac{5}{6} \Rightarrow 7\frac{10}{6} - 5\frac{5}{6} = 2\frac{5}{6}$$

138. $$76\frac{7}{15} - 50\frac{1}{12} \Rightarrow 76\frac{28}{60} - 50\frac{5}{60} = 26\frac{23}{60}$$

139. $$9\frac{9}{16} - 3\frac{5}{12} \Rightarrow 9\frac{27}{48} - 3\frac{20}{48} = 6\frac{7}{48}$$

140. $$30\frac{7}{16} - 22\frac{5}{6} \Rightarrow 30\frac{21}{48} - 22\frac{40}{48} \Rightarrow 29\frac{69}{48} - 22\frac{40}{48} = 7\frac{29}{48}$$

141. $$9\frac{8}{15} - 4\frac{17}{20} \Rightarrow 9\frac{32}{60} - 4\frac{51}{60} \Rightarrow 8\frac{92}{60} - 4\frac{51}{60} = 4\frac{41}{60}$$

142. $$33\frac{17}{30} - 25\frac{7}{9} \Rightarrow 33\frac{51}{90} - 25\frac{70}{90} \Rightarrow 32\frac{141}{90} - 25\frac{70}{90} = 7\frac{71}{90}$$

143. $$11\frac{23}{48} - \frac{23}{24} \Rightarrow 11\frac{23}{48} - \frac{46}{48} \Rightarrow 10\frac{71}{48} - \frac{46}{48} = 10\frac{25}{48}$$

144. $$8\frac{11}{20} - \frac{9}{10} \Rightarrow 8\frac{11}{20} - \frac{18}{20} \Rightarrow 7\frac{31}{20} - \frac{18}{20} = 7\frac{13}{20}$$

145. a. $$30\frac{3}{4} - 27\frac{1}{8} \Rightarrow 30\frac{6}{8} - 27\frac{1}{8} = 3\frac{5}{8}$$

During the year. $3\frac{5}{8}$ more inches of rain fall in Westport.

b. $$42\frac{1}{3} - 14\frac{5}{8} \Rightarrow 42\frac{8}{24} - 14\frac{15}{24} \Rightarrow 41\frac{32}{24} - 14\frac{15}{24} = 27\frac{17}{24}$$

$10\left(27\frac{17}{24}\right) = \frac{10}{1} \cdot \frac{665}{24} = \frac{6650}{24} = 277\frac{2}{24}$ $= 277\frac{1}{12}$; Over 10 years, Salem receives $277\frac{1}{12}$ more inches of rain.

CHAPTER 3 REVIEW EXERCISES

146. Rain in Westview $= 2\left(35\frac{1}{4}\right) = \frac{2}{1} \cdot \frac{141}{4} = \frac{282}{4} = 70\frac{2}{4} = 70\frac{1}{2}$ inches

$$\begin{array}{r} 70\frac{1}{2} \\ -\ 42\frac{1}{3} \\ \hline \end{array} \Rightarrow \begin{array}{r} 70\frac{3}{6} \\ -\ 42\frac{2}{6} \\ \hline 28\frac{1}{6} \end{array}$$

Westview would receive $28\frac{1}{6}$ more inches of rain than Salem.

147. $\frac{1}{4} + \frac{3}{8} \div \frac{1}{2} = \frac{1}{4} + \frac{3}{8} \cdot \frac{2}{1} = \frac{1}{4} + \frac{6}{8} = \frac{2}{8} + \frac{6}{8} = \frac{8}{8} = 1$

148. $\frac{1}{3} \div \frac{5}{9} + \frac{1}{6} = \frac{1}{3} \cdot \frac{9}{5} + \frac{1}{6} = \frac{3}{5} + \frac{1}{6} = \frac{18}{30} + \frac{5}{30} = \frac{23}{30}$

149. $\frac{5}{8} \div \frac{3}{4} + \frac{3}{4} = \frac{5}{8} \cdot \frac{4}{3} + \frac{3}{4} = \frac{5}{6} + \frac{3}{4} = \frac{20}{24} + \frac{18}{24} = \frac{38}{24} = 1\frac{14}{24} = 1\frac{7}{12}$

150. $\frac{1}{3} \div \frac{1}{6} + \frac{4}{9} = \frac{1}{3} \cdot \frac{6}{1} + \frac{4}{9} = \frac{2}{1} + \frac{4}{9} = 2 + \frac{4}{9} = 2\frac{4}{9}$

151. $\frac{3}{4} - \left(\frac{1}{2}\right)^2 = \frac{3}{4} - \frac{1}{2} \cdot \frac{1}{2} = \frac{3}{4} - \frac{1}{4} = \frac{2}{4} = \frac{1}{2}$

152. $\frac{4}{5} - \left(\frac{2}{5}\right)^2 = \frac{4}{5} - \frac{2}{5} \cdot \frac{2}{5} = \frac{4}{5} - \frac{4}{25} = \frac{20}{25} - \frac{4}{25} = \frac{16}{25}$

153. $\left(\frac{9}{8}\right)^2 - \left(\frac{1}{2} \div \frac{4}{5} - \frac{3}{8}\right) = \frac{9}{8} \cdot \frac{9}{8} - \left(\frac{1}{2} \cdot \frac{5}{4} - \frac{3}{8}\right) = \frac{81}{64} - \left(\frac{5}{8} - \frac{3}{8}\right) = \frac{81}{64} - \left(\frac{2}{8}\right)$

$= \frac{81}{64} - \frac{16}{64} = \frac{65}{64} = 1\frac{1}{64}$

154. $\left(\frac{1}{2}\right)^2 + \left(\frac{4}{5} \cdot \frac{5}{8} + \frac{2}{3}\right) = \frac{1}{2} \cdot \frac{1}{2} + \left(\frac{1}{2} + \frac{2}{3}\right) = \frac{1}{4} + \left(\frac{3}{6} + \frac{4}{6}\right) = \frac{1}{4} + \frac{7}{6} = \frac{6}{24} + \frac{28}{24}$

$= \frac{34}{24} = 1\frac{10}{24} = 1\frac{5}{12}$

155. $\frac{3}{8} + \frac{1}{4} + \frac{1}{2} + \frac{3}{4} = \frac{3}{8} + \frac{2}{8} + \frac{4}{8} + \frac{6}{8} = \frac{15}{8}; \frac{15}{8} \div 4 = \frac{15}{8} \cdot \frac{1}{4} = \frac{15}{32}$

156. $\frac{3}{7} + \frac{1}{4} + \frac{2}{7} + \frac{5}{28} = \frac{12}{28} + \frac{7}{28} + \frac{8}{28} + \frac{5}{28} = \frac{32}{28} = \frac{8}{7}; \frac{8}{7} \div 4 = \frac{8}{7} \cdot \frac{1}{4} = \frac{2}{7}$

157. $1\frac{2}{3} + 1\frac{5}{12} + 2\frac{1}{2} + \frac{3}{4} + \frac{5}{6} = 1 + 1 + 2 + \frac{8}{12} + \frac{5}{12} + \frac{6}{12} + \frac{9}{12} + \frac{10}{12} = 4 + \frac{38}{12}$

$= 4 + 3\frac{2}{12} = 7\frac{1}{6} = \frac{43}{6}; \frac{43}{6} \div 5 = \frac{43}{6} \cdot \frac{1}{5} = \frac{43}{30} = 1\frac{13}{30}$

158. $\frac{2}{3}+\frac{5}{12}+\frac{1}{2}+1\frac{3}{4}+1\frac{5}{6}=1+1+\frac{8}{12}+\frac{5}{12}+\frac{6}{12}+\frac{9}{12}+\frac{10}{12}=2+\frac{38}{12}$
$=2+3\frac{2}{12}=5\frac{1}{6}=\frac{31}{6};\ \frac{31}{6}\div 5=\frac{31}{6}\cdot\frac{1}{5}=\frac{31}{30}=1\frac{1}{30}$

159. $3\left(\frac{10}{10}\right)+1\left(\frac{9}{10}\right)+3\left(\frac{8}{10}\right)+5\left(\frac{7}{10}\right)+4\left(\frac{6}{10}\right)+3\left(\frac{5}{10}\right)+1\left(\frac{3}{10}\right)$
$=\frac{30}{10}+\frac{9}{10}+\frac{24}{10}+\frac{35}{10}+\frac{24}{10}+\frac{15}{10}+\frac{3}{10}=\frac{140}{10}$
$\frac{140}{10}\div 20=\frac{140}{10}\cdot\frac{1}{20}=\frac{7}{10}$; The average score was $\frac{7}{10}$.

160. The top six scores are $\frac{10}{10}, \frac{10}{10}, \frac{10}{10}, \frac{9}{10}, \frac{8}{10}$, and $\frac{8}{10}$.
$3\left(\frac{10}{10}\right)+1\left(\frac{9}{10}\right)+2\left(\frac{8}{10}\right)=\frac{30}{10}+\frac{9}{10}+\frac{16}{10}=\frac{55}{10}$
$\frac{55}{10}\div 6=\frac{55}{10}\cdot\frac{1}{6}=\frac{55}{60}=\frac{11}{12}$; The average score is $\frac{11}{12}$.

Chapter 3 True/False Concept Review (page 269)

1. true

2. false; $\frac{7}{8}=0\frac{7}{8}$

3. false; The numerator always equals the denominator, so the fraction is improper.

4. true

5. false; When a fraction is simplified, its value remains unchanged.

6. true

7. true

8. true

9. false; The reciprocal of an improper fraction is less than 1.

10. true

11. true

12. true

13. false; Like fractions have the same denominator.

14. false; You can add the whole numbers and then add the fraction parts separately.

15. true

16. true

17. true

18. true

Chapter 3 Test (page 270)

1. $\begin{array}{r} 20 \\ 3\overline{)61} \\ \underline{6} \\ 01 \\ \underline{0} \\ 1 \end{array} \Rightarrow 20\frac{1}{3}$

2. $\frac{7}{8}+\frac{5}{12}=\frac{7}{8}\cdot\frac{3}{3}+\frac{5}{12}\cdot\frac{2}{2} = \frac{21}{24}+\frac{10}{24} = \frac{31}{24}=1\frac{7}{24}$

3. $8\frac{7}{9}=\frac{9(8)+7}{9} = \frac{72+7}{9}=\frac{79}{9}$

4. LCM = 280
$\frac{2}{5}=\frac{2}{5}\cdot\frac{56}{56}=\frac{112}{280}$
$\frac{3}{8}=\frac{3}{8}\cdot\frac{35}{35}=\frac{105}{280}$
$\frac{3}{7}=\frac{3}{7}\cdot\frac{40}{40}=\frac{120}{280}$
$\frac{3}{8}<\frac{2}{5}<\frac{3}{7}$

5. $11=\frac{11}{1}$

6. $72\div 8=9$
$\frac{3}{8}=\frac{3}{8}\cdot\frac{9}{9}=\frac{27}{72}$

7. $5\frac{3}{10}+3\frac{5}{6}=5+3+\frac{3}{10}+\frac{5}{6}=8+\frac{3}{10}\cdot\frac{3}{3}+\frac{5}{6}\cdot\frac{5}{5}=8+\frac{9}{30}+\frac{25}{30}=8+\frac{34}{30}=8+1\frac{4}{30}=9\frac{2}{15}$

8. $\left(3\frac{2}{3}\right)\left(5\frac{1}{9}\right)=\frac{11}{3}\cdot\frac{46}{9}=\frac{506}{27}=18\frac{20}{27}$

9. $\frac{1}{2}-\frac{3}{8}\div\frac{3}{4}=\frac{1}{2}-\frac{3}{8}\cdot\frac{4}{3}=\frac{1}{2}-\frac{12}{24} = \frac{1}{2}-\frac{1}{2}=0$

10. $\frac{68}{102}=\frac{2\cdot \cancel{34}}{3\cdot \cancel{34}}=\frac{2}{3}$

11. $\begin{array}{r} 17\frac{4}{5} \\ -\,11\phantom{\frac{4}{5}} \\ \hline 6\frac{4}{5} \end{array}$

12. $\frac{4}{5}\cdot\frac{7}{8}\cdot\frac{15}{21}=\frac{\overset{1}{\cancel{4}}}{\underset{1}{\cancel{5}}}\cdot\frac{\overset{1}{\cancel{7}}}{\underset{2}{\cancel{8}}}\cdot\frac{\overset{3}{\cancel{15}}}{\underset{3}{\cancel{21}}}=\frac{1}{1}\cdot\frac{1}{2}\cdot\frac{\overset{1}{\cancel{3}}}{\underset{1}{\cancel{3}}}=\frac{1}{2}$

13. $\frac{2}{3}-\frac{4}{9}=\frac{2}{3}\cdot\frac{3}{3}-\frac{4}{9}=\frac{6}{9}-\frac{4}{9}=\frac{2}{9}$

14. $1\frac{2}{9}\div 3\frac{2}{3}=\frac{11}{9}\div\frac{11}{3}=\frac{11}{9}\cdot\frac{3}{11}=\frac{33}{99}=\frac{1}{3}$

15. $\frac{3}{7}\cdot\frac{4}{5}=\frac{3\cdot 4}{7\cdot 5}=\frac{12}{35}$

16. $\begin{array}{r} 11\frac{7}{12} \\ -\;4\frac{14}{15} \\ \hline \end{array} \Rightarrow \begin{array}{r} 11\frac{35}{60} \\ -\;4\frac{56}{60} \\ \hline \end{array} \Rightarrow \begin{array}{r} 10\frac{95}{60} \\ -\;4\frac{56}{60} \\ \hline 6\frac{39}{60}=6\frac{13}{20} \end{array}$

CHAPTER 3 TEST

17. $\frac{220}{352} = \frac{5 \cdot \cancel{44}}{8 \cdot \cancel{44}} = \frac{5}{8}$

18. $\frac{1}{35} + \frac{5}{14} + \frac{2}{5} = \frac{1}{35} \cdot \frac{2}{2} + \frac{5}{14} \cdot \frac{5}{5} + \frac{2}{5} \cdot \frac{14}{14} = \frac{2}{70} + \frac{25}{70} + \frac{28}{70} = \frac{55}{70} = \frac{11}{14}$

19. The reciprocal of $3\frac{1}{5} = \frac{16}{5}$ is $\frac{5}{16}$.

20. The reciprocal of $\frac{8}{21}$ is $\frac{21}{8} = 2\frac{5}{8}$.

21. proper: $\frac{7}{8}, \frac{7}{9}, \frac{8}{9}$

22. $\frac{7}{3} \div \frac{8}{9} = \frac{7}{3} \cdot \frac{9}{8} = \frac{7}{\underset{1}{\cancel{3}}} \cdot \frac{\overset{3}{\cancel{9}}}{8} = \frac{21}{8} = 2\frac{5}{8}$

23.

$$\begin{array}{r} 11\frac{7}{10} \\ -\ 3\frac{3}{8} \\ \hline \end{array} \Rightarrow \begin{array}{r} 11\frac{28}{40} \\ -\ 3\frac{15}{40} \\ \hline 8\frac{13}{40} \end{array}$$

24. The whole is divided into 12 parts.

7 of the parts are shaded. $\Rightarrow \frac{7}{12}$

25.

$$\begin{array}{r} 11 \\ -\ 3\frac{5}{11} \\ \hline \end{array} \Rightarrow \begin{array}{r} 10\frac{11}{11} \\ -\ 3\frac{5}{11} \\ \hline 7\frac{6}{11} \end{array}$$

26. $\frac{4}{15} + \frac{8}{15} = \frac{12}{15} = \frac{4}{5}$

27. $1\frac{3}{8} + \frac{1}{4} + 3\frac{1}{2} + 2\frac{3}{8} = 1 + 3 + 2 + \frac{3}{8} + \frac{2}{8} + \frac{4}{8} + \frac{3}{8} = 6 + \frac{12}{8} = 6 + 1\frac{4}{8} = 7\frac{1}{2} = \frac{15}{2}$;
$\frac{15}{2} \div 4 = \frac{15}{2} \cdot \frac{1}{4} = \frac{15}{8} = 1\frac{7}{8}$

28. $\frac{8}{25} \cdot \frac{9}{16} = \frac{\overset{1}{\cancel{8}}}{25} \cdot \frac{9}{\underset{2}{\cancel{16}}} = \frac{9}{50}$

29. LCM = 112
$\frac{5}{7} = \frac{5}{7} \cdot \frac{16}{16} = \frac{80}{112}$; $\frac{11}{16} = \frac{11}{16} \cdot \frac{7}{7} = \frac{77}{112}$
$\frac{5}{7} > \frac{11}{16} \Rightarrow$ true

30. These represent 1: $\frac{5}{5}, \frac{6}{6}, \frac{7}{7}$

31. $126\frac{1}{2} \div 5\frac{3}{4} = \frac{253}{2} \div \frac{23}{4} = \frac{253}{2} \cdot \frac{4}{23} = \frac{\overset{11}{\cancel{253}}}{\underset{1}{\cancel{2}}} \cdot \frac{\overset{2}{\cancel{4}}}{\underset{1}{\cancel{23}}} = \frac{22}{1} = 22$; There are 22 truckloads in the car.

32. $20\left(1\frac{1}{4}\right) = \frac{20}{1} \cdot \frac{5}{4} = \frac{\overset{5}{\cancel{20}}}{1} \cdot \frac{5}{\underset{1}{\cancel{4}}} = \frac{25}{1} = 25$; She needs to make 25 pounds of candy.

Exercises 4.1 (page 283)

1. 0.26 = 2 tenths + 6 hundredths; twenty-six hundredths

3. 0.267 = 2 tenths + 6 hundredths + 7 thousandths; two hundred sixty-seven thousandths

5. 7.002 = 7 ones and 2 thousandths; seven and two thousandths

7. 11.92 = 1 ten + 1 one and 9 tenths + 2 hundredths; eleven and ninety-two hundredths

9. forty-two hundredths $\Rightarrow$ 2 is in the hundredths place. $\Rightarrow$ 0.42

11. four hundred nine thousandths $\Rightarrow$ 9 is in the thousandths place. $\Rightarrow$ 0.409

13. nine and fifty-nine thousandths $\Rightarrow$ 9 is to the left of the decimal point, and 9 is in the thousandths place. $\Rightarrow$ 9.059

15. three hundred eight ten-thousandths $\Rightarrow$ 8 is in the ten-thousandths place. $\Rightarrow$ 0.0308

17. 0.805 = 8 tenths + 5 thousandths; eight hundred five thousandths

19. 80.05 = 8 tens and 5 hundredths; eighty and five hundredths

21. 61.0203 = 6 tens + 1 ones and 2 hundredths + 3 ten-thousandths; sixty-one and two hundred three ten-thousandths

23. 90.003 = 9 tens and 3 thousandths; ninety and three thousandths

25. thirty-five ten-thousandths $\Rightarrow$ 5 is in the ten-thousandths place. $\Rightarrow$ 0.0035

27. one thousand eight hundred and twenty-eight thousandths $\Rightarrow$ 1800 is to the left of the decimal point, and 8 is in the thousandths place. $\Rightarrow$ 1800.028

29. five hundred five and five thousandths $\Rightarrow$ 505 is to the left of the decimal point, and 5 is in the thousandths place. $\Rightarrow$ 505.005

31. sixty-five and sixty-five thousandths $\Rightarrow$ 65 is to the left of the decimal point, and 5 is in the thousandths place. $\Rightarrow$ 65.065

33. **Unit**
The digit in the ones place is 5. Since 7 is in the tenths place, and since 7 is greater than 5, round up. Change the 5 in the ones place to 6, and drop all digits to the right. The rounded number is 36.

Tenth
The digit in the tenths place 7. Since 7 is in the hundredths place, and since 7 is greater than 5, round up. Change the 7 in the tenths place to 8, and drop all digits to the right. The rounded number is 35.8.

Hundredth
The digit in the hundredths place is 7. Since 7 is in the thousandths place, and since 7 is greater than 5, round up. Change the 7 in the hundredths place to 8, and drop all digits to the right. The rounded number is 35.78.

35. **Unit**
The digit in the ones place is 9. Since 6 is in the tenths place, and since 6 is greater than 5, round up. Change the 9 in the ones place to 0, increase the digit in the tens position from 2 to 3, and drop all digits to the right. The rounded number is 730.

Tenth
The digit in the tenths place 6. Since 3 is in the hundredths place, and since 3 is less than 5, round down. Leave the 6 in the tenths place, and drop all digits to the right. The rounded number is 729.6.

Hundredth
The digit in the hundredths place is 3. Since 8 is in the thousandths place, and since 8 is greater than 5, round up. Change the 3 in the hundredths place to 4, and drop all digits to the right. The rounded number is 729.64.

37. **Unit**
The digit in the ones place is 0. Since 6 is in the tenths place, and since 6 is greater than 5, round up. Change the 0 in the ones place to 1, and drop all digits to the right. The rounded number is 1.

Tenth
The digit in the tenths place 6. Since 1 is in the hundredths place, and since 1 is less than 5, round down. Leave the 6 in the tenths place, and drop all digits to the right. The rounded number is 0.6.

Hundredth
The digit in the hundredths place is 1. Since 5 is in the thousandths place, round up. Change the 1 in the hundredths place to 2, and drop all digits to the right. The rounded number is 0.62.

39. The digit in the cents (hundredths) place is 8. Since 5 is in the thousandths place, round up. Change the 8 in the hundredths place to 9, and drop all digits to the right. The rounded number is $67.49.

41. The digit in the cents (hundredths) place is 2. Since 3 is in the thousandths place, and since 3 is less than 5, round down. Leave the 2 in the hundredths place, and drop all digits to the right. The rounded number is $548.72.

43. **Ten**
The digit in the tens place is 3. Since 5 is in the ones place, round up. Change the 3 in the tens place to 4, change the digit in the ones place to 0, and drop all digits to the right. The rounded number is 40.

Hundredth
The digit in the hundredths place is 8. Since 3 is in the thousandths place, and since 3 is less than 5, round down. Leave the 8 in the hundredths place, and drop all digits to the right. The rounded number is 35.78.

Thousandth
The digit in the thousandths place is 3. Since 4 is in the ten-thousandths place, and since 4 is less than 5, round down. Leave the 3 in the thousandths place, and drop all digits to the right. The rounded number is 35.783.

45. **Ten**
The digit in the tens place is 8. Since 6 is in the ones place, and since 6 is greater than 5, round up. Change the 8 in the tens place to 9, change the digit in the ones place to 0, and drop all digits to the right. The rounded number is 90.

Hundredth
The digit in the hundredths place is 2. Since 7 is in the thousandths place, and since 7 is greater than 5, round up. Change the 2 in the hundredths place to 3, and drop all digits to the right. The rounded number is 86.33.

Thousandth
The digit in the thousandths place is 7. Since 8 is in the ten-thousandths place, and since 8 is greater than 5, round up. Change the 7 in the thousandths place to 8, and drop all digits to the right. The rounded number is 86.328.

47. **Ten**
The digit in the tens place is 0. Since 0 is in the ones place, and since 0 is less than 5, round down. Leave the 0 in the tens place, change the digit in the ones place to 0, and drop all digits to the right. The rounded number is 0.

Hundredth
The digit in the hundredths place is 1. Since 4 is in the thousandths place, and since 4 is less than 5, round down. Leave the 1 in the hundredths place, and drop all digits to the right. The rounded number is 0.91.

Thousandth
The digit in the thousandths place is 4. Since 8 is in the ten-thousandths place, and since 8 is greater than 5, round up. Change the 4 in the thousandths place to 5, and drop all digits to the right. The rounded number is 0.915.

49. The digit in the dollars (ones) place is 2. Since 4 is in the tenths place, and since 4 is less than 5, round down. Leave the 2 in the ones place, and drop all digits to the right. The rounded number is \$72.

51. The digit in the dollars (ones) place is 1. Since 5 is in the tenths place, round up. Change the 1 in the ones place to 2, and drop all digits to the right. The rounded number is \$7822.

53. $22.1906 = 2$ tens $+ 2$ ones and 1 tenth $+ 9$ hundredths $+ 6$ ten-thousandths; The word name is twenty-two and one thousand nine hundred six ten-thousandths.

55. The amount is 19.8068. The digit in the hundredths place is 0. Since 6 is in the thousandths place, and since 6 is greater than 5, round up. Change the 0 in the hundredths place to 1 and drop all digits to the right. To the nearest hundredth, the tax rate is 19.81.

57. 57.79 = 5 tens + 7 ones and 7 tenths + 9 hundredths; Carlos writes "fifty-seven and seventy-nine hundredths dollars."

59. Since the arrow is between 3.74 and 3.745, the hundredths digit is 4 and the thousandths digit is less than 5. To the nearest hundredth, the position of the arrow is 3.74.

61. Since the arrow is between 3.74 and 3.745, the tenths digit is 7 and the hundredths digit is 4. To the nearest tenth, the position of the arrow is 3.7.

63. 756.7104 = 7 hundreds + 5 tens + 6 ones and 7 tenths + 1 hundredth + 4 ten-thousandths; seven hundred fifty-six and seven thousand one hundred four ten-thousandths

65. The digit in the cents (hundredths) place is 9. Since 0 is in the thousandths place, and since 0 is less than 5, round down. Leave the 9 in the hundredths place, and drop all digits to the right. To the nearest cent, the value is $3478.59.

67. two hundred thirteen and one thousand, one hundred one ten-thousandths ⇒ 213 is to the left of the decimal point, and 1 is in the ten-thousandths place. ⇒ 213.1101

69. The digit in the ones place is 0. Since 7 is in the tenths place, and since 7 is greater than 5, round up. Change the 0 in the ones place to 1, and drop all digits to the right. The density is about 131 people per square km.

71. Since the hundredths position is filled, it appears that the number has been rounded to the nearest hundredth.

73.

Thousand
The digit in the thousands place is 7. Since 5 is in the hundreds place, round up. Change the 7 in the thousands place to 8, change the hundreds, tens, and units digits to 0, and drop all digits to the right. The rounded number is 8000

Hundredth
The digit in the hundredths place is 5. Since 9 is in the thousandths place, and since 9 is greater than 5, round up. Change the 5 in the hundredths place to 6, and drop all digits to the right. The rounded number is 7564.36.

Ten-thousandth
The digit in the ten-thousandths place is 2. Since 6 is in the hundred-thousandths place, and since 6 is greater than 5, round up. Change the 2 in the ten-thousandths place 3, and drop all digits to the right. The rounded number is 7564.3593.

75. **Thousand**
The digit in the thousands place is 8. Since 0 is in the hundreds place, and since 0 is less than 5, round down. Leave the 8 in the thousands place, and drop all digits to the right. The rounded number is 78,000.

Hundredth
The digit in the hundredths place is 8. Since 8 is in the thousandths place, and since 8 is greater than 5, round up. Change the 8 in the hundredths place to 9, and drop all digits to the right. The rounded number is 78,042.39.

Ten-thousandth
The digit in the ten-thousandths place is 7. Since 5 is in the hundred-thousandths place, round up. Change the 7 in the ten-thousandths place to 8, and drop all digits to the right. The rounded number is 78,042.3888.

77. Since the thousandths position is filled, it appears that the number has been rounded to the nearest thousandth.

79. **One**
The digit in the ones place is 3. Since 0 is in the tenths place, and since 0 is less than 5, round down. Leave the 3 in the ones place, and drop all digits to the right. The speed is about 763 miles per hour.

Tenth
The digit in the tenths place is 0. Since 3 is in the hundredths place, and since 3 is less than 5, round down. Leave the 0 in the tenths place, and drop all digits to the right. The speed is about 763.0 miles per hour.

81. In each number, the "4" means that you have 4 values of a place value position. However, in 43.29, the "4" represents 4 groups of 10, while in 18.64, the "4" represents 4 groups of hundredths.

83. The digit in the thousandths place is 2. Since 8 is in the ten-thousandths place, and since 8 is greater than 5, round up. Change the 2 in the thousandths place to 3, and drop all digits to the right. The rounded number is 8.283. This rounded number is greater than the original number: $8.283 > 8.28282828$.

85. five-hundred twenty-two hundred millionths; There are no digits left of the decimal point, and 2 is in the hundred-millionths place; 0.00000522

87. **Answers may vary.**

89. $132 \div 12 = 11$
$$\frac{7}{12} = \frac{7}{12} \cdot \frac{11}{11} = \frac{77}{132}$$

91. $1000 \div 40 = 25$
$$\frac{19}{40} = \frac{19}{40} \cdot \frac{25}{25} = \frac{475}{1000}$$

93. 30,951 is larger than 30,899, so the statement "$30{,}951 > 30{,}899$" is true.

95. 6209, 6215, 6218, 6223, 6227

97. $\frac{17}{20}(40) = \frac{17}{20} \cdot \frac{40}{1} = \frac{17}{\cancel{20}_{1}} \cdot \frac{\cancel{40}^{2}}{1} = \frac{34}{1} = 34$; Kobe must make 34 out of 40 field goals.

Exercises 4.2 (page 291)

1. $0.83 = \text{eighty-three hundredths} = \dfrac{83}{100}$

3. $0.65 = \text{sixty-five hundredths} = \dfrac{65}{100} = \dfrac{13}{20}$

5. $\text{six hundred fifty-eight thousandths} = \dfrac{658}{1000} = \dfrac{329}{500}$

7. $0.82 = \text{eighty-two hundredths} = \dfrac{82}{100} = \dfrac{41}{50}$

9. $0.48 = \text{forty-eight hundredths} = \dfrac{48}{100} = \dfrac{12}{25}$

11. $10.41 = \text{ten and forty-one hundredths} = 10\dfrac{41}{100}$

13. $0.125 = \text{one hundred twenty-five thousandths} = \dfrac{125}{1000} = \dfrac{1}{8}$

15. $12.24 = \text{twelve and twenty-four hundredths} = 12\dfrac{24}{100} = 12\dfrac{6}{25}$

17. $11.344 = \text{eleven and three hundred forty-four thousandths} = 11\dfrac{344}{1000} = 11\dfrac{43}{125}$

19. $\text{seven hundred fifty thousandths} = \dfrac{750}{1000} = \dfrac{3}{4}$

21. $0.7, 0.1, 0.4 \Rightarrow 0.1, 0.4, 0.7$

23. $0.17, 0.06, 0.24 \Rightarrow 0.06, 0.17, 0.24$

25. $3.26, 3.185, 3.179 \Rightarrow 3.260, 3.185, 3.179$
$\Rightarrow 3.179, 3.185, 3.260$
$\Rightarrow 3.179, 3.185, 3.26$

27. $0.38, 0.3 \Rightarrow 0.38, 0.30$
$0.38 < 0.3 \Rightarrow \text{false}$

29. $10.48 > 10.84 \Rightarrow \text{false}$

31. $0.0477, 0.047007, 0.047, 0.046, 0.047015 \Rightarrow 0.047700, 0.047007, 0.047000, 0.046000, 0.047015$
$\Rightarrow 0.046000, 0.047000, 0.047007, 0.047015, 0.047700$
$\Rightarrow 0.046, 0.047, 0.047007, 0.047015, 0.0477$

33. $0.555, 0.55699, 0.5552, 0.55689 \Rightarrow 0.55500, 0.55699, 0.55520, 0.55689$
$\Rightarrow 0.55500, 0.55520, 0.55689, 0.55699$
$\Rightarrow 0.555, 0.5552, 0.55689, 0.55699$

35. $25.005, 25.051, 25.0059, 25.055 \Rightarrow 25.0050, 25.0510, 25.0059, 25.0550$
$\Rightarrow 25.0050, 25.0059, 25.0510, 25.0550$
$\Rightarrow 25.005, 25.0059, 25.051, 25.055$

37. $3.1231 < 3.1213 \Rightarrow \text{false}$

39. $74.6706 < 74.7046 \Rightarrow \text{true}$

41. $0.0625 =$ six hundred twenty-five ten-thousandths $= \frac{625}{10{,}000} = \frac{1}{16}$;
As a reduced fraction, the probability is $\frac{1}{16}$.

43. $2.675, 2.6351, 2.636 \Rightarrow 2.6750, 2.6351, 2.6360 \Rightarrow 2.6351, 2.6360, 2.6750$
The best (lowest) bid is \$2.6351, made by Tillamook Dairy.

45. $0.832 =$ eight hundred thirty-two thousandths $= \frac{832}{1000} = \frac{104}{125}$; The highest percentage of free throws made in a season is $\frac{104}{125}$.

47. $0.405 =$ four hundred five thousandths $= \frac{405}{1000} = \frac{81}{200}$; The lowest percentage of free throws made in a game by both teams is $\frac{81}{200}$.

49. $0.8375 =$ eight thousand, three hundred seventy-five ten-thousandths $= \frac{8375}{10{,}000} = \frac{67}{80}$

51. $25.025 =$ twenty-five and twenty-five thousandths $= 25\frac{75}{1000} = 25\frac{1}{40}$

53. $0.547, 0.55 \Rightarrow 0.547, 0.550 \Rightarrow 0.547 < 0.550 \Rightarrow$ Norado needs less acid.

55. $3.0007, 3.002, 3.00077, 3.00092, 3.00202 \Rightarrow 3.00070, 3.00200, 3.00077, 3.00092, 3.00202$
$\Rightarrow 3.00070, 3.00077, 3.00092, 3.00200, 3.00202$
$\Rightarrow 3.0007, 3.00077, 3.00092, 3.002, 3.00202$

57. $82.86, 83.01, 82.85, 82.58, 83.15, 83.55, 82.80, 82.78$
$\Rightarrow 82.58, 82.78, 82.80, 82.85, 82.86, 83.01. 83.15. 83.55$

59. $0.725 = \frac{725}{1000} = \frac{29}{40} = \frac{29}{40} \cdot \frac{7}{7} = \frac{203}{280}; \frac{5}{7} \cdot \frac{40}{40} = \frac{200}{280} \Rightarrow$ Maria should choose 0.725 yard.

61. $341.408 = 341.4080; 341.1622 < 341.4080$; Belgium had fewer people per square km.

63. $7.95 < 8.9 < 10.85; 7.95 = 7\frac{95}{100} = 7\frac{19}{20}; 8.9 = 8\frac{9}{10}; 10.85 = 10\frac{85}{100} = 10\frac{17}{20}$
The fat grams are $7\frac{19}{20}$, $8\frac{9}{10}$, and $10\frac{17}{20}$. Frozen plain hash browns have the lowest amount of fat.

65. $0.609 = \frac{609}{1000}$. He made $\frac{609}{1000}$ of his field goals and missed $\frac{391}{1000}$ of them.

67.

National League

Year	Name	Team	Average
2005	Derek Lee	Chicago	0.335
2007	Matt Holiday	Colorado	0.340
2006	Freddy Sanchez	Pittsburgh	0.344
2004	Barry Bonds	San Francisco	0.362
2008	Chipper Jones	Atlanta	0.364

American League

Year	Name	Team	Average
2008	Joe Mauer	Minnesota	0.328
2005	Michael Young	Texas	0.331
2006	Joe Mauer	Minnesota	0.347
2007	Maglio Ordonez	Detroit	0.363
2004	Ichiro Suzuki	Seattle	0.372

69. $0.44 = \dfrac{44}{100} = \dfrac{11}{25}; 0.404 = \dfrac{404}{1000} = \dfrac{101}{250}; 0.04044 = \dfrac{4044}{100{,}000} = \dfrac{1011}{25{,}000}$

71.
$$\begin{array}{rrrrrr} & & {\scriptstyle 1} & {\scriptstyle 2} & {\scriptstyle 1} & \\ & & & 4 & 7 & 9 \\ & & 3 & 7 & 1 & 2 \\ & & & & 9 & 3 \\ + & & & 7 & 2 & 2 \;5 \\ \hline & 1 & 1 & 5 & 0 & 9 \end{array}$$

73.
$$\begin{array}{rrrrrr} & & {\scriptstyle 13} & {\scriptstyle 16} & & \\ & {\scriptstyle 2} & {\scriptstyle \not{3}} & {\scriptstyle \not{6}} & {\scriptstyle 14} & \\ & \not{3} & \not{4} & 7 & \not{4} & 8 \\ - & 2 & 7 & 9 & 6 & 3 \\ \hline & & 6 & 7 & 8 & 5 \end{array}$$

75. $\dfrac{1}{2} + \dfrac{3}{4} + \dfrac{1}{8} = \dfrac{1}{2} \cdot \dfrac{4}{4} + \dfrac{3}{4} \cdot \dfrac{2}{2} + \dfrac{1}{8}$
$= \dfrac{4}{8} + \dfrac{6}{8} + \dfrac{1}{8} = \dfrac{11}{8} = 1\dfrac{3}{8}$

77. $\dfrac{25}{64} - \dfrac{3}{8} = \dfrac{25}{64} - \dfrac{3}{8} \cdot \dfrac{8}{8} = \dfrac{25}{64} - \dfrac{24}{64} = \dfrac{1}{64}$

79. $456 + 389 + 1034 + 672 + 843 + 467 + 732 = 4593$; The attendance was 4593.

Exercises 4.3 (page 299)

1.
$$\begin{array}{r} {\scriptstyle 1} \\ 0.7 \\ +\,0.7 \\ \hline 1.4 \end{array}$$

3.
$$\begin{array}{r} 3.7 \\ +\,2.2 \\ \hline 5.9 \end{array}$$

5.
$$\begin{array}{r} {\scriptstyle 1} \\ 1.6 \\ 5.5 \\ +\;\;8.7 \\ \hline 15.8 \end{array}$$

7.
$$\begin{array}{r} {\scriptstyle 1\,1} \\ 34.80 \\ +\;\;5.29 \\ \hline 40.09 \end{array}$$

9. four

11.
$$\begin{array}{r} 21.300 \\ +\,34.567 \\ \hline 55.867 \end{array}$$

13.
$$\begin{array}{r} {\scriptstyle 2\,1\;\;1} \\ 5.24 \\ 0.66 \\ 19.70 \\ +\;\;6.08 \\ \hline 31.68 \end{array}$$

15.
$$\begin{array}{r} {\scriptstyle 1\;\;1\,1} \\ 2.337 \\ 0.672 \\ +\,4.056 \\ \hline 7.065 \end{array}$$

17.
$$\begin{array}{r} 0.0017 \\ 1.0070 \\ 7.0000 \\ +\,1.0710 \\ \hline 9.0797 \end{array}$$

19.
$$\begin{array}{r} {\scriptstyle 2\;\;\;1\,1} \\ 67.062 \\ 74.007 \\ 7.160 \\ +\;\;\;9.256 \\ \hline 157.485 \end{array}$$

21.
$$\begin{array}{r} {\scriptstyle 1\,3\,1\,2} \\ 0.07810 \\ 0.00932 \\ 0.07639 \\ +\,0.00759 \\ \hline 0.17140 \end{array}$$

23.
$$\begin{array}{r} {\scriptstyle 2\,2} \\ 0.0670 \\ 0.4560 \\ 0.0964 \\ 0.5321 \\ +\,0.1120 \\ \hline 1.2635 \end{array}$$

SECTION 4.3

25.
$$\begin{array}{r} {\scriptstyle 2\,2} \\ 7.8000 \\ 35.6640 \\ +\ \ 76.9236 \\ \hline 120.3876 \end{array}$$

27.
$$\begin{array}{r} {\scriptstyle 1\,1\,2\ \ 1\,1} \\ 75.995 \\ 24.900 \\ +\ 694.447 \\ \hline 795.342 \end{array}$$

29.
$$\begin{array}{r} {\scriptstyle 1\,1\ \ 1} \\ 23.070 \\ 6.700 \\ 0.468 \\ +\ \ 8.030 \\ \hline 38.268 \end{array}$$

31.
$$\begin{array}{r} 0.7 \\ -\ 0.4 \\ \hline 0.3 \end{array}$$

33.
$$\begin{array}{r} 8.6 \\ -\ 2.5 \\ \hline 6.1 \end{array}$$

35.
$$\begin{array}{r} 6.45 \\ -\ 2.35 \\ \hline 4.10 \end{array}$$

37.
$$\begin{array}{r} {\scriptstyle 3\,15\ \ 3\,12} \\ 45.42 \\ -\ 27.38 \\ \hline 18.04 \end{array}$$

39.
$$\begin{array}{r} {\scriptstyle 9} \\ {\scriptstyle 1\ \ 10\,11} \\ 32.01 \\ -\ 11.14 \\ \hline 20.87 \end{array}$$

41.
$$\begin{array}{r} {\scriptstyle 11} \\ {\scriptstyle 6\,1\ \ 13} \\ 0.723 \\ -\ 0.457 \\ \hline 0.266 \end{array}$$

43.
$$\begin{array}{r} {\scriptstyle 11} \\ {\scriptstyle 5\,1\ \ 13} \\ 4.623 \\ -\ 2.379 \\ \hline 2.244 \end{array}$$

45.
$$\begin{array}{r} {\scriptstyle 12} \\ {\scriptstyle 7\,2\ \ 11} \\ 0.831 \\ -\ 0.462 \\ \hline 0.369 \end{array}$$

47.
$$\begin{array}{r} {\scriptstyle 12\ \ 13\,14} \\ {\scriptstyle 2\,2\ \ 3\ 4\ 16} \\ 33.456 \\ -\ 29.457 \\ \hline 3.999 \end{array}$$

49.
$$\begin{array}{r} {\scriptstyle 2\,12\,6\ \ 15\,7\,10} \\ 327.580 \\ -\ 245.674 \\ \hline 81.906 \end{array}$$

51.
$$\begin{array}{r} {\scriptstyle 11} \\ {\scriptstyle 4\,1\ \ 10} \\ 0.09520 \\ -\ 0.06434 \\ \hline 0.03086 \end{array}$$

53.
$$\begin{array}{r} {\scriptstyle 13} \\ {\scriptstyle 3\,11\ \ 2\,3\ \ 11} \\ 41.8341 \\ -\ 34.6152 \\ \hline 7.2189 \end{array}$$

55.
$$\begin{array}{r} {\scriptstyle 4\,10} \\ 0.0750 \\ -\ 0.0023 \\ \hline 0.0727 \end{array}$$

57.
$$\begin{array}{r} {\scriptstyle 10\ \ 9} \\ {\scriptstyle 5\,0\ \ 10\,12} \\ 61.02 \\ -\ 56.78 \\ \hline 4.24 \end{array}$$

59.
$$\begin{array}{r} {\scriptstyle 16} \\ {\scriptstyle 8\,6\ \ 18} \\ 11.978 \\ -\ 11.789 \\ \hline 0.189 \end{array}$$

61. $0.0643 + 0.8143 + 0.513 - (0.4083 + 0.7114) = 1.3916 - 1.1197 = 0.2719$

63. $9.056 - (5.55 + 2.62) - 0.0894 = 9.056 - 8.17 - 0.0894 = 0.886 - 0.0894 = 0.7966$

65.
$$\begin{array}{r} {\scriptstyle 1\,2} \\ 19.2 \\ 21.9 \\ 20.4 \\ +\ 23.7 \\ \hline 85.2 \end{array}$$
Manuel bought 85.2 gallons of gas on the trip.

67.
$$\begin{array}{r} {\scriptstyle 1\,2\,3\ \ 2\,1} \\ 457.386 \\ 423.900 \\ 606.777 \\ 29.420 \\ +\ \ 171.874 \\ \hline 1689.357 \end{array}$$
To the nearest tenth, the sum is 1689.4.

69. $1.847 + 1.22 + 1.144 + 0.744 + 0.63 + 0.553 = 6.138$; The total gross state products for the 6 states was \$6.138 trillion.

SECTION 4.3

71. $1.144 + 0.744 + 0.553 = 2.441$; The total gross state product for the 3 states on the east coast was \$2.441 trillion.

73. $7.98 + 5.29 + 27.85 = 41.12$; $72.00 - 41.12 = 30.88$; He spent \$30.88 on gas.

75. $3.09 + 1.49 + 2.50 + 4.39 + 7.99 + 9.27 = 28.73$; The total cost is \$28.73.

77. $53.91 + 22.53 = 76.44$; The total length of the Japanese tunnels is 76.44 km.

79. 1995-2000: $38.6 - 35.8 = 2.8$; 2000-2005: $42.0 - 38.6 = 3.4$; 2005-2010: $45.7 - 42.0 = 3.7$
The period from 2005-2010 is projected for the largest increase, 3.7 million families.

81. $46.8 + 35.7 = 82.5$; The height is about 83 feet.

83. $1.4375 + 0.3125 = 1.75$; $8.34375 - 1.75 = 6.59375$; The distance is 6.59375 in.

85. $12.16 - 11.382 = 0.778$;
Sera is 0.778 seconds faster.

87. $9.35 + 9.91 + 10.04 + 9.65 = 38.95$;
The race was completed in 38.95 seconds.

89. $12.83 + 13.22 + 13.56 = 39.61$; $52.78 - 39.61 = 13.17$; The time must be 13.16 seconds or less.

91. To subtract fractions, you need to get common denominators, subtract the numerators, then write the difference over the common denominator. To subtract decimals, align the decimal points so that place values line up. Then subtract each column, borrowing when needed.

93. A total of twenty-six 5.83s must be added.

95. The pattern is to subtract 0.01, then 0.002, then 0.0003, and so on.... $0.188 - 0.0003 = 0.1877$; The missing number is 0.1877.

97. $\frac{7}{16} = \frac{7}{16} \cdot \frac{625}{625} = \frac{4375}{10{,}000} = 0.4375$; $6\frac{7}{16} = 6.4375$; $6.4375 - 5.99 = 0.4475$ is the difference.

99. $2.25 = 2\frac{1}{4}$; $3.5 = 3\frac{1}{2}$; $4.8 = 4\frac{8}{10} = 4\frac{4}{5}$

$$2\frac{1}{2} + 3\frac{1}{8} + 4\frac{3}{4} + 2\frac{1}{4} + 3\frac{1}{2} + 4\frac{4}{5} = 2 + 3 + 4 + 2 + 3 + 4 + \frac{1}{2} + \frac{1}{8} + \frac{3}{4} + \frac{1}{4} + \frac{1}{2} + \frac{4}{5}$$

$$= 18 + \frac{2}{2} + \frac{1}{8} + \frac{4}{4} + \frac{4}{5} = 18 + 1 + 1 + \frac{5}{40} + \frac{32}{40} = \boxed{20\frac{37}{40}}$$

$$2\frac{1}{2} = 2\frac{5}{10} = 2.5;\ 3\frac{1}{8} = 3\frac{125}{1000} = 3.125;\ 4\frac{3}{4} = 4\frac{75}{100} = 4.75$$

$$2.5 + 3.125 + 4.75 + 2.25 + 3.5 + 4.8 = \boxed{20.925}$$

101.

$$\begin{array}{rrrrrrr} & & & & 1 & & \\ & & & & \cancel{1} & & \\ & & & & \cancel{2} & & \\ & & & 7 & 0 & 3 \\ \times & & & 5 & 5 & 7 \\ \hline & & 4 & 9 & 2 & 1 \\ & 3 & 5 & 1 & 5 & 0 \\ 3 & 5 & 1 & 5 & 0 & 0 \\ \hline 3 & 9 & 1 & 5 & 7 & 1 \end{array}$$

103. $(83)(27)(19) = (2241)(19)$
$= 42{,}579$

105. $\dfrac{36}{75} \cdot \dfrac{15}{16} \cdot \dfrac{40}{27} = \dfrac{\overset{4}{\cancel{36}}}{\underset{15}{\cancel{75}}} \cdot \dfrac{\overset{3}{\cancel{15}}}{\underset{2}{\cancel{16}}} \cdot \dfrac{\overset{5}{\cancel{40}}}{\underset{3}{\cancel{27}}}$
$= \dfrac{\overset{2}{\cancel{4}}}{\underset{3}{\cancel{15}}} \cdot \dfrac{\overset{1}{\cancel{3}}}{\underset{1}{\cancel{2}}} \cdot \dfrac{\overset{1}{\cancel{5}}}{\underset{1}{\cancel{3}}} = \dfrac{2}{3}$

107. $\left(4\frac{1}{2}\right)\left(5\frac{3}{5}\right) = \dfrac{9}{\underset{1}{\cancel{2}}} \cdot \dfrac{\overset{14}{\cancel{28}}}{5} = \dfrac{126}{5} = 25\frac{1}{5}$

109. $24(345) = 8280$; It will take 8280 pears to fill the order.

Getting Ready for Algebra (page 307)

1.
$$\begin{aligned} 16.3 &= x + 5.2 \\ 16.3 - 5.2 &= x + 5.2 - 5.2 \\ 11.1 &= x \end{aligned}$$

3.
$$\begin{aligned} y - 0.64 &= 13.19 \\ y - 0.64 + 0.64 &= 13.19 + 0.64 \\ y &= 13.83 \end{aligned}$$

5.
$$\begin{aligned} t + 0.03 &= 0.514 \\ t + 0.03 - 0.03 &= 0.514 - 0.03 \\ t &= 0.484 \end{aligned}$$

7.
$$\begin{aligned} x - 7.3 &= 5.21 \\ x - 7.3 + 7.3 &= 5.21 + 7.3 \\ x &= 12.51 \end{aligned}$$

9.
$$\begin{aligned} 7.33 &= w + 0.13 \\ 7.33 - 0.13 &= w + 0.13 - 0.13 \\ 7.2 &= w \end{aligned}$$

11.
$$\begin{aligned} t - 8.37 &= 0.08 \\ t - 8.37 + 8.37 &= 0.08 + 8.37 \\ t &= 8.45 \end{aligned}$$

13.
$$\begin{aligned} 5.78 &= a + 1.94 \\ 5.78 - 1.94 &= a + 1.94 - 1.94 \\ 3.84 &= a \end{aligned}$$

15.
$$\begin{aligned} 6.6 &= x - 9.57 \\ 6.6 + 9.57 &= x - 9.57 + 9.57 \\ 16.17 &= x \end{aligned}$$

17.
$$\begin{aligned} a + 82.3 &= 100 \\ a + 82.3 - 82.3 &= 100 - 82.3 \\ a &= 17.7 \end{aligned}$$

19.
$$\begin{aligned} s - 2.5 &= 4.773 \\ s - 2.5 + 2.5 &= 4.773 + 2.5 \\ s &= 7.273 \end{aligned}$$

21.

$$\begin{aligned} c + 432.8 &= 1029.16 \\ c + 432.8 - 432.8 &= 1029.16 - 432.8 \\ c &= 596.36 \end{aligned}$$

23. Let $p =$ the price 2 years ago.

$$\boxed{\text{Price 2 yrs ago}} - \boxed{\text{Decrease}} = \boxed{\text{Current price}}$$

$$\begin{aligned} p - 52.75 &= 374.98 \\ p - 52.75 + 52.75 &= 374.98 + 52.75 \\ p &= 427.73 \end{aligned}$$

Two years ago, the price was \$427.73.

25. Let $m =$ the amount of markup.

$$\boxed{\text{Cost}} + \boxed{\text{Markup}} = \boxed{\text{Price}}$$

$$\begin{aligned} 875.29 + m &= 1033.95 \\ 875.29 - 875.29 + m &= 1033.95 - 875.29 \\ m &= 158.66 \Rightarrow \text{The markup is \$158.66.} \end{aligned}$$

27. Let $C =$ the cost of the groceries.

$$\boxed{\text{Cost of bus pass}} + \boxed{\text{Cost of groceries}} = \boxed{\text{Total cost}}$$

$$\begin{aligned} 24 + C &= 61 \\ 24 - 24 + C &= 61 - 24 \\ C &= 37 \Rightarrow \text{The shopper can spend \$37 on groceries.} \end{aligned}$$

Exercises 4.4 (page 310)

1. Put one decimal place in the product.

$$\begin{array}{r} 0.5 \\ \times\quad 9 \\ \hline 4.5 \end{array}$$

3. Put one decimal place in the product.

$$\begin{array}{r} {\scriptstyle 4} \\ 1.9 \\ \times\quad 5 \\ \hline 9.5 \end{array}$$

5. Put two decimal places in the product.

$$\begin{array}{r} 0.09 \\ \times\quad 8 \\ \hline 0.72 \end{array}$$

7. Put two decimal places in the product.

$$\begin{array}{r} 0.8 \\ \times\quad 0.8 \\ \hline 0.64 \end{array}$$

9. Put three decimal places in the product.

$$\begin{array}{r} 0.06 \\ \times\quad 0.6 \\ \hline 0.036 \end{array}$$

11. Put three decimal places in the product.

$$\begin{array}{r} {\scriptstyle 5} \\ 0.18 \\ \times\quad 0.7 \\ \hline 0.126 \end{array}$$

SECTION 4.4

13. four

15. Put five decimal places in the product.

$$\begin{array}{r} {\scriptstyle 5\ 1} \\ 7.72 \\ \times \quad 0.008 \\ \hline 0.06176 \end{array}$$

17. Put three decimal places in the product.

$$\begin{array}{r} {\scriptstyle 1\ 1} \\ \not{3}\ \not{2} \\ 2.44 \\ \times \quad 4.7 \\ \hline 1708 \\ 9760 \\ \hline 11.468 \end{array}$$

19. Put four decimal places in the product.

$$\begin{array}{r} {\scriptstyle 3\ 1} \\ \not{1} \\ 6.84 \\ \times \quad 0.42 \\ \hline 1368 \\ 27360 \\ \hline 2.8728 \end{array}$$

21. Put four decimal places in the product.

$$\begin{array}{r} {\scriptstyle 1} \\ \not{3} \\ 5.92 \\ \times \quad 2.04 \\ \hline 2368 \\ 0000 \\ 118400 \\ \hline 12.0768 \end{array}$$

23. Put three decimal places in the product.

$$\begin{array}{r} {\scriptstyle 1\ 3} \\ \not{2} \\ 42.7 \\ \times \quad 0.53 \\ \hline 1281 \\ 21350 \\ \hline 22.631 \end{array}$$

25. Put six decimal places in the product.

$$\begin{array}{r} {\scriptstyle 3\ 3} \\ \not{3}\ \not{4} \\ 0.356 \\ \times \quad 0.067 \\ \hline 2492 \\ 21360 \\ \hline 0.023852 \end{array}$$

27. Put six decimal places in the product

$$\begin{array}{r} {\scriptstyle 2} \\ \not{1} \\ 0.0416 \\ \times \quad 4.02 \\ \hline 832 \\ 0000 \\ 166400 \\ \hline 0.167232 \end{array}$$

29. Put 7 decimal places in the product.

$$\begin{array}{r} {\scriptstyle 1\ 2} \\ \not{1}\ \not{2} \\ 0.825 \\ \times \quad 0.0054 \\ \hline 3300 \\ 41250 \\ \hline 0.0044550 \end{array}$$

31. Put 4 decimal places in the product.

$$\begin{array}{r} {\scriptstyle 1\ 1} \\ \not{3}\ \not{2} \\ \not{4}\ \not{3} \\ 8.54 \\ \times \quad 3.78 \\ \hline 6832 \\ 59780 \\ 256200 \\ \hline 32.2812 \end{array}$$

33. Put 2 decimal places in the product.

$$\begin{array}{r} {\scriptstyle 2} \\ 4.4 \\ \times \quad 0.6 \\ \hline 2.64 \end{array}$$

Put 4 decimal places in the product.

$$\begin{array}{r} {\scriptstyle 2\ 1} \\ \not{5}\ \not{3} \\ 2.64 \\ \times \quad 0.48 \\ \hline 2112 \\ 10560 \\ \hline 1.2672 \end{array}$$

35. Put 3 decimal places in the product.

$$\begin{array}{r} 11 \\ \cancel{1} \\ 0.846 \\ \times \quad 32 \\ \hline 1\,692 \\ 25\,380 \\ \hline 27.072 \end{array}$$

$27.072 \approx 27.1$

37. Put 4 decimal places in the product.

$$\begin{array}{r} 1\,2\;\,1 \\ \cancel{1}\,\cancel{3}\;\cancel{2} \\ \cancel{1}\;\cancel{1} \\ \cancel{2}\,\cancel{5}\;\cancel{3} \\ 64.85 \\ \times \quad 34.26 \\ \hline 3\,89\,10 \\ 12\,97\,00 \\ 259\,40\,00 \\ 1945\,50\,00 \\ \hline 2221.76\,10 \end{array}$$

$2221.761 \approx 2221.8$

39. Put 4 decimal places in the product.

$$\begin{array}{r} 2\,2\;\;\, \\ \cancel{2}\,\cancel{3}\;\cancel{1} \\ \cancel{4}\,\cancel{6}\;\cancel{2} \\ 16.93 \\ \times \quad 31.47 \\ \hline 1\,18\,51 \\ 6\,77\,20 \\ 16\,93\,00 \\ 507\,90\,00 \\ \hline 532.78\,71 \end{array}$$

$532.7871 \approx 532.79$

41. Put 4 decimal places in the product.

$$\begin{array}{r} 1 \\ \cancel{1}\;\;\cancel{1} \\ \cancel{2}\;\;\cancel{4} \\ \cancel{2}\;\;\cancel{3} \\ 34.06 \\ \times \quad 23.75 \\ \hline 1\,70\,30 \\ 23\,84\,20 \\ 102\,18\,00 \\ 681\,20\,00 \\ \hline 808.92\,50 \end{array}$$

Put 6 decimal places in the product.

$$\begin{array}{r} 2\,\cancel{2}\;\;\cancel{1} \\ 3\,\cancel{3}\;\cancel{1}\,\cancel{2} \\ 808.925 \\ \times \quad 0.134 \\ \hline 3\,235\,700 \\ 24\,267\,750 \\ 80\,892\,500 \\ \hline 108.395\,950 \end{array}$$

$108.39595 \approx 108.40$

43. $19.7 + 21.4 + 18.6 + 20.9 + 18.4 = 99$
Grant purchased 99 gallons of gas.

45. $18.4(2.629) = 48.3736$; To the nearest cent, Grant paid \$48.37 for the fifth fill-up.

47. 1st: $19.7(2.399) \approx 47.26$; 2nd: $21.4(2.419) \approx 51.77$; 3rd: $18.6(2.559) \approx 47.60$;
4th: $20.9(2.399) \approx 50.14$; 5th: $18.4(2.629) \approx 48.37$; Grant paid least for the \$2.399 fill-up.

SECTION 4.4

49. Put four decimal places in the product.

$$\begin{array}{r} 2\;2\;\;1\;\;\;\;\; \\ \cancel{3}\;\cancel{3}\;\;\cancel{3}\;\;\;\;\; \\ \cancel{4}\;\cancel{4}\;\;\cancel{2}\;\;\;\;\; \\ 78.95 \\ \times \qquad 3.65 \\ \hline 3\;94\;75 \\ 47\;37\;00 \\ 236\;85\;00 \\ \hline 288.16\;75 \end{array}$$

51. Put three decimal places in the product.

$$\begin{array}{r} 4\;\;\;\;\;2\;\; \\ \cancel{6}\;\;\;\;\;\cancel{3}\;\; \\ 58\;0.0\;4 \\ \times \qquad 1\;5.8 \\ \hline 464\;0\;3\;2 \\ 2900\;2\;0\;0 \\ 5800\;4\;0\;0 \\ \hline 9164.6\;3\;2 \end{array}$$

53. $(4.57)(234.7)(21.042) = (1072.579)(21.042) = 22569.207318 \approx 22{,}569.207$

55. $(20.4)(0.48)(8.02)(50.4) = (9.792)(8.02)(50.4) = (78.53184)(50.4) = 3958.004736 \approx 3958.00$

57. $6.99(4.35) = 30.4065$; Sonya paid \$30.41 for the steak.

59. $3(100) = 300$; $710 - 300 = 410$ extra miles; $3(26.95) = 80.85$; $410(0.24) = 98.40$; $80.85 + 98.40 = 179.25$; It costs \$179.25 to rent the car.

61. $3(100) = 300$; $425 - 300 = 125$ extra miles; $3(30.00) = 90.00$; $125(0.275) = 34.375 \approx 34.38$; $90.00 + 34.38 = 124.38$; It costs \$124.38 to rent the full-size car.
$2(100) = 200$; $625 - 200 = 425$ extra miles; $2(26.95) = 53.90$; $425(0.24) = 102$;
$53.90 + 102.00 = 155.90$; $155.90 - 124.38 = 31.52$. It costs \$31.52 less to rent the full-size car.

63. Store 1: $80 + 91.95(18) = 80 + 1655.10 = 1735.10$
Store 2: $125 + 67.25(24) = 125 + 1614 = 1739$
Store 3: $350 + 119.55(12) = 350 + 1434.60 = 1784.60$; Store 1 has the lowest cost.

65. Calories for 25 minutes (per pound) $= 25(0.102) = 2.55$; Calories burned for 25 minutes (187 pounds) $= 187(2.55) = 476.85$; Calories burned in 1 week $= 5(476.85) = 2384.25$ calories

67. $43(17.325) = 744.975$; The total number of linear feet is 744.975 feet.

69. 1995: $4(79.4) = 317.6$; 2005: $4(63.7) = 254.8$; 2015: $4(59) = 236$; In 1995, 2005, and 2015, the total weight of meat consumed was 317.6 lb, 254.8 lb, and 236 lb, respectively. One reason for the decrease could be increased consumer health concerns associated with red meat.

71. # flushes $= 5(41{,}782) = 208{,}910$; pre-1970: $5.5(208{,}910) = 1{,}149{,}005$;
1970s: $3.5(208{,}910) = 731{,}185$; low-flow: $1.55(208{,}910) = 323{,}810.5$;
$1{,}149{,}005 - 323{,}810.5 = 825{,}194.5$; The number of gallons used per day are 1,149,005 for the older models, 731,185 for the 1970s models, and 323,810.5 for the low-flow models. The total number of gallons saved using the low-flow model is 825,194.5 gallons.

73. $1{,}461{,}200 \div 1000 = 1461.2$; $1461.2(2.775) = 4054.83$; The tax bill is about \$4055.

75. $19.19 - 9.58 = 9.61$; He ran the second 100 meters in 9.61 seconds.

77. Find the number of decimal places in the two factors and add them together. This is the number of decimal places in the product.

79. The smallest whole number is 111..

81. The pattern is to multiply by 2 and divide by 10, then multiply by 3 and divide by 10, and so on...
$0.0504(5) \div 10 = 0.252 \div 10 = 0.0252$; The missing number is 0.0252.

83. $0.3(?) = 0.36 \Rightarrow 0.3(1.2) = 0.36$; $1.5 + 2.7 - ? = 1.2$; $4.2 - ? = 1.2$; $? = 3$
The missing number is 3.

85. $82(10{,}000) = 820{,}000$

87. $22{,}000{,}000 \div 10{,}000 = 2200$

89. $692 \times 10^3 = 692{,}000$

91. $4{,}760{,}000 \div 10^4 = 476$

93. $4{,}210{,}000{,}000 \div 10^6 = 4210$

Exercises 4.5 (page 319)

1. $19.3 \div 10 = 1.93$

3. $(92.6)(100) = 9260$

5. $(1.3557)(1000) = 1355.7$

7. $\frac{58.18}{100} = 0.5818$

9. $\frac{8325}{100} = 83.25$

11. $0.107 \times 10^4 = 1070$

13. right

15. $(78.324)(1000) = 78{,}324$

17. $99.7 \div 10 = 9.97$

19. $57.9(1000) = 57{,}900$

21. $\frac{9077.5}{10{,}000} = 0.90775$

23. $28.73(100{,}000) = 2{,}873{,}000$

25. $6056.32 \div 100 = 60.5632$

27. $32.76 \div 100{,}000 = 0.0003276$

29. $750{,}000 = 7.5 \times 10^5$

31. $0.000091 = 9.1 \times 10^{-5}$

33. $4195.3 = 4.1953 \times 10^3$

35. $12 \times 10^5 = 1{,}200{,}000$

37. $4 \times 10^{-3} = 0.004$

39. $9.43 \times 10^5 = 943{,}000$

41. $43{,}700 = 4.37 \times 10^4$

43. $0.000000587 = 5.87 \times 10^{-7}$

45. $0.0000000000684 = 6.84 \times 10^{-11}$

47. $64.004 = 6.4004 \times 10^1$

49. $7.341 \times 10^{-5} = 0.00007341$

51. $1.77 \times 10^9 = 1{,}770{,}000{,}000$

53. $3.11 \times 10^{-8} = 0.0000000311$

55. $1.48 \times 10^{-8} = 0.0000000148$

SECTION 4.5

57. $100(39.68) = 3968$; The total cost is \$3968.

59. $100(3100) = 310{,}000$; The total cost is \$310,000.

61. $52{,}000{,}000 = 5.2 \times 10^7$; The total land area is 5.2×10^7 square miles.

63. $0.000000001 = 1 \times 10^{-9}$; One nanometer is 1×10^{-9} m.

65. $0.000054 = 5.4 \times 10^{-5}$; Light travels 1 mile in approximately 5.4×10^{-5} second.

67. $4 \times 10^{-5} = 0.00004$; The shortest wavelength of visible light is 0.00004 cm.

69. $3.467 \times 10^8 = 346{,}700{,}000$; $3.521 \times 10^8 = 352{,}100{,}000$; $2.783 \times 10^8 = 278{,}300{,}000$; $2.552 \times 10^8 = 255{,}200{,}000$; The total use is 1,232,300,000 BTUs.

71. $3{,}060{,}000 \div 100{,}000 = 30.6$; The per capita consumption was 30.6 gallons.

73. They multiply the batting average by 1000.

75. The average is 0.424.

77. $206{,}265(93{,}000{,}000) = 19{,}182{,}645{,}000{,}000 \approx 1.92 \times 10^{13}$ miles

79. $$\frac{(3.25 \times 10^{-3})(2.4 \times 10^{3})}{(4.8 \times 10^{-4})(2.5 \times 10^{-3})} = \frac{(0.00325)(2400)}{(0.00048)(0.0025)} = \frac{7.8}{0.0000012} = 6{,}500{,}000 = 6.5 \times 10^6$$

81. $$\begin{aligned}(5.5 \times 10^{-7})(8.1 \times 10^{12})(2 \times 10^{5})(1.5 \times 10^{-9}) &= (0.00000055)(8{,}100{,}000{,}000{,}000)(200000)(0.0000000015)\\ &= (4{,}455{,}000)(200000)(0.0000000015)\\ &= (4{,}455{,}000)(0.0003) = 1336.5 = 1.3365 \times 10^3\end{aligned}$$

83.
$$\begin{array}{r} 312\\ 59\overline{)18408}\\ \underline{177}\\ 70\\ \underline{59}\\ 118\\ \underline{118}\\ 0 \end{array}$$

85.
$$\begin{array}{r} 62\text{ R}1\\ 12\overline{)745}\\ \underline{72}\\ 25\\ \underline{24}\\ 1 \end{array}$$

87.
$$\begin{array}{r} 2596\\ 243\overline{)630828}\\ \underline{486}\\ 1448\\ \underline{1215}\\ 2332\\ \underline{2187}\\ 1458\\ \underline{1458}\\ 0 \end{array}$$

89.
$$\begin{array}{r} 9202 \\ 711\overline{)6542851} \\ \underline{6399} \\ 1438 \\ \underline{1422} \\ 165 \\ \underline{0} \\ 1651 \\ \underline{1422} \\ 229 \end{array}$$
$9202 \approx 9200$

91. 4 feet = 48 inches
$1008 \div 48 = 21$
The width is 21 inches, or 1.75 feet.

Exercises 4.6 (page 330)

1.
$$\begin{array}{r} 0.8 \\ 8\overline{)6.4} \\ \underline{6\ 4} \\ 0 \end{array}$$

3.
$$\begin{array}{r} 4.9 \\ 4\overline{)19.6} \\ \underline{16} \\ 3\ 6 \\ \underline{3\ 6} \\ 0 \end{array}$$

5.
$$\begin{array}{r} 0.1\overline{)32.67} \\ \Downarrow \\ 326.7 \\ 1.\overline{)326.7} \\ \underline{3} \\ 0\,2 \\ \underline{2} \\ 0\,6 \\ \underline{6} \\ 0\ 7 \\ \underline{7} \\ 0 \end{array}$$

7.
$$\begin{array}{r} 0.13\overline{)393.9} \\ \Downarrow \\ 3030. \\ 13\overline{)39390.} \\ \underline{39} \\ 0\,0\,3 \\ \underline{0} \\ 3\,9 \\ \underline{3\,9} \\ 0\,0 \\ \underline{0} \\ 0 \end{array}$$

9.
$$\begin{array}{r} 5.53 \\ 60\overline{)331.80} \\ \underline{300} \\ 31\,8 \\ \underline{30\,0} \\ 1\,8\,0 \\ \underline{1\,8\,0} \\ 0 \end{array}$$

11.
$$\begin{array}{r} 0.7875 \\ 36\overline{)28.3500} \\ \underline{25\ 2} \\ 3\ 1\,5 \\ \underline{2\ 8\,8} \\ 2\,7\,0 \\ \underline{2\,5\,2} \\ 1\,8\,0 \\ \underline{1\,8\,0} \\ 0 \end{array}$$

13. whole number

15.
$$\begin{array}{r} 1.20 \\ 6\overline{)7.23} \\ \underline{6} \\ 1\ 2 \\ \underline{1\ 2} \\ 0\,3 \\ \underline{0} \end{array}$$
≈ 1.2

SECTION 4.6

17. $1.6\overline{)10.551}$

$\Downarrow$

$$\begin{array}{r} 6.59 \\ 16\overline{)105.51} \\ 96 \\ \hline 9\,5 \\ 8\,0 \\ \hline 1\,5\,1 \\ 1\,4\,4 \\ \hline \end{array}$$

$6.59 \approx 6.6$

19.

$$\begin{array}{r} 0.098 \\ 6\overline{)0.5934} \\ 54 \\ \hline 53 \\ 48 \\ \hline \end{array}$$

$0.098 \approx 0.10$

21. $0.7\overline{)5.687}$

$\Downarrow$

$$\begin{array}{r} 8.124 \\ 7\overline{)56.870} \\ 56 \\ \hline 0\,8 \\ 7 \\ \hline 1\,7 \\ 1\,4 \\ \hline 3\,0 \\ 2\,8 \\ \hline \end{array}$$

$8.124 \approx 8.12$

23. $0.413\overline{)0.793}$

$\Downarrow$

$$\begin{array}{r} 1.920 \\ 413\overline{)793.000} \\ 413 \\ \hline 3800 \\ 3717 \\ \hline 830 \\ 826 \\ \hline 40 \\ 0 \\ \hline \end{array}$$

$1.920 \approx 1.92$

25. $0.0552\overline{)25}$

$\Downarrow$

$$\begin{array}{r} 452.898 \\ 552\overline{)250000.000} \\ 2208 \\ \hline 2920 \\ 2760 \\ \hline 1600 \\ 1104 \\ \hline 4960 \\ 4416 \\ \hline 5440 \\ 4968 \\ \hline 4720 \\ 4416 \\ \hline \end{array}$$

$452.898 \approx 452.90$

27. $2.15\overline{)19.68}$

$\Downarrow$

$$\begin{array}{r} 9.1534 \\ 215\overline{)1968.0000} \\ 1935 \\ \hline 330 \\ 215 \\ \hline 1150 \\ 1075 \\ \hline 750 \\ 645 \\ \hline 1050 \\ 860 \\ \hline \end{array}$$

$9.1534 \approx 9.153$

29. $4.16\overline{)0.06849}$

$\Downarrow$

$$\begin{array}{r} 0.0164 \\ 416\overline{)6.8490} \\ 416 \\ \hline 2689 \\ 2496 \\ \hline 1930 \\ 1664 \\ \hline 266 \end{array}$$

$0.0164 \approx 0.016$

SECTION 4.6

31. $0.009\overline{)0.13}$

$\Downarrow$

$$\begin{array}{r} 14.4444 \\ 9\overline{)130.0000} \\ \underline{9} \\ 40 \\ \underline{36} \\ 40 \\ \underline{36} \\ 40 \\ \underline{36} \\ 40 \\ \underline{36} \\ 40 \\ \underline{36} \end{array}$$

$14.4444 \approx 14.444$

33. $8.3 + 5.9 = 14.2;\ 14.2 \div 2 = 7.1$

35. $5.7 + 10.2 = 15.9;\ 15.9 \div 2 = 7.95$

37. $10.6 + 8.4 + 2.9 = 21.9;\ 21.9 \div 3 = 7.3$

39. $12.1 + 12.5 + 12.6 = 37.2;\ 37.2 \div 3 = 12.4$

41. $12.5 + 7.1 + 16.7 + 2.8 = 39.1;$
$39.1 \div 4 = 9.775$

43. **Average**
$7.8 + 9.08 + 3.9 + 5.7 = 26.48$
$26.48 \div 4 = 6.62$

Median
In order: 3.9, 5.7, 7.8, 9.08
$(5.7 + 7.8) \div 2 = 13.5 \div 2 = 6.75$

Mode
none

45. **Average**
$15.8 + 23.64 + 22.46 + 23.64 + 18.7 = 104.24$
$104.24 \div 5 = 20.848$

Median
In order: 15.8, 18.7, 22.46, 23.64, 23.64
22.46

Mode
23.64

47. **Average**
$14.3 + 15.4 + 7.6 + 17.4 + 21.6 = 76.3$
$76.3 \div 5 = 15.26$

Median
In order: 7.6, 14.3, 15.4, 17.4, 21.6
15.4

Mode
none

49. **Average**
$0.675 + 0.431 + 0.662 + 0.904 = 2.672$
$2.672 \div 4 = 0.668$

Median
In order: 0.431, 0.662, 0.675, 0.904
$0.662 + 0.675 = 1.337$
$1.337 \div 2 = 0.6685$

Mode
none

51. **Average**
$0.5066 + 0.6055 + 0.5506 + 0.5066 + 0.6505 = 2.8198;\ 2.8198 \div 5 = 0.56396$

Median
In order: 0.5066, 0.5066, 0.5506, 0.6055, 0.6505; 0.5506

Mode: 0.5066

SECTION 4.6

53. $24.64 + 24.64 + 24.55 + 24.69 + 24.68 = 123.2$; $123.2 \div 5 = 24.64$
The average closing price was \$24.64.

55. $0.19\overline{)17.43}$

$\Downarrow$

$$\begin{array}{r} 91.736 \\ 19\overline{)1743.000} \\ \underline{171} \\ 33 \\ \underline{19} \\ 140 \\ \underline{133} \\ 70 \\ \underline{57} \\ 130 \\ \underline{114} \end{array}$$

$91.736 \approx 91.74$

57. $\$3.79 \div 4$

$$\begin{array}{r} 0.9475 \\ 4\overline{)3.7900} \\ \underline{3\,6} \\ 19 \\ \underline{16} \\ 30 \\ \underline{28} \\ 20 \\ \underline{20} \end{array}$$

The unit price is about \$0.948, or 94.8¢, per lb.

59. $\$6.97 \div 3.5$

$3.5\overline{)6.97}$

$\Downarrow$

$$\begin{array}{r} 1.9914 \\ 35\overline{)69.7000} \\ \underline{35} \\ 347 \\ \underline{315} \\ 320 \\ \underline{315} \\ 50 \\ \underline{35} \\ 150 \\ \underline{140} \end{array}$$

The unit price is about \$1.991, or 199.1¢, per lb.

61. $\$10.38 \div 2.6$

$2.6\overline{)10.38}$

$\Downarrow$

$$\begin{array}{r} 3.992 \\ 26\overline{)103.800} \\ \underline{78} \\ 258 \\ \underline{234} \\ 240 \\ \underline{234} \\ 60 \\ \underline{52} \end{array}$$

$1.5(3.992) \approx 5.988$
The cod will cost \$5.99.

63. $245{,}610 \div 256$

$$\begin{array}{r} 959.414 \\ 256\overline{)245610.000} \\ \underline{2304} \\ 1521 \\ \underline{1280} \\ 2410 \\ \underline{2304} \\ 1060 \\ \underline{1024} \\ 360 \\ \underline{256} \\ 1040 \\ \underline{1024} \end{array}$$

$959.414 \approx 959.41$
The average donation was \$959.41.

65. $72 + 73 + 78 + 84 + 88 + 91 + 92 + 92 + 90 + 84 + 78 + 73 = 995$; $995 \div 12 \approx 82.9167$
In order: 72, 73, 73, 78, 78, 84, 84, 88, 90, 91, 92, 92; $(84 + 84) \div 2 = 168 \div 2 = 84$
The average temperature is 82.9° F. The median temperature is 84° F. The mode temperatures are 73° F, 78° F, 84° F, and 92° F.

SECTION 4.6

67. $2.489 + 2.599 + 2.409 + 2.489 + 2.619 + 2.599 + 2.329 + 2.479 = 20.012$; $20.012 \div 8 = 2.5015$
In order: 2.329, 2.409, 2.479, 2.489, 2.489, 2.599, 2.599, 2.619
$(2.489 + 2.489) \div 2 = 4.978 \div 2 = 2.489$
The average price is about \$2.502 per gallon. The median price is \$2.489 per gallon. The mode prices are \$2.489 and \$2.599 per gallon.

69-71.

City	High	Low	Range
Detroit	37	26	11
Cincinnati	43	31	12
Chicago	35	30	5
St. Louis	44	28	16
Kansas City	33	24	9
Minneapolis	27	17	10
Milwaukee	33	28	5
Rapid City	40	18	22

69. $26 + 31 + 30 + 28 + 24 + 17 + 28 + 18 = 202$; $202 \div 8 \approx 25.3$; The average low temperature was 25.3° F.

71. The mode for the high temperatures was 33° F.

73. $766.5 \div 15.8$

$15.8\overline{)766.5}$

⇓

$$\begin{array}{r} 48.5 \\ 158\overline{)7665.0} \\ \underline{632} \\ 1345 \\ \underline{1264} \\ 810 \\ \underline{790} \end{array}$$

June got about 49 miles per gallon.

75. $867 \div 2.75$

$2.75\overline{)867}$

⇓

$$\begin{array}{r} 315.2 \\ 275\overline{)86700.0} \\ \underline{825} \\ 420 \\ \underline{275} \\ 1450 \\ \underline{1375} \\ 750 \\ \underline{550} \end{array}$$

About 315 feet of cable are on the spool.

77. $630.6 \div 32.7$

$32.7\overline{)630.6}$

⇓

$$\begin{array}{r} 19.28 \\ 327\overline{)6306.00} \\ \underline{327} \\ 3036 \\ \underline{2943} \\ 930 \\ \underline{654} \\ 2760 \\ \underline{2616} \end{array}$$

The length is about 19.3 feet.

79. Denmark has the smallest area and Norway has the smallest population.

81. Based on #80, Denmark has the smallest area but is over 6 times more crowded than Sweden. Finland and Norway have about the same density.

83. $110 \div 9 \approx 12.22$; $20 \div 12.22 \approx 1.64$
The ERA is about 1.64.

85. The average is $\frac{7}{10}$ or higher. This means that the runner succeeds at least 7 out of 10 times.

87. The decimal point in a quotient is placed directly over the decimal point in the dividend when the divisor is a whole number. When the divisor is a decimal, move the decimal point to make the divisor a whole number. Then move the decimal point in the dividend the same amount. Place the decimal point in the quotient above the new location of the decimal point in the dividend. The procedure works because a division problem can be thought of as the fraction $\frac{\text{dividend}}{\text{divisor}}$. Moving the decimal point the same number of places in each is equivalent to multiplying the numerator and denominator of this fraction by the same power of 10.

89. **Answers may vary.**

91. $8.23 \div 0.56 \approx 14.6964$;
$14.6964 \div 2.47 \approx 5.95$

93. $\frac{95}{114} = \frac{5 \cdot \cancel{19}}{6 \cdot \cancel{19}} = \frac{5}{6}$

95. $4\frac{8}{11} = \frac{11(4) + 8}{11} = \frac{52}{11}$

97. $\frac{215}{12} = 17\frac{11}{12}$

99. $100 \div 25 = 4$; $\frac{17}{25} \cdot \frac{4}{4} = \frac{68}{100}$

101. $24 \div 40 = \frac{24}{40} = \frac{3 \cdot \cancel{8}}{5 \cdot \cancel{8}} = \frac{3}{5}$

Getting Ready for Algebra (page 337)

1.
$$2.7x = 18.9$$
$$\frac{2.7x}{2.7} = \frac{18.9}{2.7}$$
$$x = 7$$

3.
$$0.04y = 12.34$$
$$\frac{0.04y}{0.04} = \frac{12.34}{0.04}$$
$$y = 308.5$$

5.
$$0.9476 = 4.12t$$
$$\frac{0.9476}{4.12} = \frac{4.12t}{4.12}$$
$$0.23 = t$$

7.
$$3.3m = 0.198$$
$$\frac{3.3m}{3.3} = \frac{0.198}{3.3}$$
$$m = 0.06$$

9.
$$0.016q = 9$$
$$\frac{0.016q}{0.016} = \frac{9}{0.016}$$
$$q = 562.5$$

11.
$$9 = 0.32h$$
$$\frac{9}{0.32} = \frac{0.32h}{0.32}$$
$$28.125 = h$$

13.
$$\frac{y}{9.5} = 0.28$$
$$9.5\left(\frac{y}{9.5}\right) = 9.5(0.28)$$
$$y = 2.66$$

15.
$$0.312 = \frac{c}{0.65}$$
$$0.65(0.312) = 0.65\left(\frac{c}{0.65}\right)$$
$$0.2028 = c$$

17.
$$0.0325 = \frac{x}{32}$$
$$32(0.0325) = 32\left(\frac{x}{32}\right)$$
$$1.04 = x$$

19.
$$\frac{s}{0.07} = 0.345$$
$$0.07\left(\frac{s}{0.07}\right) = 0.07(0.345)$$
$$s = 0.02415$$

21.
$$\frac{z}{21.02} = 4.08$$
$$21.02\left(\frac{z}{21.02}\right) = 21.02(4.08)$$
$$z = 85.7616$$

23.
$$T = sC$$
$$3617.9 = s(157.3)$$
$$\frac{3617.9}{157.3} = \frac{s(157.3)}{157.3}$$
$$23 = s$$
There are 23 servings.

25.
$$E = IR$$
$$209 = I(22)$$
$$\frac{209}{22} = \frac{I(22)}{22}$$
$$9.5 = I$$
The current is 9.5 amps.

27.
$$A = lw$$
$$250.24 = l(13.6)$$
$$\frac{250.24}{13.6} = \frac{l(13.6)}{13.6}$$
$$18.4 = l$$
The length is 18.4 feet.

29. Let S represent the number of students.
$$20S = 3500$$
$$\frac{20S}{20} = \frac{3500}{20}$$
$$S = 175$$
There are 175 students.

Exercises 4.7 (page 342)

1. $\frac{3}{4} = 3 \div 4 = 0.75$

```
   0. 7 5
 4|3. 0 0
   2 8
   ---
     2 0
     2 0
     ---
       0
```

3. $\frac{3}{20} = 3 \div 20 = 0.15$

```
     0. 1 5
 2 0|3. 0 0
     2 0
     -----
     1 0 0
     1 0 0
     -----
         0
```

SECTION 4.7

5. $\dfrac{13}{16} = 13 \div 16 = 0.8125$

$$\begin{array}{r}
0.8125 \\
16\overline{)13.0000} \\
\underline{12\,8} \\
20 \\
\underline{16} \\
40 \\
\underline{32} \\
80 \\
\underline{80} \\
0
\end{array}$$

7. $6\dfrac{5}{8} = 6 + 5 \div 8 = 6.625$

$$\begin{array}{r}
0.625 \\
8\overline{)5.000} \\
\underline{4\,8} \\
20 \\
\underline{16} \\
40 \\
\underline{40} \\
0
\end{array}$$

9. $56\dfrac{73}{125} = 56 + 73 \div 125 = 56.584$

$$\begin{array}{r}
0.584 \\
125\overline{)73.000} \\
\underline{62\,5} \\
10\,50 \\
\underline{10\,00} \\
500 \\
\underline{500} \\
0
\end{array}$$

11. **Tenth**

$\dfrac{3}{7} = 3 \div 7 = 0.4$

$$\begin{array}{r}
0.42 \\
7\overline{)3.00} \\
\underline{2\,8} \\
20 \\
\underline{14}
\end{array}$$

Hundredth

$\dfrac{3}{7} = 3 \div 7 = 0.43$

$$\begin{array}{r}
0.428 \\
7\overline{)3.000} \\
\underline{2\,8} \\
20 \\
\underline{14} \\
60 \\
\underline{56}
\end{array}$$

13. **Tenth**

$\dfrac{5}{12} = 5 \div 12 = 0.4$

$$\begin{array}{r}
0.41 \\
12\overline{)5.00} \\
\underline{4\,8} \\
20 \\
\underline{12}
\end{array}$$

Hundredth

$\dfrac{5}{12} = 5 \div 12 = 0.42$

$$\begin{array}{r}
0.416 \\
12\overline{)5.000} \\
\underline{4\,8} \\
20 \\
\underline{12} \\
80 \\
\underline{72}
\end{array}$$

15. **Tenth**

$\dfrac{11}{13} = 11 \div 13 = 0.8$

$$\begin{array}{r}
0.84 \\
13\overline{)11.00} \\
\underline{10\,4} \\
60 \\
\underline{52}
\end{array}$$

Hundredth

$\dfrac{11}{13} = 11 \div 13 = 0.85$

$$\begin{array}{r}
0.846 \\
13\overline{)11.000} \\
\underline{10\,4} \\
60 \\
\underline{52} \\
80 \\
\underline{78}
\end{array}$$

SECTION 4.7

17. **Tenth**

$\frac{2}{15} = 2 \div 15 = 0.1$

$$\begin{array}{r} 0.13 \\ 15\overline{)2.00} \\ \underline{1\,5} \\ 50 \\ \underline{45} \end{array}$$

Hundredth

$\frac{2}{15} = 2 \div 15 = 0.13$

$$\begin{array}{r} 0.133 \\ 15\overline{)2.000} \\ \underline{1\,5} \\ 50 \\ \underline{45} \\ 50 \\ \underline{45} \end{array}$$

19. **Tenth**

$\frac{7}{18} = 7 \div 18 = 0.4$

$$\begin{array}{r} 0.38 \\ 18\overline{)7.00} \\ \underline{5\,4} \\ 160 \\ \underline{144} \end{array}$$

$7\frac{7}{18} = 7.4$

Hundredth

$\frac{7}{18} = 7 \div 18 = 0.39$

$$\begin{array}{r} 0.388 \\ 18\overline{)7.000} \\ \underline{5\,4} \\ 160 \\ \underline{144} \\ 160 \\ \underline{144} \end{array}$$

$7\frac{7}{18} = 7.39$

21. $\frac{9}{11} = 9 \div 11 = 0.\overline{81}$

$$\begin{array}{r} 0.818 \\ 11\overline{)9.000} \\ \underline{8\,8} \\ 20 \\ \underline{11} \\ 90 \\ \underline{88} \end{array}$$

23. $\frac{1}{12} = 1 \div 12 = 0.08\overline{3}$

$$\begin{array}{r} 0.0833 \\ 12\overline{)1.0000} \\ \underline{9\,6} \\ 40 \\ \underline{36} \\ 40 \\ \underline{36} \end{array}$$

25. **Hundredth**

$\frac{9}{79} = 9 \div 79$

$= 0.11$

$$\begin{array}{r} 0.113 \\ 79\overline{)9.000} \\ \underline{7\,9} \\ 110 \\ \underline{79} \\ 310 \\ \underline{237} \end{array}$$

Thousandth

$\frac{9}{79} = 9 \div 79$

$= 0.114$

$$\begin{array}{r} 0.1139 \\ 79\overline{)9.0000} \\ \underline{7\,9} \\ 110 \\ \underline{79} \\ 310 \\ \underline{237} \\ 730 \\ \underline{711} \end{array}$$

27. **Hundredth**

$\frac{45}{46} = 45 \div 46$

$= 0.98$

$$\begin{array}{r} 0.978 \\ 46\overline{)45.000} \\ \underline{41\,4} \\ 360 \\ \underline{322} \\ 380 \\ \underline{368} \end{array}$$

Thousandth

$\frac{45}{46} = 45 \div 46$

$= 0.978$

$$\begin{array}{r} 0.9782 \\ 46\overline{)45.0000} \\ \underline{41\,4} \\ 360 \\ \underline{322} \\ 380 \\ \underline{368} \\ 120 \\ \underline{92} \end{array}$$

29. $\frac{5}{13} = 5 \div 13 = 0.\overline{384615}$

$$\begin{array}{r} 0.3846153 \\ 13\,\overline{)\,5.0000000} \\ \underline{39} \\ 110 \\ \underline{104} \\ 60 \\ \underline{52} \\ 80 \\ \underline{78} \\ 20 \\ \underline{13} \\ 70 \\ \underline{65} \\ 50 \\ \underline{39} \end{array}$$

31. $\frac{6}{7} = 6 \div 7 = 0.\overline{857142}$

$$\begin{array}{r} 0.8571428 \\ 7\,\overline{)\,6.0000000} \\ \underline{56} \\ 40 \\ \underline{35} \\ 50 \\ \underline{49} \\ 10 \\ \underline{7} \\ 30 \\ \underline{28} \\ 20 \\ \underline{14} \\ 60 \\ \underline{56} \end{array}$$

33. $2\frac{3}{8} = 2 + 3 \div 8 = 2 + 0.375 = 2.375$; The reading will be 2.375 in.

$$\begin{array}{r} 0.375 \\ 8\,\overline{)\,3.000} \\ \underline{24} \\ 60 \\ \underline{56} \\ 40 \\ \underline{40} \\ 0 \end{array}$$

35. $\frac{3}{8} = 3 \div 8 = 0.375$ $\quad 1\frac{1}{4} = 1 + 1 \div 4 = 1 + 0.25 = 1.25$ $\quad \frac{1}{2} = 1 \div 2 = 0.5$

$$\begin{array}{r} 0.375 \\ 8\,\overline{)\,3.000} \\ \underline{24} \\ 60 \\ \underline{56} \\ 40 \\ \underline{40} \\ 0 \end{array} \qquad \begin{array}{r} 0.25 \\ 4\,\overline{)\,1.00} \\ \underline{8} \\ 20 \\ \underline{20} \\ 0 \end{array} \qquad \begin{array}{r} 0.5 \\ 2\,\overline{)\,1.0} \\ \underline{10} \\ 0 \end{array}$$

The measurements are 0.375 in., 1.25 in., and 0.5 in.

SECTION 4.7

37. **Hundredth**

$\frac{21}{52} = 21 \div 52 = 0.40$

$$\begin{array}{r} 0.403 \\ 52\overline{)21.000} \\ \underline{20\,8} \\ 20 \\ \underline{0} \\ 200 \\ \underline{156} \end{array}$$

Thousandth

$\frac{21}{52} = 21 \div 52 = 0.404$

$$\begin{array}{r} 0.4038 \\ 52\overline{)21.0000} \\ \underline{20\,8} \\ 20 \\ \underline{0} \\ 200 \\ \underline{156} \\ 440 \\ \underline{416} \end{array}$$

Ten-thousandth

$\frac{21}{52} = 21 \div 52 = 0.4038$

$$\begin{array}{r} 0.40384 \\ 52\overline{)21.00000} \\ \underline{20\,8} \\ 20 \\ \underline{0} \\ 200 \\ \underline{156} \\ 440 \\ \underline{416} \\ 240 \\ \underline{208} \end{array}$$

39. **Hundredth**

$16\frac{15}{101} = 16 + 15 \div 101$
$= 16 + 0.15 = 16.15$

$$\begin{array}{r} 0.148 \\ 101\overline{)15.000} \\ \underline{10\,1} \\ 4\,90 \\ \underline{4\,04} \\ 860 \\ \underline{808} \end{array}$$

Thousandth

$16\frac{15}{101} = 16 + 15 \div 101$
$= 16 + 0.149 = 16.149$

$$\begin{array}{r} 0.1485 \\ 101\overline{)15.0000} \\ \underline{10\,1} \\ 4\,90 \\ \underline{4\,04} \\ 860 \\ \underline{808} \\ 520 \\ \underline{505} \end{array}$$

Ten-thousandth

$16\frac{15}{101} = 16 + 15 \div 101$
$= 16 + 0.1485 = 16.1485$

$$\begin{array}{r} 0.14851 \\ 101\overline{)15.00000} \\ \underline{10\,1} \\ 4\,90 \\ \underline{4\,04} \\ 860 \\ \underline{808} \\ 520 \\ \underline{505} \\ 150 \\ \underline{101} \end{array}$$

41. $7.14 \div 1\frac{3}{4} = 7\frac{7}{50} \div 1\frac{3}{4} = \frac{357}{50} \div \frac{7}{4} = \frac{357}{50} \cdot \frac{4}{7} = \frac{102}{25} = 4\frac{2}{25}$

$7.14 \div 1\frac{3}{4} = 7.14 \div 1.75 = 4.08$; The cost is $\$4\frac{2}{25}$, or \$4.08 per yard. You may find decimals easier to use because you can avoid fractions with large numerators and denominators.

43. $6.45 = 6\frac{45}{100} = 6\frac{9}{20} = 6\frac{27}{60}$
He can run the mile in 6 minutes and 27 seconds.

45. $19\frac{14}{60} = 19\frac{7}{30} = 19 + 7 \div 30$
$\approx 19 + 0.23 = 19.23$
There were 19.23 hours of daylight.

47. 2 hours, 25 minutes, 15 seconds $= 2 + \frac{25}{60} + \frac{15}{3600} = 2 + \frac{1500}{3600} + \frac{15}{3600} = 2\frac{1515}{3600} = 2\frac{101}{240}$

$101 \div 240 = 0.4208\overline{3}$; The time was $2\frac{101}{240}$ hours, or about 2.42 hours.

49. $7 \div 625 = 0.0112$

The larger is $\frac{7}{625}$.

51. $\frac{1.23}{80} = \frac{123}{8000}$; 123 is less than $\frac{1}{10}$ of 8000, so the fraction is less than 0.1. $1.23 \div 80 = 0.015375$.

53. $\frac{5}{\frac{16}{15}} = 5 \div (16 \div 15) = 5 \div 1.0667 \approx 4.69;\ \frac{\frac{5}{16}}{15} = (5 \div 16) \div 15 = 0.3125 \div 15 \approx 0.02$

55. $(9 - 5) \cdot 5 - 14 + 6 \div 2 = 4 \cdot 5 - 14 + 3 = 20 - 14 + 3 = 6 + 3 = 9$

57. $(7 - 4)^2 - 9 \div 3 + 7 = 3^2 - 9 \div 3 + 7 = 9 - 3 + 7 = 6 + 7 = 13$

59. $345 - 271 \approx 350 - 270 = 80$

61. $265 \times 732 \approx 300 \times 700 = 210{,}000$

63. $$\begin{aligned}(145{,}780 + 234{,}700 + 195{,}435 + 389{,}500 + 275{,}000 + 305{,}677) \div 6 &= 1{,}546{,}092 \div 6 \\ &= 257{,}682\end{aligned}$$

The average sale price was $257,682.

Exercises 4.8 (page 349)

1. $0.9 - 0.7 + 0.3 = 0.2 + 0.3 = 0.5$

3. $0.36 \div 9 - 0.02 = 0.04 - 0.02 = 0.02$

5. $2.4 - 3(0.7) = 2.4 - 2.1 = 0.3$

7. $$\begin{aligned}6(2.7) + 3(4.4) &= 16.2 + 3(4.4) \\ &= 16.2 + 13.2 = 29.4\end{aligned}$$

9. $0.19 + (0.7)^2 = 0.19 + 0.49 = 0.68$

11. $9.35 - 2.54 + 6.91 - 3.65 = 6.81 + 6.91 - 3.65 = 13.72 - 3.65 = 10.07$

13. $9.6 \div 2.4(12.7) = 4(12.7) = 50.8$

15. $2.28 \div 0.38(0.37) = 6(0.37) = 2.22$

17. $(4.6)^2 - 2.6(4.1) = 21.16 - 2.6(4.1) = 21.16 - 10.66 = 10.5$

19. $(6.7)(1.4)^3 \div 0.7 = (6.7)(2.744) \div 0.7 = 18.3848 \div 0.7 = 26.264$

21. $$\begin{array}{r} 0.0749 \\ 0.0861 \\ +\,0.0392 \\ \hline \end{array} \Rightarrow \begin{array}{r} 0.07 \\ 0.09 \\ +\,0.04 \\ \hline 0.20 \end{array}$$

23. $$\begin{array}{r} 0.838 \\ -\,0.369 \\ \hline \end{array} \Rightarrow \begin{array}{r} 0.8 \\ -\,0.4 \\ \hline 0.4 \end{array}$$

25. $$\begin{array}{r} 6.299 \\ 3.0055 \\ 0.67 \\ +\,0.0048 \\ \hline \end{array} \Rightarrow \begin{array}{r} 6. \\ 3. \\ 1. \\ +\ \ 0. \\ \hline 10. \end{array}$$

27. $$\begin{array}{r} 7.972 \\ -\,6.7234 \\ \hline \end{array} \Rightarrow \begin{array}{r} 8. \\ -\,7. \\ \hline 1. \end{array}$$

SECTION 4.8

29. $0.00922(0.237) \approx 0.009(0.2) = 0.0018$

31. $11.876(4.368) \approx 10(4) = 40$

33. $2.88 \div 0.0462 \approx 3 \div 0.04 = 300 \div 4 \approx 75 \Rightarrow$ ten

35. $0.0000891 \div 3.78 \approx 0.0000891 \div 4 \approx 0.000022 \Rightarrow$ hundred-thousandth

37.

$$\begin{array}{r} 0.0494 \\ 0.0663 \\ +\,0.07425 \\ \hline \end{array} \Rightarrow \begin{array}{r} 0.05 \\ 0.07 \\ +\,0.07 \\ \hline 0.19 \end{array}$$

The answer is reasonable.

39. $0.00576(0.0491) \approx 0.006(0.05) = 0.0003 \Rightarrow$ The answer is reasonable.

41. $3(0.89) + 4(1.09) + 2(2.49) + 6(0.59) = 2.67 + 4.36 + 4.98 + 3.54 = 15.55$;
Elmer spends \$15.55.

43. $0.0023452 \div 0.572 \approx 0.002 \div 0.6 = 0.02 \div 6 \approx 0.0033 \Rightarrow$ The largest nonzero place value in the quotient is thousandths, so answer (d) is most reasonable.

45.

$$\begin{aligned} (9.9)(4.3) - (5.6)(5.1) + (2.3)^2 &= 42.57 - (5.6)(5.1) + (2.3)^2 \\ &= 42.57 - 28.56 + (2.3)^2 \\ &= 42.57 - 28.56 + 5.29 = 14.01 + 5.29 = 19.3 \end{aligned}$$

47.

$$\begin{aligned} 32.061 - \left[(1.1)^3(1.5) + 4.25\right] &= 32.061 - [(1.331)(1.5) + 4.25] \\ &= 32.061 - [1.9965 + 4.25] = 32.061 - [6.2465] = 25.8145 \end{aligned}$$

49.

$$\begin{aligned} 3.8(3.46 + 6.89 - 1.27) - 2.25(3.54) &= 3.8(10.35 - 1.27) - 2.25(3.54) \\ &= 3.8(9.08) - 2.25(3.54) \\ &= 34.504 - 2.25(3.54) = 34.504 - 7.965 = 26.539 \end{aligned}$$

51. $0.00762(0.215) \approx 0.008(0.2) = 0.0016 \Rightarrow$ The answer is reasonable.

53. $6(23.45) \approx 6(20) = \$120$; The estimate is \$120, but this was obtained by rounding the prices down. She will not have enough money to buy the shirts.

55. $31.8 + 31.8 + 46.8 \approx 32 + 32 + 47 = 111$; The perimeter is about 111 yards.

57. $2.00 + 0.75 + 0.33 = 3.08$; $7.50 - 3.08 = 4.42$; $48(4.42) = 212.16$; $3(212.16) = 636.48$
Rosalie will save \$636.48 per year by bringing her lunch from home.

59.

$$\begin{aligned} &14.88 + 4(2) + 12.99 + 16.88 + 3(12.97) + 6(0.88) + 3(4.88) + 4(4.88) - 14.50 \\ &\quad = 14.88 + 8 + 12.99 + 16.88 + 38.91 + 5.28 + 14.64 + 19.52 - 14.50 = 116.60 \end{aligned}$$

Matthew paid \$116.60 for the items.

61. $1{,}350{,}000 + 810{,}000 + 2(435{,}000) + 300{,}000 + 4(233{,}000) = 4{,}262{,}000$;
$4{,}262{,}000 \div 9 \approx 473{,}555.56$; The average earnings were \$473,555.56.

63. $30 + 26 + 29 + 39 + 29 + 26 + 24 = 203$; $203 \div 7 = 29$;
$32 + 32 + 21 + 31 + 22 + 31 + 18 = 187$; $187 \div 7 = 26.7$;
The Red Wings averaged 29 shots per game, while the Penguins averaged 26.7 shots per game.

65. In the first expression, you must divide $8.3 \div 5$ before doing any addition. In the second expression, you must perform the addition before dividing by 5. The brackets [and] indicate a group of expressions that must be evaluated before combining with any numbers outside of the brackets.

67. $3.62 \div (0.02 + 72.3 \cdot 0.2) = 0.25$

69. $(1.4^2 - 0.8)^2 = 1.3456$

71. **Answers may vary.**

73. $\dfrac{27}{32} = 27 \div 32 = 0.84375$

75. $\dfrac{58}{25} = 58 \div 25 = 2.32$

77. $0.408 = \dfrac{408}{1000} = \dfrac{51}{125}$

79. $6.84 = 6\dfrac{84}{100} = 6\dfrac{21}{25}$

81. $588.88 - 98.50 = 490.38$;
The sale price should be \$490.38.

Getting Ready for Algebra (page 356)

1.
$$\begin{aligned} 2.5x - 7.6 &= 12.8 \\ 2.5x - 7.6 + 7.6 &= 12.8 + 7.6 \\ 2.5x &= 20.4 \\ \frac{2.5x}{2.5} &= \frac{20.4}{2.5} \\ x &= 8.16 \end{aligned}$$

3.
$$\begin{aligned} 1.8x + 6.7 &= 12.1 \\ 1.8x + 6.7 - 6.7 &= 12.1 - 6.7 \\ 1.8x &= 5.4 \\ \frac{1.8x}{1.8} &= \frac{5.4}{1.8} \\ x &= 3 \end{aligned}$$

5.
$$\begin{aligned} 4.115 &= 2.15t + 3.9 \\ 4.115 - 3.9 &= 2.15t + 3.9 - 3.9 \\ 0.215 &= 2.15t \\ \frac{0.215}{2.15} &= \frac{2.15t}{2.15} \\ 0.1 &= t \end{aligned}$$

7.
$$\begin{aligned} 0.03x - 18.7 &= 3.53 \\ 0.03x - 18.7 + 18.7 &= 3.53 + 18.7 \\ 0.03x &= 22.23 \\ \frac{0.03x}{0.03} &= \frac{22.23}{0.03} \\ x &= 741 \end{aligned}$$

9.
$$\begin{aligned} 7x + 0.06 &= 2.3 \\ 7x + 0.06 - 0.06 &= 2.3 - 0.06 \\ 7x &= 2.24 \\ \frac{7x}{7} &= \frac{2.24}{7} \\ x &= 0.32 \end{aligned}$$

11.
$$\begin{aligned} 3.65m - 122.2 &= 108.115 \\ 3.65m - 122.2 + 122.2 &= 108.115 + 122.2 \\ 3.65m &= 230.315 \\ \frac{3.65m}{3.65} &= \frac{230.315}{3.65} \\ m &= 63.1 \end{aligned}$$

13.
$$\begin{aligned} 5000 &= 125y + 2055 \\ 5000 - 2055 &= 125y + 2055 - 2055 \\ 2945 &= 125y \\ \frac{2945}{125} &= \frac{125y}{125} \\ 23.56 &= y \end{aligned}$$

15.
$$\begin{aligned} 60p - 253 &= 9.5 \\ 60p - 253 + 253 &= 9.5 + 253 \\ 60p &= 262.5 \\ \frac{60p}{60} &= \frac{262.5}{60} \\ p &= 4.375 \end{aligned}$$

17.
$$\begin{aligned} 8.551 &= 4.42 + 0.17x \\ 8.551 - 4.42 &= 4.42 - 4.42 + 0.17x \\ 4.131 &= 0.17x \\ \frac{4.131}{0.17} &= \frac{0.17x}{0.17} \\ 24.3 &= x \end{aligned}$$

19.
$$\begin{aligned} 45 &= 1.75h - 1.9 \\ 45 + 1.9 &= 1.75h - 1.9 + 1.9 \\ 46.9 &= 1.75h \\ \frac{46.9}{1.75} &= \frac{1.75h}{1.75} \\ 26.8 &= h \end{aligned}$$

21.
$$\begin{aligned} 1375 &= 80c + 873 \\ 1375 - 873 &= 80c + 873 - 873 \\ 502 &= 80c \\ \frac{502}{80} &= \frac{80c}{80} \\ 6.275 &= c \end{aligned}$$

23.
$$\begin{aligned} F &= 1.8C + 32 \\ 248 &= 1.8C + 32 \\ 248 - 32 &= 1.8C + 32 - 32 \\ 216 &= 1.8C \\ \frac{216}{1.8} &= \frac{1.8C}{1.8} \\ 120 &= C \end{aligned}$$

The Celsius temperature is 120°C.

25.
$$\begin{aligned} D + NP &= B \\ 661.5 + N(73.5) &= 1764 \\ 661.5 - 661.5 + N(73.5) &= 1764 - 661.5 \\ N(73.5) &= 1102.5 \\ \frac{N(73.5)}{73.5} &= \frac{1102.5}{73.5} \\ N &= 15; \end{aligned}$$

Gina has made 15 payments.

27. Let H represent the hours of labor.
$$\begin{aligned} 36H + 137.5 &= 749.5 \\ 36H + 137.5 - 137.5 &= 749.5 - 137.5 \\ 36H &= 612 \\ \frac{36H}{36} &= \frac{612}{36} \\ H &= 17 \end{aligned}$$

She put in 17 hours of labor.

Chapter 4 Review Exercises (page 360)

1. 6.12 = 6 ones and 1 tenth + 2 hundredths; six and twelve hundredths

2. 0.843 = 8 tenths + 4 hundredths + 3 thousandths; eight hundred forty-three thousandths

3. 15.058 = 1 ten + 5 ones and 5 hundredths + 8 thousandths; fifteen and fifty-eight thousandths

4. 0.0000027 = 2 millionths + 7 ten-millionths; twenty-seven ten-millionths

5. twenty-one and five hundredths = 21.05

6. four hundred nine ten-thousandths ⇒ 9 is in the ten-thousandths place ⇒ 0.0409

CHAPTER 4 REVIEW EXERCISES

7. four hundred and four hundredths = 400.04

8. one hundred twenty-five and forty-five thousandths = 125.045

9. **Tenth**
The digit in the tenths place is 7. Since 6 is in the hundredths place, and since 6 is greater than 5, round up. Change the 7 in the tenths place to 8, and drop all digits to the right. The rounded number is 34.8.

Hundredth
The digit in the hundredths place is 6. Since 4 is in the thousandths place, and since 4 is less than 5, round down. Leave the 6 in the hundredths place, and drop all digits to the right. The rounded number is 34.76.

Thousandth
The digit in the thousandths place is 4. Since 8 is in the ten-thousandths place, and since 8 is greater than 5, round up. Change the 4 in the thousandths place to 5, and drop all digits to the right. The rounded number is 34.765.

10. **Tenth**
The digit in the tenths place is 8. Since 7 is in the hundredths place, and since 7 is greater than 5, round up. change the 8 in the tenths place to 9, and drop all digits to the right. The rounded number is 7.9.

Hundredth
The digit in the hundredths place is 7. Since 3 is in the thousandths place, and since 3 is less than 5, round down. Leave the 7 in the hundredths place, and drop all digits to the right. The rounded number is 7.87.

Thousandth
The digit in the thousandths place is 3. Since 6 is in the ten-thousandths place, and since 6 is greater than 5, round up. Change the 3 in the thousandths place to 4, and drop all digits to the right. The rounded number is 7.874.

11. **Tenth**
The digit in the tenths place is 4. Since 6 is in the hundredths place, and since 6 is greater than 5, round up. Change the 4 in the tenths place to 5, and drop all digits to the right. The rounded number is 0.5.

Hundredth
The digit in the hundredths place is 6. Since 7 is in the thousandths place, and since 7 is greater than 5, round up. Change the 6 in the hundredths place to 7, and drop all digits to the right. The rounded number is 0.47.

Thousandth
The digit in the thousandths place is 7. Since 2 is in the ten-thousandths place, and since 2 is less than 5, round down. Leave the 7 in the thousandths place, and drop all digits to the right. The rounded number is 0.467.

12. The digit in the thousandths place is 7. Since 9 is in the ten-thousandths place, and since 9 is greater than 5, round up. Change the 7 in the thousandths place to 8, and drop all digits to the right. The rounded number is 91.458.

13. $0.76 = \frac{76}{100} = \frac{19}{25}$

14. $7.035 = 7\frac{35}{1000} = 7\frac{7}{200}$

15. $0.00256 = \frac{256}{100{,}000} = \frac{8}{3125}$

16. $0.0545 = \frac{545}{10{,}000} = \frac{109}{2000}$

CHAPTER 4 REVIEW EXERCISES

17. 0.89, 0.95, 1.01

18. 0.0900, 0.0930, 0.0899 $\Rightarrow$ 0.0899, 0.09, 0.093

19. 0.717, 7.017, 7.022, 7.108

20. 34.0230, 34.1030, 34.0204, 34.0239 $\Rightarrow$ 34.0204, 34.023, 34.0239, 34.103

21. $6.1774 < 6.1780 \Rightarrow$ true

22. $87.0309 > 87.0319 \Rightarrow$ false

23.

$$\begin{array}{r} {\scriptstyle 11\;2} \\ 11.356 \\ 0.670 \\ 13.082 \\ +\;\;9.600 \\ \hline 34.708 \end{array}$$

24.

$$\begin{array}{r} {\scriptstyle 1\;\;111} \\ 12.0678 \\ 7.0120 \\ 56.0921 \\ +\;\;0.0045 \\ \hline 75.1764 \end{array}$$

25.

$$\begin{array}{r} {\scriptstyle 11\quad\quad 10} \\ {\scriptstyle 1\not{1}\;\; 10\,7\,\not{0}\;\, 16} \\ 22.\not{0}\,\not{8}\,\not{1}\,\not{6} \\ -\;\;8.3629 \\ \hline 13.7187 \end{array}$$

26.

$$\begin{array}{r} {\scriptstyle 139} \\ {\scriptstyle 3\;\;10\,7\,\not{3}\;\;\not{10}\,10} \\ 54.0\not{8}\,\not{4}\,\not{0}\,\not{0} \\ -\;23.64936 \\ \hline 30.43464 \end{array}$$

27.

$$\begin{array}{r} {\scriptstyle 12\;\;121} \\ 3.4050 \\ 8.1200 \\ 0.0098 \\ 0.3456 \\ 11.3000 \\ +\;24.9345 \\ \hline 48.1149 \end{array}$$

28.

$$\begin{array}{r} {\scriptstyle 6\;10\,7\;13} \\ 56.7\not{0}\,\not{8}\,\not{3} \\ -\;21.6249 \\ \hline 35.0834 \end{array}$$

29. $6.1 + 7.2 + 4.9 + 6.3 + 8.1 = 32.6$ in.

30. $4 + 6.4 + 5 = 15.4$ m

31. $1295 + 582.75 + 356.12 + 82.24 + 323.75 + 45 + 325.45 = 3010.31$
$6475 - 3010.31 = 3464.69$; Hilda's take-home pay is \$3464.69 per month.

32. $785.95 - 615.55 = 170.4$; Mary saves \$170.40.

33. Put three decimal places in the product:

$$\begin{array}{r} 8.07 \\ \times\quad 3.5 \\ \hline 4035 \\ 24210 \\ \hline 28.245 \end{array}$$

34. Put three decimal places in the product:

$$\begin{array}{r} {\scriptstyle 1} \\ {\scriptstyle \not{1}\;\not{2}} \\ 11.24 \\ \times\quad 3.5 \\ \hline 5620 \\ 33720 \\ \hline 39.340 \end{array}$$

CHAPTER 4 REVIEW EXERCISES

35. Put seven decimal places in the product:

$$\begin{array}{rr} & 2\ 2 \\ & \not{3}\ \not{4} \\ & 7\ 7 \\ & 0.006\ 78 \\ \times & 3.59 \\ \hline & 0\ 061\ 02 \\ & 00\ 339\ 00 \\ & 0\ 02\ 034\ 00 \\ \hline & 0.02\ 434\ 02 \end{array}$$

36. Put five decimal places in the product:

$$\begin{array}{rr} & 1\ \ \ 4\ 5 \\ & \not{1}\ \ \ \not{4}\ \not{5} \\ & 12.0\ 57 \\ \times & 8.08 \\ \hline & 96\ 4\ 56 \\ & 0\ 00\ 0\ 00 \\ & 96\ 45\ 6\ 00 \\ \hline & 97.42\ 0\ 56 \end{array}$$

37. Put six decimal places in the product:

$$\begin{array}{rr} & 2\ \ \\ & \not{1}\ \ \\ & 2.004 \\ \times & 0.074 \\ \hline & 8\ 016 \\ & 140\ 280 \\ \hline & 0.148\ 296 \end{array} \approx 0.148$$

38. Put five decimal places in the product:

$$\begin{array}{rr} & 2\ 6\ \ \\ & \not{2}\ \not{5}\ \ \\ & 42.7 \\ \times & 0.009\ 8 \\ \hline & 341\ 6 \\ & 3\ 843\ 0 \\ \hline & 0.4\ 184\ 6 \end{array} \approx 0.42$$

39. $(0.03)(4.12)(0.015) = (0.1236)(0.015)$
$= 0.001854 \approx 0.0019$

40. $\text{Area} = \text{Length} \cdot \text{Width} = (7.84)(3.5)$
$= 27.44\ \text{m}^2$

41. $52.35(23.75) \approx 1243.31$; Millie will pay \$1243.31.

42. Method 1: $850 + 5(12)(401.64) = 850 + 60(401.64) = 850 + 24{,}098.40 = 24{,}948.40$
Method 2: $475 + 54(443.10) = 475 + 23{,}927.40 = 24{,}402.40$
Method 3: $600 + 4(12)(495.30) = 600 + 48(495.30) = 600 + 23{,}744.40 = 24{,}374.40$
The third method results in the lowest overall cost.

43. $13.765 \div 10^3 = 0.013765$

44. $7.023 \times 10^6 = 7{,}023{,}000$

45. $0.7321(100{,}000) = 73{,}210$

46. $9.503 \div 100 = 0.09503$

47. $0.0078 = 7.8 \times 10^{-3}$

48. $34.67 = 3.467 \times 10$

49. $0.0000143 = 1.43 \times 10^{-5}$

50. $65{,}700.8 = 6.57008 \times 10^4$

51. $7 \times 10^7 = 70{,}000{,}000$

52. $8.13 \times 10^{-6} = 0.00000813$

53. $6.41 \times 10^{-2} = 0.0641$

54. $3.505 \times 10^3 = 3505$

55. $37{,}350 \div 1000 = 37.35$; The average price is \$37.35.

56. $50{,}000{,}000{,}000 = 5 \times 10^{10}$; The stock market lost $\$5 \times 10^{10}$.

CHAPTER 4 REVIEW EXERCISES

57. $0.3\overline{)0.0111}$

$\Downarrow$

$$\begin{array}{r} 0.037 \\ 3\overline{)0.111} \\ 9 \\ \hline 21 \\ 21 \\ \hline 0 \end{array}$$

58.

$$\begin{array}{r} 0.54 \\ 75\overline{)40.50} \\ 375 \\ \hline 300 \\ 300 \\ \hline 0 \end{array}$$

59. $0.32\overline{)56.7}$

$\Downarrow$

$$\begin{array}{r} 177.1875 \\ 32\overline{)5670.0000} \\ 32 \\ \hline 247 \\ 224 \\ \hline 230 \\ 224 \\ \hline 60 \\ 32 \\ \hline 280 \\ 256 \\ \hline 240 \\ 224 \\ \hline 160 \\ 160 \\ \hline 0 \end{array}$$

60. $0.17\overline{)0.01003}$

$\Downarrow$

$$\begin{array}{r} 0.059 \\ 17\overline{)1.003} \\ 85 \\ \hline 153 \\ 153 \\ \hline 0 \end{array}$$

61. $0.456\overline{)0.38304}$

$\Downarrow$

$$\begin{array}{r} 0.84 \\ 456\overline{)383.04} \\ 3648 \\ \hline 1824 \\ 1824 \\ \hline 0 \end{array}$$

62. $2.015\overline{)6.3271}$

$\Downarrow$

$$\begin{array}{r} 3.14 \\ 2015\overline{)6327.10} \\ 6045 \\ \hline 2821 \\ 2015 \\ \hline 8060 \\ 8060 \\ \hline 0 \end{array}$$

63. $4.7\overline{)332.618}$

$\Downarrow$

$$\begin{array}{r} 70.769 \\ 47\overline{)3326.180} \\ 329 \\ \hline 36 \\ 0 \\ \hline 361 \\ 329 \\ \hline 328 \\ 282 \\ \hline 460 \\ 423 \\ \hline \end{array}$$

$70.769 \approx 70.77$

64. $0.068\overline{)0.01956}$

$\Downarrow$

$$\begin{array}{r} 0.287 \\ 68\overline{)19.560} \\ 136 \\ \hline 596 \\ 544 \\ \hline 520 \\ 476 \\ \hline \end{array}$$

$0.287 \approx 0.29$

CHAPTER 4 REVIEW EXERCISES

65. $13{,}745.50 \div 210 \approx 65.45$
The average donation was about \$65.45.

66. $375.9 \div 12.8 \approx 29$; Carol got about 29 miles per gallon.

67. **Average**
$4.56 + 11.93 + 13.4 + 1.58 + 8.09 = 39.56$
$39.56 \div 5 = 7.912$

Median
In order: 1.58, 4.56, 8.09, 11.93, 13.4
8.09

68. **Average**
$61.78 + 50.32 + 86.3 + 95.04 = 293.44$
$293.44 \div 4 = 73.36$

Median
In order: 50.32, 61.78, 86.3, 95.04
$(61.78 + 86.3) \div 2 = 148.08 \div 2 = 74.04$

69. **Average**
$0.5672 + 0.6086 + 0.3447 + 0.5555 = 2.076$
$2.076 \div 4 = 0.519$

Median
In order: 0.3447, 0.5555, 0.5672, 0.6086
$(0.5555 + 0.5672) \div 2 = 1.1227 \div 2 = 0.56135$

70. **Average**
$14.6 + 18.95 + 12.9 + 23.5 + 16.75 = 86.7$
$86.7 \div 5 = 17.34$

Median
In order: 12.9, 14.6, 16.75, 18.95, 23.5
16.75

71. $4.69 \div 3 \approx 1.5633 \approx 1.563$
The unit cost is about \$1.563 per ounce.

72. $12 + 21 + 5 + 18 + 46 + 67 + 17 = 186$
$186 \div 7 \approx 26.57 \approx 26.6$; About 26.6 robberies per day were reported.

73. $\dfrac{9}{16} = 9 \div 16 = 0.5625$

$$\begin{array}{r}
0.5625 \\
16\overline{)9.0000} \\
\underline{8\,0} \\
1\,00 \\
\underline{96} \\
40 \\
\underline{32} \\
80 \\
\underline{80} \\
0
\end{array}$$

74. $\dfrac{7}{20} = 7 \div 20 = 0.35$

$$\begin{array}{r}
0.35 \\
20\overline{)7.00} \\
\underline{6\,0} \\
1\,00 \\
\underline{1\,00} \\
0
\end{array}$$

75. $17\dfrac{47}{125} = 17 + 47 \div 125$
$= 17.376$

$$\begin{array}{r}
0.376 \\
125\overline{)47.000} \\
\underline{37\,5} \\
9\,50 \\
\underline{8\,75} \\
750 \\
\underline{750} \\
0
\end{array}$$

CHAPTER 4 REVIEW EXERCISES

76. $\frac{11}{37} = 11 \div 37 \approx 0.3$

$$\begin{array}{r} 0.29 \\ 37\overline{)11.00} \\ \underline{7\,4} \\ 3\,60 \\ \underline{3\,33} \end{array}$$

77. $\frac{57}{93} = 57 \div 93 \approx 0.61$

$$\begin{array}{r} 0.612 \\ 93\overline{)57.000} \\ \underline{55\,8} \\ 1\,20 \\ \underline{93} \\ 270 \\ \underline{186} \end{array}$$

78. $\frac{54}{61} = 54 \div 61 \approx 0.885$

$$\begin{array}{r} 0.8852 \\ 61\overline{)54.0000} \\ \underline{48\,8} \\ 5\,20 \\ \underline{4\,88} \\ 320 \\ \underline{305} \\ 150 \\ \underline{122} \end{array}$$

79. $\frac{9}{13} = 9 \div 13 = 0.\overline{692307}$

$$\begin{array}{r} 0.6923076 \\ 13\overline{)9.0000000} \\ \underline{7\,8} \\ 1\,20 \\ \underline{1\,17} \\ 30 \\ \underline{26} \\ 40 \\ \underline{39} \\ 10 \\ \underline{0} \\ 100 \\ \underline{91} \\ 90 \\ \underline{78} \end{array}$$

80. $\frac{7}{48} = 7 \div 48 = 0.1458\overline{3}$

$$\begin{array}{r} 0.145833 \\ 48\overline{)7.000000} \\ \underline{4\,8} \\ 2\,20 \\ \underline{1\,92} \\ 280 \\ \underline{240} \\ 400 \\ \underline{384} \\ 160 \\ \underline{144} \\ 160 \\ \underline{144} \end{array}$$

81. $24\frac{9}{32} = 24 + 9 \div 32 \approx 24.28$

$$\begin{array}{r} 0.281 \\ 32\overline{)9.000} \\ \underline{6\,4} \\ 2\,60 \\ \underline{2\,56} \\ 40 \\ \underline{32} \end{array}$$

The value is about \$24.28 per share.

82. $60\frac{11}{66} = 60\frac{1}{6} = 60 + 1 \div 6 \approx 60.17$

$$\begin{array}{r} 0.166 \\ 6\overline{)1.000} \\ \underline{6} \\ 40 \\ \underline{36} \\ 40 \\ \underline{36} \end{array}$$

Idaho had the winning (longer) shot put.

83. $0.65 + 4.29 - 2.71 + 3.04 = 4.94 - 2.71 + 3.04 = 2.23 + 3.04 = 5.27$

84. $13.8 \div 0.12 \times 4.03 = 115 \times 4.03 = 463.45$

85. $(6.7)^2 - (4.4)(2.93) = (6.7)(6.7) - (4.4)(2.93) = 44.89 - 12.892 = 31.998$

86. $(5.5)(2.4)^3 \div 9.9 = (5.5)(13.824) \div 9.9 = 76.032 \div 9.9 = 7.68$

87. $(6.3)(5.08) - (2.6)(0.17) + 2.42 = 32.004 - 0.442 + 2.42 = 31.562 + 2.42 = 33.982$

88. $6.2(3.45 - 2.07 + 0.98) - 3.1(1.45) = 6.2(1.38 + 0.98) - 3.1(1.45)$
$= 6.2(2.36) - 3.1(1.45) = 14.632 - 4.495 = 10.137$

89.
$$\begin{array}{r} 3.67 \\ 4.874 \\ 0.0621 \\ 0.00045 \\ +\ 1.134 \\ \hline \end{array} \Rightarrow \begin{array}{r} 3.7 \\ 4.9 \\ 0.1 \\ 0.0 \\ +\ 1.1 \\ \hline 9.8 \end{array}$$

The answer is reasonable.

90.
$$\begin{array}{r} 0.0672 \\ -\ 0.037612 \\ \hline \end{array} \Rightarrow \begin{array}{r} 0.07 \\ -\ 0.04 \\ \hline 0.03 \end{array}$$

The answer is not reasonable.

91. $0.00562(4.235) \approx 0.006(4) = 0.024 \Rightarrow$ The answer is reasonable.

92. $0.678 \div 0.0032 \approx 0.7 \div 0.003 = 700 \div 3 \approx 233 \Rightarrow$ The largest nonzero place value in the quotient is hundreds, so answer (c) is most reasonable.

93. $9(31.95) \approx 9(30) = \$270$; Ron should be able to buy all 9 plaques.

94. $4(48.50) + 12.35 + 4(1.25) + 129.75 + 75 \approx 4(49) + 12 + 4(1) + 130 + 75$
$= 196 + 12 + 4 + 130 + 75 = 417$

Millie does not have enough to make all the purchases.

Chapter 4 True/False Concept Review (page 364)

1. false; The word name is "seven hundred nine thousandths."

2. true

3. false; The expanded form is $\frac{8}{10} + \frac{5}{100}$.

4. true

5. false; $0.732687 < 0.74$

6. true

7. true

8. true

9. false; Since the 7 is followed by a 4, we round down and leave the 7.

10. true

11. false; $9.7 - 0.2 = 9.5$

12. false; The product may have extra zeros to the right that can be omitted, such as in the problem $0.5(3.02) = 1.51$

CHAPTER 4 TRUE/FALSE CONCEPT REVIEW

13. true

14. false; The decimal point is moved to the left.

15. false; Move the decimal point to the left.

16. true

17. false; $\frac{1}{3}$ is not an exact decimal.

18. false; $\frac{4}{11} = 0.\overline{36}$

19. true

20. true

Chapter 4 Test (page 365)

1. $0.87\overline{)4.7441}$

$\Downarrow$

$$\begin{array}{r} 5.4529 \\ 87\overline{)474.4100} \\ \underline{435} \\ 394 \\ \underline{348} \\ 461 \\ \underline{435} \\ 260 \\ \underline{174} \\ 860 \\ \underline{783} \end{array}$$

$5.4529 \approx 5.453$

2. 0.6780, 0.6820, 0.6789, 0.6699, 0.6707 $\Rightarrow$
0.6699, 0.6707, 0.678, 0.6789, 0.682

3. $75.032 = 7$ tens $+ 5$ ones and 3 hundredths $+ 2$ thousandths; seventy-five and thirty-two thousandths

4. Put four decimal places in the product:

$$\begin{array}{r} 3\ 1 \\ 7\ \not{3} \\ \not{2}\ \not{1} \\ 6.84 \\ \times\quad 4.93 \\ \hline 20\ 52 \\ 6\ 15\ 60 \\ \underline{27\ 36\ 00} \\ 33.72\ 12 \end{array}$$

5. $\frac{23}{125} = 23 \div 125 = 0.184$

$$\begin{array}{r} 0.184 \\ 125\overline{)23.000} \\ \underline{12\ 5} \\ 10\ 50 \\ \underline{10\ 00} \\ 500 \\ \underline{500} \\ 0 \end{array}$$

CHAPTER 4 TEST

6. The digit in the hundredths place is 9. Since 6 is in the thousandths place, and since 6 is greater than 5, round up. Change the 9 in the hundredths place to 0, increase the digit in the tenths place from 8 to 9, and drop all digits to the right. The rounded number is 57.90.

7.

$$\begin{array}{r} \scriptstyle 9\ 9 \\ \scriptstyle 6\ \ \not{1}\!\not{0}\ \not{1}\!\not{0}\ 10 \\ 87.\not{0}\ \not{0}\ \not{0} \\ -\ 14.8\ 3\ 7 \\ \hline 72.1\ 6\ 3 \end{array}$$

8. $18.725 = 18\frac{725}{1000} = 18\frac{29}{40}$

9. $0.000000723 = 7.23 \times 10^{-7}$

10. $\frac{17}{23} = 17 \div 23 = 0.739$

$$\begin{array}{r} 0.\ 7\ 3\ 9\ 1 \\ 23\overline{)1\ 7.\ 0\ 0\ 0\ 0} \\ 1\ 6\ 1\ \ \ \ \ \ \ \ \ \\ \hline 9\ 0\ \ \ \ \ \ \\ 6\ 9\ \ \ \ \ \ \\ \hline 2\ 1\ 0\ \ \ \ \\ 2\ 0\ 7\ \ \ \ \\ \hline 3\ 0\ \ \\ 2\ 3\ \ \\ \hline \end{array}$$

11. The digit in the hundreds position is 9. Since the digit in the tens position is 8, and 8 is greater than 5, round up. Change the 9 in the hundreds position to 0, and increase the digit in the thousands position from 2 to 3. Replace the digits in the tens and ones position with 0, and drop all digits to the right. The rounded number is 73,000.

12. $2.227 \div 0.33 \times 1.5 + 11.47 = 6.9 \times 1.5 + 11.47 = 10.35 + 11.47 = 21.82$

13.

$$\begin{array}{r} \scriptstyle 9\ \ \ \ \ \ \ \ \ \ \ \ \ \ \\ \scriptstyle 2\ \not{1}\!\not{0}\ 15\ \ \ \ 2\ 14 \\ \not{3}0\ \not{5}\ .6\not{3}\not{4} \\ -\ 20\ 8\ .519 \\ \hline 9\ 7\ .115 \end{array}$$

14. $5.94 \times 10^{-5} = 0.0000594$

15. 9045.065

16. $0.000917(100{,}000) = 91.7$

17. $309{,}720 = 3.0972 \times 10^{5}$

18.

$$\begin{array}{r} \scriptstyle 2\ 2\ \ 2\ \ \ \ \\ 17.980 \\ 1.467 \\ 18.920 \\ +\ \ 8.370 \\ \hline 46.737 \end{array}$$

19. Put six decimal places in the product:

$$\begin{array}{r} \scriptstyle \not{2}\ \not{2}\ \ \ \ \ \\ \scriptstyle \not{2}\ \not{2}\ \ \ \ \ \\ 34.4 \\ \times\ \ \ 0.0016\ 5 \\ \hline 172\ 0 \\ 2064\ 0 \\ 3440\ 0 \\ \hline 0.0\ 5676\ 0 \end{array}$$

20.

$$\begin{array}{r} 0.\ 0\ 0\ 0\ 3\ 7 \\ 72\overline{)0.\ 0\ 2\ 6\ 6\ 4} \\ 2\ 1\ 6\ \ \ \\ \hline 5\ 0\ 4 \\ 5\ 0\ 4 \\ \hline 0 \end{array}$$

21. $124{,}658.95 + 110{,}750.50 + 134{,}897.70 + 128{,}934.55 + 141{,}863.20 + 119{,}541.10 = 760{,}646$; $760{,}646 \div 6 \approx 126{,}774.33$; The average offering was about \$126,774.33.

22.
$$\begin{array}{r} {\scriptstyle 1\,1\,2\,3\;\;2} \\ 911.840 \\ 45.507 \\ 6003.620 \\ 7.200 \\ 35.780 \\ +\;\;891.361 \\ \hline 7895.308 \end{array}$$

23. $3.48 \div 4 = 0.87$ per plant
$78(0.87) = 67.86$
He pays \$67.86.

24. $2214 \div 30.8 \approx 72$; He played in 72 games.

25. $201 \div 293 \times 1000 \approx 0.6860 \times 1000 = 686.0 \approx 686$; The slugging percentage is about 686.

26. Harold: $267.8 - 254.63 = 13.17$; Jerry: $209.4 - 196.2 = 13.2$;
$13.2 - 13.17 = 0.03$; Jerry lost 0.03 pounds more than Harold.

Exercises 5.1 (page 376)

1. $\frac{16}{60} = \frac{4}{15}$

3. $\frac{15\text{ mi}}{75\text{ mi}} = \frac{1}{5}$

5. $\frac{20¢}{25¢} = \frac{4}{5}$

7. $\frac{2\text{ dimes}}{8\text{ nickels}} = \frac{20¢}{40¢} = \frac{1}{2}$

9. $\frac{2\text{ ft}}{56\text{ in.}} = \frac{24\text{ in.}}{56\text{ in.}} = \frac{3}{7}$

11. $\frac{140\text{ min}}{5\text{ hr}} = \frac{140\text{ min}}{300\text{ min}} = \frac{7}{15}$

13. $\frac{\$4540}{40\text{ donors}} = \frac{\$227}{2\text{ donors}}$

15. $\frac{110\text{ miles}}{2\text{ hours}} = \frac{55\text{ miles}}{1\text{ hour}}$

17. $\frac{175\text{ miles}}{5\text{ gallons}} = \frac{35\text{ miles}}{1\text{ gallon}}$

19. $\frac{195\text{ rose bushes}}{26\text{ rows}} = \frac{15\text{ rose bushes}}{2\text{ rows}}$

21. $\frac{10\text{ trees}}{35\text{ ft}} = \frac{2\text{ trees}}{7\text{ ft}}$

23. $\frac{85\text{ scholarships}}{240\text{ applicants}} = \frac{17\text{ scholarships}}{48\text{ applicants}}$

25. $\frac{5340\text{ apples}}{89\text{ boxes}} = \frac{60\text{ apples}}{1\text{ box}}$

27. $\frac{345\text{ pies}}{46\text{ sales}} = \frac{15\text{ pies}}{2\text{ sales}}$

29. $\frac{600\text{ mi}}{15\text{ gal}} = 600 \div 15 = 40$ mi per gal

31. $\frac{36\text{ ft}}{9\text{ sec}} = 36 \div 9 = 4$ ft per sec

33. $\frac{\$1.60}{5\text{ lb}} = 1.60 \div 5 = \$0.32$ per lb

35. $\frac{4\text{ qt}}{500\text{ mi}} = 4 \div 500 = 0.008$ qt per mi

37. $\frac{36\text{ children}}{15\text{ families}} = 36 \div 15$
$= 2.4$ children per family

39. $\frac{1000 \text{ ft}}{12 \text{ sec}} = 1000 \div 12 \approx 83.3$ ft per sec

41. $\frac{4695 \text{ lb}}{25 \text{ in.}^2} = 4695 \div 25 = 187.8$ lb per in.^{2}

43. $\frac{2225 \text{ gal}}{3 \text{ hr}} = 2225 \div 3 \approx 741.7$ gal per hr

45. **a.** $\frac{\$56.95}{4 \text{ roses}} = 56.95 \div 4 \approx \$14.24$ per rose
The price per rose is about $14.24.

b. $4(17.95) = 71.80$;
$71.80 - 56.95 = 14.85$; This is a savings of $14.85 over buying each separately.

47. $\frac{\$0.79}{8 \text{ oz}} = 0.79 \div 8 = \$0.09875$ per oz
$\frac{\$2.19}{20 \text{ oz}} = 2.19 \div 20 = \$0.1095$ per oz
The 8-oz can costs 9.88¢ per oz, while the 20-oz can costs 10.95¢ per oz. The 8-oz can is the better buy.

49. $\frac{\$2.80}{5 \text{ lb}} = 2.80 \div 5 = \$0.56$ per lb
$\frac{\$5.49}{10 \text{ lb}} = 5.49 \div 10 = \$0.549$ per lb
The 10-lb bag is the better buy.

51. **a.** $\frac{\$10.77}{48 \text{ oz}} = 10.77 \div 48 \approx \$0.22$ per oz
The salsa costs 22¢ per oz.

b. $\frac{\$5}{32 \text{ oz}} = 5 \div 32 \approx \$0.16$ per oz
The sale price is 16¢ per oz.

c. $4(16)(0.22) = 14.08$
$4(16)(0.16) = 10.24$
$14.08 - 10.24 = 3.84$
Jerry can save $3.84.

53. **Answers may vary.**

55. **a.** $\frac{46 \text{ type O}}{100 \text{ total}} = \frac{23}{50}$
The ratio of people with type O blood to all people is $\frac{23}{50}$.

b. $\frac{4 \text{ type AB}}{39 + 11 \text{ type A or B}} = \frac{4}{50} = \frac{2}{25}$
The ratio of people with type AB blood to people with type A or type B blood is $\frac{2}{25}$.

57. $\frac{\$40}{\$70} = \frac{4}{7}$; The ratio of the sale price to the regular price is $\frac{4}{7}$.

59. $\frac{22{,}450 \text{ people}}{230 \text{ mi}^2} = 22{,}450 \div 230 \approx 97.6$
The density is about 97.6 people per mi^2.

61. **Answers may vary.**

63. **a.** $\frac{35 \text{ gal}}{100 \text{ gal}} = \frac{7}{20}$; The ratio of laundry use to toilet use is $\frac{7}{20}$.

b. $\frac{80 \text{ gal}}{15 \text{ gal}} = \frac{16}{3}$; The ratio of bath and shower use to dishwashing use is $\frac{16}{3}$.

65. $0.05 \text{ mg per L} = \frac{0.05 \text{ mg}}{1 \text{ L}} = \frac{0.05 \text{ mg}}{1 \text{ L}} \cdot \frac{25}{25} = \frac{1.25 \text{ mg}}{25 \text{ L}}$; 1.25 mg of lead will pollute 25 L of water.

67. $\dfrac{105 \text{ in.}}{5 \text{ turns}} = 105 \div 5 = 21$ in. per turn

69. $\dfrac{4020 \text{ credits}}{645 \text{ students}} = 4020 \div 645 \approx 6$ credits per student; $\dfrac{4020 \text{ credits}}{15 \text{ credits}} = 4020 \div 15 = 268$ FTE

71. This is a ratio since no units are given. It means that 1 unit (such as inches) on the map represents 150,000 units in the real world.

73. **Answers may vary.**

75. **Answers may vary.**

77. $\dfrac{5}{9} \cdot \dfrac{3}{25} = \dfrac{\overset{1}{\cancel{5}}}{\underset{3}{\cancel{9}}} \cdot \dfrac{\overset{1}{\cancel{3}}}{\underset{5}{\cancel{25}}} = \dfrac{1}{15}$

79. Put three decimal places in the product.

$$\begin{array}{r} \scriptstyle 2 \\ 1\,1.6 \\ \times \quad 0.0\,4 \\ \hline 0.4\,6\,4 \end{array}$$

81. $\dfrac{7}{8}(0.6) = 0.875(0.6) = 0.525$

83.
$$\begin{array}{r} 9 \\ -\,4\frac{7}{15} \\ \hline \end{array} \Rightarrow \begin{array}{r} 8\frac{15}{15} \\ -\,4\frac{7}{15} \\ \hline 4\frac{8}{15} \end{array}$$

85. $\dfrac{3}{4} = \dfrac{3}{4} \cdot \dfrac{125}{125} = \dfrac{375}{500}$; $\dfrac{94}{125} = \dfrac{94}{125} \cdot \dfrac{4}{4} = \dfrac{376}{500}$

$\dfrac{94}{125}$ is larger.

Exercises 5.2 (page 384)

1.
$$\begin{aligned} \frac{4}{27} &= \frac{20}{135} \\ 4(135) &\overset{?}{=} 27(20) \\ 540 &\overset{?}{=} 540 \end{aligned}$$
true

3.
$$\begin{aligned} \frac{3}{2} &= \frac{9}{4} \\ 3(4) &\overset{?}{=} 2(9) \\ 12 &\overset{?}{=} 18 \end{aligned}$$
false

5.
$$\begin{aligned} \frac{4}{10} &= \frac{5}{20} \\ 4(20) &\overset{?}{=} 10(5) \\ 80 &\overset{?}{=} 50 \end{aligned}$$
false

7.
$$\begin{aligned} \frac{18}{12} &= \frac{15}{10} \\ 18(10) &\overset{?}{=} 12(15) \\ 180 &\overset{?}{=} 180 \end{aligned}$$
true

9.
$$\begin{aligned} \frac{42}{24} &= \frac{63}{55} \\ 42(55) &\overset{?}{=} 24(63) \\ 2310 &\overset{?}{=} 1512 \end{aligned}$$
false

11.
$$\begin{aligned} \frac{30}{27} &= \frac{60}{45} \\ 30(45) &\overset{?}{=} 27(60) \\ 1350 &\overset{?}{=} 1620 \end{aligned}$$
false

13.
$$\begin{aligned} \frac{13}{4} &= \frac{9.75}{3} \\ 13(3) &\overset{?}{=} 4(9.75) \\ 39 &\overset{?}{=} 39 \end{aligned}$$
true

15.
$$\begin{aligned} \frac{5}{8} &= \frac{d}{32} \\ 5(32) &= 8d \\ 160 &= 8d \\ \frac{160}{8} &= \frac{8d}{8} \\ 20 &= d \end{aligned}$$

SECTION 5.2

17.
$$\frac{2}{6}=\frac{c}{18}$$
$$2(18)=6c$$
$$36=6c$$
$$\frac{36}{6}=\frac{6c}{6}$$
$$6=c$$

19.
$$\frac{28}{y}=\frac{14}{5}$$
$$28(5)=y(14)$$
$$140=14y$$
$$\frac{140}{14}=\frac{14y}{14}$$
$$10=y$$

21.
$$\frac{13}{39}=\frac{7}{c}$$
$$13c=39(7)$$
$$13c=273$$
$$\frac{13c}{13}=\frac{273}{13}$$
$$c=21$$

23.
$$\frac{16}{12}=\frac{3}{x}$$
$$16x=12(3)$$
$$16x=36$$
$$\frac{16x}{16}=\frac{36}{16}$$
$$x=\frac{9}{4}$$

25.
$$\frac{p}{14}=\frac{5}{42}$$
$$p(42)=14(5)$$
$$42p=70$$
$$\frac{42p}{42}=\frac{70}{42}$$
$$p=\frac{5}{3}$$

27.
$$\frac{x}{5}=\frac{23}{10}$$
$$x(10)=5(23)$$
$$10x=115$$
$$\frac{10x}{10}=\frac{115}{10}$$
$$x=11.5$$

29.
$$\frac{2}{z}=\frac{5}{11}$$
$$2(11)=z(5)$$
$$22=5z$$
$$\frac{22}{5}=\frac{5z}{5}$$
$$4.4=z$$

31.
$$\frac{13}{6}=\frac{w}{2}$$
$$13(2)=6w$$
$$26=6w$$
$$\frac{26}{6}=\frac{6w}{6}$$
$$4\frac{1}{3}=w$$

33.
$$\frac{15}{16}=\frac{12}{a}$$
$$15a=16(12)$$
$$15a=192$$
$$\frac{15a}{15}=\frac{192}{15}$$
$$a=12.8$$

35.
$$\frac{1.5}{d}=\frac{0.03}{2}$$
$$1.5(2)=d(0.03)$$
$$3=0.03d$$
$$\frac{3}{0.03}=\frac{0.03d}{0.03}$$
$$100=d$$

37.
$$\frac{\frac{3}{5}}{b}=\frac{8}{5}$$
$$\frac{3}{5}(5)=b(8)$$
$$3=8b$$
$$\frac{3}{8}=\frac{8b}{8}$$
$$\frac{3}{8}=b$$

39.
$$\frac{0.9}{4.5}=\frac{0.09}{x}$$
$$0.9x=4.5(0.09)$$
$$0.9x=0.405$$
$$\frac{0.9x}{0.9}=\frac{0.405}{0.9}$$
$$x=0.45$$

41.
$$\frac{9}{1.8}=\frac{w}{0.15}$$
$$9(0.15)=1.8w$$
$$1.35=1.8w$$
$$\frac{1.35}{1.8}=\frac{1.8w}{1.8}$$
$$0.75=w$$

43.
$$\frac{y}{5}=\frac{10}{\frac{1}{5}}$$
$$y\left(\frac{1}{5}\right)=5(10)$$
$$\frac{1}{5}y=50$$
$$5\cdot\frac{y}{5}=5(50)$$
$$y=250$$

45.
$$\frac{t}{24}=\frac{3\frac{1}{2}}{10\frac{1}{2}}$$
$$\frac{t}{24}=\frac{\frac{7}{2}}{\frac{21}{2}}$$
$$t\cdot\frac{21}{2}=24\cdot\frac{7}{2}$$
$$\frac{21t}{2}=\frac{168}{2}$$
$$2\cdot\frac{21t}{2}=2\cdot\frac{168}{2}$$
$$21t=168$$
$$\frac{21t}{21}=\frac{168}{21}$$
$$t=8$$

47.
$$\frac{3}{11}=\frac{w}{5}$$
$$3(5)=11w$$
$$15=11w$$
$$\frac{15}{11}=\frac{11w}{11}$$
$$1.4\approx w$$

49.
$$\frac{16}{25}=\frac{5}{y}$$
$$16y=25(5)$$
$$16y=125$$
$$\frac{16y}{16}=\frac{125}{16}$$
$$y\approx 7.8$$

51.
$$\frac{2.5}{4.5}=\frac{a}{0.6}$$
$$2.5(0.6)=4.5a$$
$$1.5=4.5a$$
$$\frac{1.5}{4.5}=\frac{4.5a}{4.5}$$
$$0.33\approx a$$

53.
$$\frac{\frac{4}{9}}{c}=\frac{5}{14}$$
$$\frac{4}{9}(14)=c(5)$$
$$\frac{56}{9}=5c$$
$$\frac{56}{9(5)}=\frac{5c}{5}$$
$$\frac{56}{45}=c$$
$$1.24\approx c$$

55.
$$\frac{\square}{120}=\frac{x}{12}$$
$$\frac{\square}{120}=\frac{1}{12}$$
$$\square\cdot 12=120(1)$$
$$\square\cdot 12=120$$
$$\frac{\square\cdot 12}{12}=\frac{120}{12}$$
$$\square=10$$

57. The error is multiplying straight across the equal sign instead of finding the cross products.

59.
$$\frac{176}{H}=\frac{4.5}{1}$$
$$176(1)=H(4.5)$$
$$176=4.5H$$
$$\frac{176}{4.5}=\frac{4.5H}{4.5}$$
$$39\approx H$$

Fran's minimum allowable level of HDL is about 39.

61.
$$\frac{3}{20}=\frac{R}{134}$$
$$3(134)=20R$$
$$402=20R$$
$$\frac{402}{20}=\frac{20R}{20}$$
$$20\approx R$$

About 20 rivers would have increased levels of pollution.

63.
$$\frac{1}{6}=\frac{g}{560}$$
$$1(560)=6g$$
$$560=6g$$
$$\frac{560}{6}=\frac{6g}{6}$$
$$93\approx g$$

About 93 girls could be expected to be on a diet.

65. $3.1\times 5280\times 12=196{,}416$, so 3.1 mi $=196{,}416$ in. The ratios $\frac{1}{196{,}416}$ and $\frac{1}{200{,}000}$ are not equal, but the values are very close.

67.
$$\frac{3.5}{\frac{1}{4}}=\frac{7}{y}$$
$$3,5y=7\cdot\frac{1}{4}$$
$$3.5y=\frac{7}{4}$$
$$3.5y=1.75$$
$$\frac{3.5y}{3.5}=\frac{1.75}{3.5}$$
$$y=0.5$$

69.
$$\frac{9+7}{15+9}=\frac{6}{a}$$
$$\frac{16}{24}=\frac{6}{a}$$
$$\frac{2}{3}=\frac{6}{a}$$
$$2a=3(6)$$
$$2a=18$$
$$\frac{2a}{2}=\frac{18}{2}$$
$$a=9$$

71.
$$\begin{aligned} \frac{7}{w} &= \frac{18.92}{23.81} \\ 7(23.81) &= w(18.92) \\ 166.67 &= 18.92w \\ \frac{166.67}{18.92} &= \frac{18.92w}{18.92} \\ 8.809 &\approx w \end{aligned}$$

73. $813.6 - 638.4196 = 175.1804$

75. $6.45 + 7.13 + 5.11 = 18.69;\ 18.69 \div 3 = 6.23$

77. $4.835 \div 10{,}000 = 0.0004835$

79. Put four decimal places in the product.

$$\begin{array}{r} {\scriptstyle 2\ 6\ 4} \\ {\scriptstyle \not{2}\ \not{6}\ \not{4}} \\ 12.75 \\ \times \quad 8.09 \\ \hline 1\ 14\ 75 \\ 0\ 00\ 00 \\ 102\ 00\ 00 \\ \hline 103.14\ 75 \end{array}$$

81. $0.86\overline{)4.\ 3}$

$$\Downarrow$$

$$\begin{array}{r} 5. \\ 86\overline{)4\ 3\ 0.} \\ 4\ 3\ 0 \\ \hline 0 \end{array}$$

Exercises 5.3 (page 391)

1. 6

3. 15

5. $\frac{6}{4} = \frac{15}{h}$

7. 30

9. h

11. $\frac{30}{18} = \frac{h}{48}$

13. 9

15. N

17. $\frac{9}{12} = \frac{N}{108}$

19. T

21. $\frac{3}{65} = \frac{T}{910}$

23. From #22, 42 teachers are needed. This is 4 more.

25. $\frac{18}{2} = \frac{75}{x}$

27. Let m = the monthly payment.

$$\begin{aligned} \frac{\$4000 \text{ loan}}{\$95.75 \text{ pmt}} &= \frac{\$10{,}000 \text{ loan}}{m} \\ 4000m &= 95.75(10{,}000) \\ 4000m &= 957{,}500 \\ \frac{4000m}{4000} &= \frac{957{,}500}{4000} \\ m &\approx 239.38 \end{aligned}$$

The monthly payment will be about \$239.38.

SECTION 5.3

29. **a.** Let m = the number males.

$$\frac{55 \text{ males}}{100 \text{ people}} = \frac{m \text{ males}}{727{,}785 \text{ people}}$$
$$55(727{,}785) = 100m$$
$$40{,}028{,}175 = 100m$$
$$\frac{40{,}028{,}175}{100} = \frac{100m}{100}$$
$$400{,}282 \approx m$$

There were about 400,282 males.

b. $727{,}785 - 400{,}282 = 327{,}503$ females

$$\frac{400{,}282}{327{,}503} = 400{,}282 \div 327{,}503 \approx 1.22$$

There are about 1.22 males for every female.

31. Let f = amount spent on food.

$$\frac{3 \text{ food}}{10 \text{ total}} = \frac{f \text{ food}}{30{,}000 \text{ total}}$$
$$3(30{,}000) = 10f$$
$$90{,}000 = 10f$$
$$\frac{90{,}000}{10} = \frac{10f}{10}$$
$$9000 = f$$

The family would spend $9000 on food.

33. Let A = area covered.

$$\frac{30 \text{ lb}}{1500 \text{ ft}^2} = \frac{50 \text{ lb}}{A \text{ ft}^2}$$
$$30A = 1500(50)$$
$$30A = 75{,}000$$
$$\frac{30A}{30} = \frac{75{,}000}{30}$$
$$A = 2500$$

50 lb of fertilizer will cover 2500 ft^2.

35. $\frac{11}{15} = \frac{x}{30}$

37. Let r = number who can read.

$$\frac{909 \text{ read}}{1000 \text{ people}} = \frac{r}{1{,}338{,}612{,}968}$$
$$909(1{,}338{,}612{,}968) = 1000r$$
$$1{,}216{,}799{,}187{,}912 = 1000r$$
$$\frac{1{,}216{,}799{,}187{,}912}{1000} = \frac{1000r}{1000}$$
$$1{,}216{,}799{,}188 \approx r$$

$1{,}338{,}612{,}968 - 1{,}216{,}799{,}188$
$= 121{,}813{,}780$

About 121,813,780 people in China cannot read.

39. Let d = dosage.

$$\frac{60 \text{ lb}}{400 \text{ mg}} = \frac{42 \text{ lb}}{d}$$
$$60d = 400(42)$$
$$60d = 16{,}800$$
$$\frac{60d}{60} = \frac{16{,}800}{60}$$
$$d = 280$$

The dosage is 280 mg.

41. Let t = driving time.

$$\frac{451 \text{ miles}}{8.2 \text{ hr}} = \frac{935 \text{ miles}}{t}$$
$$451t = 8.2(935)$$
$$451t = 7667$$
$$\frac{451t}{451} = \frac{7667}{451}$$
$$t = 17$$

It will take 17 hours to drive 935 miles.

43. Let c = cost of carpet.

$$\frac{33 \text{ yd}^2}{\$526.35} = \frac{22 \text{ yd}^2}{c}$$
$$33c = 526.35(22)$$
$$33c = 11{,}579.7$$
$$\frac{33c}{33} = \frac{11{,}579.7}{33}$$
$$c = 350.9$$

The 2nd room will cost $350.90 to carpet.

45. Let $a =$ # of cubic centimeters needed.

$$\frac{20 \text{ mg}}{1 \text{ cc}} = \frac{8 \text{ mg}}{a \text{ cc}}$$
$$20a = 1(8)$$
$$20a = 8$$
$$\frac{20a}{20} = \frac{8}{20}$$
$$a = 0.4$$

Ida needs 0.4 cm^3 of the solution.

47. Let $w =$ her waist measurement.

$$\frac{4 \text{ waist}}{5 \text{ hip}} = \frac{w}{35}$$
$$4(35) = 5w$$
$$140 = 5w$$
$$\frac{140}{5} = \frac{5w}{5}$$
$$28 = w$$

Her waist size should be no more than 28".

49. Let $n =$ number of pounds of cashews.

$$\frac{3 \text{ lb cashews}}{10 \text{ lb total}} = \frac{n}{40 \text{ lb total}}$$
$$3(40) = 10n$$
$$120 = 10n$$
$$\frac{120}{10} = \frac{10n}{10}$$
$$12 = n$$

$40 - 12 = 28$; Betty needs 12 lb of cashews and 28 lb of peanuts.

51. Let $n =$ number of quarts of blue paint.

$$\frac{3 \text{ qt blue}}{7 \text{ qt green}} = \frac{n}{98 \text{ qt green}}$$
$$3(98) = 7n$$
$$294 = 7n$$
$$\frac{294}{7} = \frac{7n}{7}$$
$$42 = n$$

Debra needs 42 quarts of blue paint.

53. Let $n =$ number of lb of ground beef.

$$\frac{10 \text{ lb beef}}{13 \text{ lb total}} = \frac{n}{84.5 \text{ lb total}}$$
$$10(84.5) = 13n$$
$$845 = 13n$$
$$\frac{845}{13} = \frac{13n}{13}$$
$$65 = n$$

Lucia needs 65 lb of ground round.

55. Let $c =$ the cost of the computer in pounds.

$$\frac{\$1}{£0.6175} = \frac{\$899}{c}$$
$$1c = 0.6175(899)$$
$$c \approx 555.13$$

The computer costs about £555.13.

57. Let $p =$ the price of the battery

$$\frac{\$84.99}{36 \text{ months}} = \frac{p}{60 \text{ months}}$$
$$84.99(60) = 36p$$
$$5099.4 = 36p$$
$$\frac{5099.4}{36} = \frac{36p}{36}$$
$$141.65 = p$$

The battery should cost \$141.65.

59. Let $q =$ quantity of ozone.

$$\frac{1 \text{ m}^3}{235 \text{ mg}} = \frac{12 \text{ m}^3}{q}$$
$$1q = 235(12)$$
$$q = 2820$$

12 m^3 of air must have less than 2820 mg of ozone.

61. Let $c =$ cost of grocery store box.

$$\frac{\text{4 batches}}{\$8.90} = \frac{\text{1 batch}}{c}$$
$$4c = 8.90(1)$$
$$4c = 8.90$$
$$\frac{4c}{4} = \frac{8.90}{4}$$
$$c = 2.225$$

The store needs to charge \$2.22, a reduction of 17¢, to beat the warehouse.

63. Let $x =$ distance in the real world.

$$\frac{1\frac{3}{8}\text{ in.}}{0.5\text{ mi}} = \frac{8\frac{1}{2}\text{ in.}}{x\text{ mi}}$$
$$1\frac{3}{8} \cdot x = 0.5\left(8\frac{1}{2}\right)$$
$$1.375x = 0.5(8.5)$$
$$\frac{1.375x}{1.375} = \frac{4.25}{1.375}$$
$$x = 3.09$$

The bridges are 3.09 mi apart.

65. Answers may vary.

67. Answers may vary.

69. Let $d =$ drive shaft speed.

$$\frac{\text{2.5 engine}}{\text{1 drive shaft}} = \frac{2800}{d}$$
$$2.5d = 1(2800)$$
$$\frac{2.5d}{2.5} = \frac{2800}{2.5}$$
$$d = 1120$$

The drive shaft speed is 1120 rpm.

71. **Hundredth**

The digit in the hundredths place is 5. Since 1 is in the thousandths place, and since 1 is less than 5, round down. Leave the 5 in the hundredths place, and drop all digits to the right. The rounded number 167.85.

Hundred

The digit in the hundreds place is 1. Since 6 is in the tens place, and 6 is greater than 5, round up. Change the 1 in the hundreds place to 2, and drop all digits to the right. The rounded number is 200.

73. $0.01399, 0.01100 \Rightarrow 0.01399 > 0.011$

75. $12(1.49) = 17.88$;
The ground beef costs \$17.88.

77. $0.635 = \frac{635}{1000} = \frac{127}{200}$

79. $\frac{345}{561} = 345 \div 561 \approx 0.615$

Chapter 5 Review Exercises (page 398)

1. $\frac{18}{90} = \frac{1}{5}$

2. $\frac{9}{54} = \frac{1}{6}$

3. $\frac{12\text{ m}}{10\text{ m}} = \frac{6}{5}$

4. $\frac{12\text{ km}}{9\text{ km}} = \frac{4}{3}$

5. $\frac{\text{3 dollars}}{\text{80 nickels}} = \frac{\text{60 nickels}}{\text{80 nickels}} = \frac{3}{4}$

6. $\frac{\text{660 feet}}{\text{1 mile}} = \frac{\text{660 feet}}{\text{5280 feet}} = \frac{1}{8}$

7. $\frac{\text{16 in.}}{\text{2 ft}} = \frac{\text{16 in.}}{\text{24 in.}} = \frac{2}{3}$

8. $\frac{\text{3 ft}}{\text{3 yd}} = \frac{\text{3 ft}}{\text{9 ft}} = \frac{1}{3}$

9. $\frac{\text{9 people}}{\text{10 chairs}}$

10. $\frac{\text{23 miles}}{\text{3 hikes}}$

CHAPTER 5 REVIEW EXERCISES

11. $\dfrac{40 \text{ applicants}}{15 \text{ jobs}} = \dfrac{8 \text{ applicants}}{3 \text{ jobs}}$

12. $\dfrac{10 \text{ cars}}{6 \text{ households}} = \dfrac{5 \text{ cars}}{3 \text{ households}}$

13. $\dfrac{210 \text{ books}}{45 \text{ students}} = \dfrac{14 \text{ books}}{3 \text{ students}}$

14. $\dfrac{36 \text{ buttons}}{24 \text{ bows}} = \dfrac{3 \text{ buttons}}{2 \text{ bows}}$

15. $\dfrac{765 \text{ people}}{27 \text{ committees}} = \dfrac{85 \text{ people}}{3 \text{ committees}}$

16. $\dfrac{8780 \text{ households}}{6 \text{ cable companies}} = \dfrac{4390 \text{ households}}{3 \text{ cable companies}}$

17. $\dfrac{50 \text{ mi}}{2 \text{ hr}} = 50 \div 2 = 25 \text{ mi per hr}$

18. $\dfrac{60 \text{ mi}}{4 \text{ min}} = 60 \div 4 = 15 \text{ mi per min}$

19. $\dfrac{90¢}{10 \text{ lb}} = 90 \div 10 = 9¢ \text{ per lb}$

20. $\dfrac{\$1.17}{3 \text{ lb}} = 1.17 \div 3 = \$0.39 \text{ per lb}$

21. $\dfrac{825 \text{ mi}}{22 \text{ gal}} = 825 \div 22 = 37.5 \text{ mpg}$

22. $\dfrac{13{,}266 \text{ km}}{220 \text{ gal}} = 13{,}266 \div 220$
$= 60.3 \text{ km per gal}$

23. $\dfrac{\$2.10}{6 \text{ croissants}} = 2.10 \div 6$
$= \$0.35 \text{ per croissant}$

24. $\dfrac{\$3.75}{15 \text{ oz}} = 3.75 \div 15 \approx \$0.25 \text{ per oz}$

25. $\dfrac{3500 \text{ sets}}{1000 \text{ households}} = 3500 \div 1000 = 3.5 \text{ sets per household}$
$\dfrac{500 \text{ sets}}{150 \text{ households}} = 500 \div 150 \approx 3.33$ sets per household; The rates are not the same.

26. $\dfrac{5000 \text{ automobiles}}{3750 \text{ households}} = 5000 \div 3750 \approx 1.33 \text{ sets per household}$
$\dfrac{6400 \text{ automobiles}}{4800 \text{ households}} = 6400 \div 4800 \approx 1.33$ sets per household; The rates are the same.

27. $\dfrac{15}{7} = \dfrac{75}{35}$
$15(35) \stackrel{?}{=} 7(75)$
$525 \stackrel{?}{=} 525$
true

28. $\dfrac{2}{3} = \dfrac{26}{39}$
$2(39) \stackrel{?}{=} 3(26)$
$78 \stackrel{?}{=} 78$
true

29. $\dfrac{25}{9} = \dfrac{8}{3}$
$25(3) \stackrel{?}{=} 9(8)$
$75 \stackrel{?}{=} 72$
false

30. $\dfrac{16}{25} = \dfrac{10}{15}$
$16(15) \stackrel{?}{=} 25(10)$
$240 \stackrel{?}{=} 250$
false

31. $\dfrac{31}{35} = \dfrac{6.125}{7}$
$31(7) \stackrel{?}{=} 35(6.125)$
$217 \stackrel{?}{=} 214.375$
false

32. $\dfrac{9.375}{3} = \dfrac{25}{8}$
$9.375(8) \stackrel{?}{=} 3(25)$
$75 \stackrel{?}{=} 75$
true

CHAPTER 5 REVIEW EXERCISES

33.
$$\frac{1}{4}=\frac{r}{44}$$
$$1(44)=4r$$
$$44=4r$$
$$\frac{44}{4}=\frac{4r}{4}$$
$$11=r$$

34.
$$\frac{1}{3}=\frac{s}{18}$$
$$1(18)=3s$$
$$18=3s$$
$$\frac{18}{3}=\frac{3s}{3}$$
$$6=s$$

35.
$$\frac{14}{t}=\frac{42}{27}$$
$$14(27)=t(42)$$
$$378=42t$$
$$\frac{378}{42}=\frac{42t}{42}$$
$$9=t$$

36.
$$\frac{8}{v}=\frac{2}{5}$$
$$8(5)=v(2)$$
$$40=2v$$
$$\frac{40}{2}=\frac{2v}{2}$$
$$20=v$$

37.
$$\frac{f}{9}=\frac{3}{45}$$
$$f(45)=9(3)$$
$$45f=27$$
$$\frac{45f}{45}=\frac{27}{45}$$
$$f=\tfrac{3}{5}$$

38.
$$\frac{g}{2}=\frac{2}{12}$$
$$g(12)=2(2)$$
$$12g=4$$
$$\frac{12g}{12}=\frac{4}{12}$$
$$g=\tfrac{1}{3}$$

39.
$$\frac{16}{24}=\frac{r}{16}$$
$$16(16)=24r$$
$$256=24r$$
$$\frac{256}{24}=\frac{24r}{24}$$
$$10\tfrac{2}{3}=r$$

40.
$$\frac{s}{10}=\frac{15}{16}$$
$$s(16)=10(15)$$
$$16s=150$$
$$\frac{16s}{16}=\frac{150}{16}$$
$$s=9\tfrac{3}{8}$$

41.
$$\frac{21}{25}=\frac{t}{7}$$
$$21(7)=25t$$
$$147=25t$$
$$\frac{147}{25}=\frac{25t}{25}$$
$$5\tfrac{22}{25}=t$$

42.
$$\frac{7}{5}=\frac{w}{7}$$
$$7(7)=5w$$
$$49=5w$$
$$\frac{49}{5}=\frac{5w}{5}$$
$$9\tfrac{4}{5}=w$$

43.
$$\frac{9}{11}=\frac{a}{13}$$
$$9(13)=11a$$
$$117=11a$$
$$\frac{117}{11}=\frac{11a}{11}$$
$$10.6\approx a$$

44.
$$\frac{7}{6}=\frac{6}{b}$$
$$7b=6(6)$$
$$7b=36$$
$$\frac{7b}{7}=\frac{36}{7}$$
$$b=5.1$$

45.
$$\frac{16}{7}=\frac{c}{12}$$
$$16(12)=7c$$
$$192=7c$$
$$\frac{192}{7}=\frac{7c}{7}$$
$$27.4=c$$

46.
$$\frac{16}{5}=\frac{12}{d}$$
$$16d=5(12)$$
$$16d=60$$
$$\frac{16d}{16}=\frac{60}{16}$$
$$d\approx 3.8$$

47.
$$\frac{9.99}{80}=\frac{c}{50}$$
$$9.99(50)=80c$$
$$499.5=80c$$
$$\frac{499.5}{80}=\frac{80c}{80}$$
$$6.24\approx c$$

The store brand must be cheaper than \$6.24.

48.
$$\frac{18{,}000{,}000}{3{,}000{,}000} = \frac{L}{50{,}000}$$
$$18{,}000{,}000(50{,}000) = 3{,}000{,}000L$$
$$900{,}000{,}000{,}000 = 3{,}000{,}000L$$
$$\frac{900{,}000{,}000{,}000}{3{,}000{,}000} = \frac{3{,}000{,}000L}{3{,}000{,}000}$$
$$300{,}000 = L$$
It will take 300,000 lawn mowers.

49. Let h = hours studying.
$$\frac{2 \text{ hr in class}}{5 \text{ hr studying}} = \frac{15 \text{ hr}}{h}$$
$$2h = 5(15)$$
$$2h = 75$$
$$\frac{2h}{2} = \frac{75}{2}$$
$$h = 37.5$$
Merle will study 37.5 hours per week.

50. Let f = pounds of fertilizer.
$$\frac{16 \text{ lb}}{1500 \text{ ft}^2} = \frac{f}{2500 \text{ ft}^2}$$
$$16(2500) = 1500f$$
$$40{,}000 = 1500f$$
$$\frac{40{,}000}{1500} = \frac{1500f}{1500}$$
$$26\tfrac{2}{3} = f$$
$26\frac{2}{3}$ lb of fertilizer is needed.

51. Let h = hours of work needed.
$$\frac{36 \text{ hr}}{3 \text{ credits}} = \frac{h}{15 \text{ credits}}$$
$$36(15) = 3h$$
$$540 = 3h$$
$$\frac{540}{3} = \frac{3h}{3}$$
$$180 = h$$
Juan must work 180 hours.

52. $180 \div 40 = 4.5$; He will need to work 5 weeks.

53. Let E = amount earned
$$\frac{\$120 \text{ sales}}{\$15 \text{ earned}} = \frac{\$350}{E}$$
$$120E = 15(350)$$
$$120E = 5250$$
$$\frac{120E}{120} = \frac{5250}{120}$$
$$E = 43.75$$
Larry will make $43.75.

54. Let w = gallons of water needed.
$$\frac{1.5 \text{ lb}}{1 \text{ gal}} = \frac{12 \text{ lb}}{w}$$
$$1.5w = 1(12)$$
$$1.5w = 12$$
$$\frac{1.5w}{1.5} = \frac{12}{1.5}$$
$$w = 8$$
The solution will need 8 gallons of water.

Chapter 5 True/False Concept Review (page 400)

1. true

2. true

3. false; $\frac{18 \text{ miles}}{1 \text{ gallon}} = \frac{54 \text{ miles}}{3 \text{ gallons}}$ or $\frac{18 \text{ miles}}{1 \text{ hour}} = \frac{54 \text{ miles}}{3 \text{ hours}}$

4. false; Three of the values must be known.

5. false; $t = \frac{16}{5} = 3\frac{1}{5}$

6. false; They are *stated* to be equal but it is possible that they are not *actually* equal.

7. true

8. true

9. true

10. false; The correct table is below:

	First Tree	Second Tree
Height	18	x
Shadow	17	25

Chapter 5 Test (page 400)

1. $\dfrac{12\text{ yd}}{15\text{ yd}} = \dfrac{4}{5}$

2. Let c = number answered correctly.

$$\frac{20\text{ correct}}{32\text{ total}} = \frac{c}{72}$$
$$20(72) = 32c$$
$$1440 = 32c$$
$$\frac{1440}{32} = \frac{32c}{32}$$
$$45 = c$$

Ken would need to answer 45 correctly.

3.
$$\frac{4.8}{12} = \frac{0.36}{w}$$
$$4.8w = 12(0.36)$$
$$4.8w = 4.32$$
$$\frac{4.8w}{4.8} = \frac{4.32}{4.8}$$
$$w = 0.9$$

4.
$$\frac{16}{35} = \frac{24}{51}$$
$$16(51) \stackrel{?}{=} 35(24)$$
$$816 \stackrel{?}{=} 840$$
false

5.
$$\frac{13}{36} = \frac{y}{18}$$
$$13(18) = 36y$$
$$234 = 36y$$
$$\frac{234}{36} = \frac{36y}{36}$$
$$6.5 = y$$

6.
$$\frac{9\text{ in.}}{2\text{ ft}} = \frac{6\text{ in.}}{16\text{ in.}}$$
$$\frac{9\text{ in.}}{24\text{ in.}} \stackrel{?}{=} \frac{6\text{ in.}}{16\text{ in.}}$$
$$9(16) \stackrel{?}{=} 24(6)$$
$$144 \stackrel{?}{=} 144$$
true

7. Let p = amount paid.
$$\frac{\$49.14}{7\text{ hr}} = \frac{p}{12}$$
$$49.14(12) = 7p$$
$$589.68 = 7p$$
$$\frac{589.68}{7} = \frac{7p}{7}$$
$$84.24 = p$$

Mary should be paid \$84.24.

8. $\dfrac{8\text{ hr}}{3\text{ days}} = \dfrac{8\text{ hr}}{72\text{ hr}} = \dfrac{1}{9}$

9. Let p = price of the cans.
$$\frac{24\text{ cans}}{\$19.68} = \frac{10}{p}$$
$$24p = 19.68(10)$$
$$24p = 196.8$$
$$\frac{24p}{24} = \frac{196.8}{24}$$
$$p = 8.2$$

The price is \$8.20.

10. Let b = pounds of bones.
$$\frac{40\text{ lb beef}}{7\text{ lb bones}} = \frac{100\text{ lb beef}}{b}$$
$$40b = 7(100)$$
$$40b = 700$$
$$\frac{40b}{40} = \frac{700}{40}$$
$$b = 17.5$$

There should be 17.5 pounds of bones.

11. $\dfrac{0.4}{0.5} = \dfrac{0.5}{x}$

$0.4x = 0.5(0.5)$

$0.4x = 0.25$

$\dfrac{0.4x}{0.4} = \dfrac{0.25}{0.4}$

$x = 0.625$

12. Let f = number of fish caught.

$\dfrac{3 \text{ salmon}}{4 \text{ people}} = \dfrac{f}{32 \text{ people}}$

$3(32) = 4f$

$96p = 4f$

$\dfrac{96}{4} = \dfrac{4f}{4}$

$24 = f$

They can expect to catch 24 fish.

13. Let g = gallons of gas needed.

$\dfrac{12.5 \text{ gal}}{295 \text{ mi}} = \dfrac{g}{236 \text{ mi}}$

$12.5(236) = 295g$

$2950 = 295g$

$\dfrac{2950}{295} = \dfrac{295g}{295}$

$10 = g$

Jennie will need 10 gallons of gas.

14. $\dfrac{a}{8} = \dfrac{4.24}{6.4}$

$a(6.4) = 8(4.24)$

$6.4a = 33.92$

$\dfrac{6.4a}{6.4} = \dfrac{33.92}{6.4}$

$a = 5.3$

15. yes

16. Let s = length of shadow.

$\dfrac{20 \text{ ft tree}}{15 \text{ ft shadow}} = \dfrac{14 \text{ ft tree}}{s}$

$20s = 15(14)$

$20s = 210$

$\dfrac{20s}{20} = \dfrac{210}{20}$

$s = 10.5$

The shadow will be 10.5 feet long.

17. $5580 \div 150 = 37.2$; The population density is 37.2 people per square mile.

18. $\dfrac{4.78}{y} = \dfrac{32.5}{11.2}$

$4.78(11.2) = y(32.5)$

$53.536 = 32.5y$

$\dfrac{53.536}{32.5} = \dfrac{32.5y}{32.5}$

$1.65 \approx y$

19. Let j = number of jobs.

$$\frac{1 \text{ job}}{4\frac{1}{2} \text{ hr}} = \frac{j}{117}$$
$$\frac{1}{\frac{9}{2}} = \frac{j}{117}$$
$$1(117) = \frac{9}{2}j$$
$$2(117) = 2 \cdot \frac{9}{2}j$$
$$234 = 9j$$
$$\frac{234}{9} = \frac{9j}{9}$$
$$26 = j$$

They could complete 26 jobs.

20. Let f = number of females.

$$\frac{5 \text{ females}}{8 \text{ total}} = \frac{f}{48 \text{ total}}$$
$$5(48) = 8f$$
$$240 = 8f$$
$$\frac{240}{8} = \frac{8f}{8}$$
$$30 = f$$

There will be 30 female students.

Exercises 6.1 (page 411)

1. The whole = 100 squares. 29 squares are shaded. $\frac{29}{100} = 29\%$

3. The whole = 100 squares. 72 squares are shaded. $\frac{72}{100} = 72\%$

5. The whole = 4 regions. 5 regions are shaded. $\frac{5}{4} \cdot \frac{25}{25} = \frac{125}{100} = 125\%$

7. $\frac{62}{100} = 62\%$

9. $\frac{32}{100} = 32\%$

11. $\frac{28}{50} \cdot \frac{2}{2} = \frac{56}{100} = 56\%$

13. $\frac{12}{25} \cdot \frac{4}{4} = \frac{48}{100} = 48\%$

15. $\frac{11}{20} \cdot \frac{5}{5} = \frac{55}{100} = 55\%$

17. $\frac{13}{10} \cdot \frac{10}{10} = \frac{130}{100} = 130\%$

19. $\frac{313}{313} = \frac{100}{100} = 100\%$

21. $\frac{30}{12} = \frac{5}{2} \cdot \frac{50}{50} = \frac{250}{100} = 250\%$

23.
$$\frac{85}{200} = \frac{R}{100}$$
$$85(100) = 200R$$
$$8500 = 200R$$
$$\frac{8500}{200} = R$$
$$42.5 = R \Rightarrow 42.5\%$$

25.
$$\frac{15}{40} = \frac{R}{100}$$
$$15(100) = 40R$$
$$1500 = 40R$$
$$\frac{1500}{40} = R$$
$$37.5 = R \Rightarrow 37.5\%$$

27.
$$\frac{70}{80} = \frac{R}{100}$$
$$70(100) = 80R$$
$$7000 = 80R$$
$$\frac{7000}{80} = R$$
$$87.5 = R \Rightarrow 87.5\%$$

29.
$$\frac{180}{480} = \frac{R}{100}$$
$$180(100) = 480R$$
$$18000 = 480R$$
$$\frac{18000}{480} = R$$
$$37.5 = R \Rightarrow 37.5\%$$

31.
$$\frac{68}{102} = \frac{R}{100}$$
$$68(100) = 102R$$
$$6800 = 102R$$
$$\frac{6800}{102} = R$$
$$66\tfrac{2}{3} = R \Rightarrow 66\tfrac{2}{3}\%$$

33. 2

35. The percent of eligible voters who voted was 82%.

37.
$$\frac{129}{400} = \frac{R}{100}$$
$$129(100) = 400R$$
$$12{,}900 = 400R$$
$$\frac{12{,}900}{400} = R$$
$$32\tfrac{1}{4} = R \Rightarrow 32\tfrac{1}{4}\%$$

39.
$$\frac{175}{50} = \frac{R}{100}$$
$$175(100) = 50R$$
$$17{,}500 = 50R$$
$$\frac{17{,}500}{50} = R$$
$$350 = R \Rightarrow 350\%$$

41.
$$\frac{115}{15} = \frac{R}{100}$$
$$115(100) = 15R$$
$$11{,}500 = 15R$$
$$\frac{11{,}500}{15} = R$$
$$766\tfrac{2}{3} = R \Rightarrow 766\tfrac{2}{3}\%$$

43. $\frac{11}{100} \Rightarrow$ The tax is 11%.

45. $\frac{1.95}{100} \Rightarrow$ The rate is 1.95%.

47. $\frac{1.48}{100} \Rightarrow$ The rate is 1.48%.

49.
$$\frac{800}{100{,}000} = \frac{R}{100}$$
$$800(100) = 100{,}000R$$
$$80{,}000 = 100{,}000R$$
$$\frac{80{,}000}{100{,}000} = R$$
$$0.8 = R \Rightarrow 0.8\%$$

0.8% were arrested for burglary in 1980.

$$\frac{225}{100{,}000} = \frac{R}{100}$$
$$225(100) = 100{,}000R$$
$$22{,}500 = 100{,}000R$$
$$\frac{22{,}500}{100{,}000} = R$$
$$0.225 = R \Rightarrow 0.225\%$$

0.225% were arrested for burglary in 2007.

51.
$$\frac{1}{3} = \frac{R}{100}$$
$$1(100) = 3R$$
$$100 = 3R$$
$$\frac{100}{3} = R$$
$$33\tfrac{1}{3} = R \Rightarrow 33\tfrac{1}{3}\%$$

In 2007, $33\tfrac{1}{3}$% of women 25 to 29 had a bachelor's degree.

53. $\frac{82}{100} \Rightarrow$ She spent 82% of her money.

55. $\frac{180}{900} = \frac{20}{100} \Rightarrow$ The interest was 20%.

57. Percent is the amount per 100. It is the numerator of a fraction with 100 in the denominator. It is the number of hundredths in a decimal.

59. $\frac{109}{500} = \frac{R}{100}$
$109(100) = 500R$
$10{,}900 = 500R$
$\frac{10{,}900}{500} = R$
$21.8 = R$
$\frac{109}{500}$; 21.8%

61. $\frac{776}{500} = \frac{R}{100}$
$776(100) = 500R$
$77{,}600 = 500R$
$\frac{77{,}600}{500} = R$
$155.2 = R$
$\frac{776}{500} = \frac{194}{125} = 1\frac{69}{125} = 155.2\%$

63. $47.335 \times 100 = 4733.5$

65. $207.8 \times 1000 = 207{,}800$

67. $0.0672 \div 1000 = 0.0000672$

69. $0.9003 \div 10^2 = 0.009003$

71. $35 + 30.25 + 25 + 36.75 + 6 = 133$; $133(23.85) = 3172.05$; She made \$3172.05.

Exercises 6.2 (page 418)

1. $0.47 = 47\%$

3. $2.32 = 232\%$

5. $0.08 = 8\%$

7. $4.96 = 496\%$

9. $19 = 1900\%$

11. $0.0083 = 0.83\%$

13. $0.952 = 95.2\%$

15. $0.592 = 59.2\%$

17. $0.0731 = 7.31\%$

19. $20 = 2000\%$

21. $17.81 = 1781\%$

23. $0.00044 = 0.044\%$

25. $7.1 = 710\%$

27. $0.8867 = 88.67\%$

29. $0.811\overline{6} = 81.1\overline{6}\%$ or $81\frac{1}{6}\%$

31. $0.2409 = 24.09\%$

33. $96\% = 0.96$

35. $73\% = 0.73$

37. $1.35\% = 0.0135$

39. $908\% = 9.08$

41. $652.5\% = 6.525$

43. $0.0062\% = 0.000062$

45. $0.071\% = 0.00071$

47. $3940\% = 39.4$

49. $0.092\% = 0.00092$

51. $100\% = 1$

53. $662\% = 6.62$

55. $\frac{1}{2}\% = 0.5\% = 0.005$

57. $\frac{3}{16}\% = 0.1875\%$
$= 0.001875$

59. $73\frac{3}{4}\% = 73.75\%$
$= 0.7375$

61. $\frac{1}{8}\% = 0.125\% = 0.00125$

63. $413.773\% = 4.13773$

65. $0.0463 = 4.63\%$; The tax on income is 4.63%.

67. $0.36 = 36\%$; She sold 36% of her items.

69. $3.15\% = 0.0315$; The employees will get a raise of 0.0315.

71. $6.34\% = 0.0634$; The Credit Union will use 0.0634 to compute the interest.

73. $\frac{7}{15}\% \approx 0.466\% \approx 0.005$

75. $88\frac{7}{12}\% \approx 88.583\% \approx 0.886$

77. $60\% = 0.6$; The river pollution due to agricultural runoff is 0.6 of the total.

79. $0.875 = 87.5\%$; The team won 87.5% of its matches.

81. $0.67 = 67\%$; His pass completion rate was 67%.

83. $1.272 = 127.2\%$; In 2016, the number of physical therapy jobs will be 127.2% of the number of jobs in 2006.

85. One mile is about 1.609 kilometers.

87. $40.6\% = 0.406$; The New York system has 0.406 times the number of riders of the Moscow system.

89. $27.4\% = 0.274$; The diameter of Pluto is about 0.274 times the diameter of Earth.

91. $0.062 + 0.029 = 0.091 = 9.1\%$; The percent of income paid to Social Security and Medicare is 9.1%.

93. $0.89 = 89\%$; The sale price is 89% of the regular price. The sale price is 11% off the regular price.

95. $125\% = 1.25$; The new box is 1.25 times the size of the old box.

97. The decimal will be greater than 1 when the percent is greater than 100%.

99. $18{,}000 = 1{,}800{,}000\%$

101. $0.425 = 42.5\%$; $0.42\overline{5} = 42\frac{5}{9}\%$

103. $56\frac{11}{12}\% \approx 56.916667\% \Rightarrow 0.6, 0.569$

105. Answers may vary.

107. $\frac{9}{64} = 9 \div 64 = 0.140625$

109. $\frac{117}{65} = 117 \div 65 = 1.8$

111. $0.1025 = \frac{1025}{10{,}000} = \frac{41}{400}$

113. $345{,}891.62479 \approx 350{,}000$

115. $340(0.695) = 236.3$; $340(0.629) = 213.86$; $236.3 - 213.86 = 22.44$; \$22.44 is saved.

Exercises 6.3 (page 425)

1. $\frac{67}{100} = 0.67 = 67\%$

3. $\frac{37}{50} = 0.74 = 74\%$

5. $\frac{17}{20} = 0.85 = 85\%$

7. $\frac{1}{2} = 0.5 = 50\%$

9. $\frac{17}{20} = 0.85 = 85\%$

11. $\frac{21}{20} = 1.05 = 105\%$

13. $\frac{15}{8} = 1.875 = 187.5\%$

15. $\frac{63}{1000} = 0.063 = 6.3\%$

17. $4\frac{3}{5} = \frac{23}{5} = 4.6 = 460\%$

SECTION 6.3

19. $\frac{1}{3} = 0.333\overline{3} = 33.\overline{3}\% = 33\frac{1}{3}\%$

21. $\frac{29}{6} = 4.8\overline{3} = 483.\overline{3}\% = 483\frac{1}{3}\%$

23. $4\frac{11}{12} = 4.91\overline{6} = 491.\overline{6}\% = 491\frac{2}{3}\%$

25. $\frac{8}{13} \approx 0.615 = 61.5\%$

27. $\frac{5}{9} \approx 0.556 = 55.6\%$

29. $\frac{11}{14} \approx 0.786 = 78.6\%$

31. $32\frac{11}{29} \approx 32.379 = 3237.9\%$

33. $12\% = 12 \cdot \frac{1}{100} = \frac{12}{100} = \frac{3}{25}$

35. $85\% = 85 \cdot \frac{1}{100} = \frac{85}{100} = \frac{17}{20}$

37. $130\% = 130 \cdot \frac{1}{100} = \frac{130}{100} = \frac{13}{10}$, or $1\frac{3}{10}$

39. $200\% = 200 \cdot \frac{1}{100} = \frac{200}{100} = 2$

41. $84\% = 84 \cdot \frac{1}{100} = \frac{84}{100} = \frac{21}{25}$

43. $25\% = 25 \cdot \frac{1}{100} = \frac{25}{100} = \frac{1}{4}$

45. $45\% = 45 \cdot \frac{1}{100} = \frac{45}{100} = \frac{9}{20}$

47. $150\% = 150 \cdot \frac{1}{100} = \frac{150}{100} = \frac{3}{2}$, or $1\frac{1}{2}$

49. $45.5\% = 45\frac{1}{2} \cdot \frac{1}{100} = \frac{91}{2} \cdot \frac{1}{100} = \frac{91}{200}$

51. $6.8\% = 6\frac{4}{5} \cdot \frac{1}{100} = \frac{34}{5} \cdot \frac{1}{100} = \frac{34}{500} = \frac{17}{250}$

53. $60.5\% = 60\frac{1}{2} \cdot \frac{1}{100} = \frac{121}{2} \cdot \frac{1}{100} = \frac{121}{200}$

55. $\frac{1}{4}\% = \frac{1}{4} \cdot \frac{1}{100} = \frac{1}{400}$

57. $2\frac{1}{2}\% = \frac{5}{2} \cdot \frac{1}{100} = \frac{5}{200} = \frac{1}{40}$

59. $\frac{2}{11}\% = \frac{2}{11} \cdot \frac{1}{100} = \frac{2}{1100} = \frac{1}{550}$

61. $44\frac{1}{4}\% = \frac{177}{4} \cdot \frac{1}{100} = \frac{177}{400}$

63. $331\frac{2}{3}\% = \frac{995}{3} \cdot \frac{1}{100} = \frac{995}{300} = \frac{199}{60}$, or $3\frac{19}{60}$

65. $\frac{17}{20} = 0.85 = 85\%$; Kobe Bryant made 85% of his free throws.

67. $\frac{69{,}456{,}897}{131{,}257{,}328} \approx 0.529 = 52.9\%$; He received about 52.9% of the votes.

69. $\frac{365}{538} \approx 0.678 = 67.8\%$; He received about 67.8% of the electoral college votes.

71. $\frac{5732}{9570} \approx 0.599 = 59.9\%$; He made about 59.9% of his field goals.

73. $\frac{67}{360} \approx 0.1861 = 18.61\%$

75. $1\frac{25}{66} \approx 1.3788 = 137.88\%$

77. $4\frac{4}{9}\% = \frac{40}{9} \cdot \frac{1}{100} = \frac{40}{900} = \frac{2}{45}$

SECTION 6.3

79. $0.275\% = \frac{275}{1000} \cdot \frac{1}{100} = \frac{275}{100{,}000} = \frac{11}{4000}$

81. $13 \cdot \frac{1}{8} = \frac{13}{8} = 1.625 = 162.5\%$; Miquel is taking 162.5% of the recommended level.

83. $126\% = 126 \cdot \frac{1}{100} = \frac{126}{100} = \frac{63}{50}$; It represents $\frac{63}{50}$, or $1\frac{13}{50}$, of last year's enrollment.

85. $45\% = 45 \cdot \frac{1}{100} = \frac{45}{100} = \frac{90}{200}$; $37\frac{1}{2}\% = \frac{75}{2} \cdot \frac{1}{100} = \frac{75}{200}$; $\frac{90}{200} + \frac{75}{200} = \frac{165}{200} = \frac{33}{40}$;
$1 - \frac{33}{40} = \frac{40}{40} - \frac{33}{40} = \frac{7}{40}$; About $\frac{7}{40}$ of the residents are between 25 and 40.

87. $\frac{84}{140} = \frac{3}{5} = 0.6 = 60\%$; St. Croix is $\frac{3}{5}$, or 60%, of the land area of the Virgin Islands.

89. $\frac{211}{365} \approx 0.578 = 57.8\%$; In 2004, about 57.8% of the days were smoggy. $\frac{167}{365} \approx 0.458 = 45.8\%$; In 2009, about 45.8% of the days were smoggy.
Reasons for the decrease could include better emission controls on cars and factories.

91. $42\% = 42 \cdot \frac{1}{100} = \frac{42}{100} = \frac{21}{50}$; $1 - \frac{21}{50} = \frac{50}{50} - \frac{21}{50} = \frac{29}{50}$; During the last 10 years, $\frac{29}{50}$ of the salmon run was lost.

93. $0.17\% = \frac{17}{100} \cdot \frac{1}{100} = \frac{17}{10{,}000}$; In Kenya, $\frac{17}{10{,}000}$ of the population is white. This means that 17 out of 10,000 people are white.

95. $\frac{1}{3} + \frac{1}{7} \approx 0.333 + 0.143 = 0.476 \approx 0.48$; About 48% of the price is taken off.

97. $1 - \frac{2}{3} = \frac{1}{3}$; $\frac{1}{3} \approx 0.33 = 33\%$
The grill was 33% off.

99. **Answers may vary.**

101. $2\frac{4}{13}\% \approx 2.3\%$

103. $0.00025 = \frac{25}{100{,}000} = \frac{1}{4000}$

105. $180.04\% = 180\frac{4}{100}\% = \frac{18{,}004}{100} \cdot \frac{1}{100} = \frac{18{,}004}{10{,}000} = \frac{4501}{2500} = 1\frac{2001}{2500}$

107. **Answers may vary.**

109. $0.84 = \frac{84}{100} = \frac{21}{25}$

111. $4.065 = 4\frac{65}{1000} = 4\frac{13}{200}$

113. $\frac{33}{40} = 33 \div 40 = 0.825$

115. $0.567 = 56.7\%$

117. $8.13\% = 0.0813$

Exercises 6.4 (page 433)

1. $10\% = 10 \cdot \frac{1}{100} = \frac{10}{100} = \frac{1}{10}$
$10\% = 0.10 = 0.1$

3. $0.75 = \frac{75}{100} = \frac{3}{4}$
$0.75 = 75\%$

5. $1\frac{3}{20} = \frac{23}{20} = 23 \div 20 = 1.15$
$1\frac{3}{20} = 1.15 = 115\%$

7. $0.001 = \frac{1}{1000}$
$0.001 = 0.1\%$

9. $4.25 = 4\frac{25}{100} = 4\frac{1}{4}$
$4.25 = 425\%$

11. $5\frac{1}{2}\% = \frac{11}{2} \cdot \frac{1}{100} = \frac{11}{200}$
$5\frac{1}{2}\% = 5.5\% = 0.055$

13. $13\frac{1}{2}\% = \frac{27}{2} \cdot \frac{1}{100} = \frac{27}{200}$
$13\frac{1}{2}\% = 13.5\% = 0.135$

15. $62\frac{1}{2}\% = \frac{125}{2} \cdot \frac{1}{100} = \frac{125}{200} = \frac{5}{8}$
$62\frac{1}{2}\% = 62.5\% = 0.625$

17. $\frac{17}{50} = 17 \div 50 = 0.34$
$\frac{17}{50} = 0.34 = 34\%$

19. $0.08 = \frac{8}{100} = \frac{2}{25}$
$0.08 = 8\%$

21. $0.96 = \frac{96}{100} = \frac{24}{25}$
$0.96 = 96\%$

23. $0.35 = \frac{35}{100} = \frac{7}{20}$
$0.35 = 35\%$

25. $0.125 = \frac{125}{1000} = \frac{1}{8}$
$0.125 = 12.5\%$

27. $\frac{1}{4} = 1 \div 4 = 0.25 = 25\%$; The swim suit is 25% off the regular price.

29. $\frac{1}{6} = 1 \div 6 = 0.1\overline{6}$; $0.25 > 0.1\overline{6}$, so Hank's has the better offer.

31. $96\% = 0.96$; $\frac{23}{25} = 23 \div 25 = 0.92$;
$0.92 < 0.96 < 0.965$, so Seller B has the better deal and both offers fall within the authorized amount.

33. $\frac{11}{30} = 11 \div 30 \approx 0.367$; $35\% = 0.35$;
$0.35 < 0.361 < 0.367$
Stephanie had the highest average.

35. $\frac{851}{1000} = 851 \div 1000 = 0.851$;
$0.851 < 0.89$; so Peru has the higher rate.

37. $40\% = 40 \cdot \frac{1}{100} = \frac{40}{100} = \frac{2}{5}$; The TV sells for $\frac{2}{5}$ off the regular price.

39. $\frac{1}{8} = 1 \div 8 = 0.125$; $12\% = 0.120$; $0.12 < 0.125 < 0.13$; Melinda gets the best deal from Machines Etc.

41. **Answers may vary.**

43.

Fraction	Decimal	Percent
$\frac{23}{7}$	3.3	328.6%
$\frac{4}{5}$	0.8	80%
$\frac{421}{200}$	2.1	210.5%

45.
$$\begin{aligned}\frac{45}{81} &= \frac{23}{y}\\ 45y &= 81(23)\\ 45y &= 1863\\ \frac{45y}{45} &= \frac{1863}{45}\\ y &= 41.4\end{aligned}$$

47.
$$\begin{aligned}\frac{x}{8.3} &= \frac{117}{260}\\ x(260) &= 8.3(117)\\ 260x &= 971.1\\ \frac{260x}{260} &= \frac{971.1}{260}\\ x &= 3.735\end{aligned}$$

49.
$$\begin{aligned}\frac{85}{100} &= \frac{170}{B}\\ 85B &= 100(170)\\ 85B &= 17{,}000\\ \frac{85B}{85} &= \frac{17{,}000}{85}\\ B &= 200\end{aligned}$$

51.
$$\begin{aligned}\frac{23}{100} &= \frac{A}{41.3}\\ 23(41.3) &= 100A\\ 949.9 &= 100A\\ \frac{949.9}{100} &= \frac{100A}{100}\\ 9.5 &\approx A\end{aligned}$$

53. $\frac{1}{3} \cdot 235{,}500 = 78{,}500$ in damage

$\frac{4}{5} \cdot 78{,}500 = 62{,}800$

They should collect $62,800.

Exercises 6.5 (page 441)

note: Some exercises are solved using the formula, while the others are solved using a proportion.

1.
$$\begin{aligned}A &= R \times B\\ 27 &= 0.50B\\ \frac{27}{0.50} &= \frac{0.50B}{0.50}\\ 54 &= B\end{aligned}$$

3.
$$\begin{aligned}A &= R \times B\\ A &= 0.60(125)\\ A &= 75\end{aligned}$$

5.
$$\begin{aligned}A &= R \times B\\ 12 &= R \times 4\\ \frac{12}{4} &= \frac{R \times 4}{4}\\ 3 &= R \Rightarrow R = 300\%\end{aligned}$$

7.
$$\begin{aligned}\frac{x}{100} &= \frac{150}{200}\\ x(200) &= 100(150)\\ 200x &= 15000\\ \frac{200x}{200} &= \frac{15000}{200}\\ x &= 75 \Rightarrow 75\%\end{aligned}$$

9.
$$\begin{aligned}\frac{80}{100} &= \frac{32}{x}\\ 80x &= 100(32)\\ 80x &= 3200\\ \frac{80x}{80} &= \frac{3200}{80}\\ x &= 40\end{aligned}$$

11.
$$\begin{aligned}\frac{30}{100} &= \frac{x}{91}\\ 30(91) &= 100x\\ 2730 &= 100x\\ \frac{2730}{100} &= \frac{100x}{100}\\ 27.3 &= x\end{aligned}$$

SECTION 6.5

13.
$$\begin{aligned} A &= R \times B \\ 96 &= R \times 120 \\ \frac{96}{120} &= \frac{R \times 120}{120} \\ 0.8 &= R \Rightarrow R = 80\% \end{aligned}$$

15.
$$\begin{aligned} A &= R \times B \\ 39 &= 0.39B \\ \frac{39}{0.39} &= \frac{0.39B}{0.39} \\ 100 &= B \end{aligned}$$

17.
$$\begin{aligned} A &= R \times B \\ A &= \frac{1}{300}(600) \\ A &= 2 \end{aligned}$$

19.
$$\begin{aligned} \frac{130}{100} &= \frac{x}{90} \\ 130(90) &= 100x \\ 11{,}700 &= 100x \\ \frac{11{,}700}{100} &= \frac{100x}{100} \\ 117 &= x \end{aligned}$$

21.
$$\begin{aligned} A &= R \times B \\ A &= 0.175(70) \\ A &= 12.25 \end{aligned}$$

23.
$$\begin{aligned} \frac{x}{100} &= \frac{3.33}{60} \\ x(60) &= 100(3.33) \\ 60x &= 333 \\ \frac{60x}{60} &= \frac{333}{60} \\ x &= 5.55 \Rightarrow 5.55\% \end{aligned}$$

25.
$$\begin{aligned} \frac{76}{100} &= \frac{497.8}{x} \\ 76x &= 100(497.8) \\ 76x &= 49{,}780 \\ \frac{76x}{76} &= \frac{49{,}780}{76} \\ x &= 655 \end{aligned}$$

27.
$$\begin{aligned} \frac{45.5}{100} &= \frac{x}{80} \\ 45.5(80) &= 100x \\ 3640 &= 100x \\ \frac{3640}{100} &= \frac{100x}{100} \\ 36.4 &= x \end{aligned}$$

29.
$$\begin{aligned} A &= R \times B \\ 105.3 &= 0.39B \\ \frac{105.3}{0.39} &= \frac{0.39B}{0.39} \\ 270 &= B \end{aligned}$$

31.
$$\begin{aligned} \frac{124}{100} &= \frac{328.6}{x} \\ 124x &= 100(328.6) \\ 124x &= 32{,}860 \\ \frac{124x}{124} &= \frac{32{,}860}{124} \\ x &= 265 \end{aligned}$$

33.
$$\begin{aligned} A &= R \times B \\ 96 &= R \times 125 \\ \frac{96}{125} &= \frac{R \times 125}{125} \\ 0.768 &= R \Rightarrow R = 76.8\% \end{aligned}$$

35.
$$\begin{aligned} A &= R \times B \\ A &= 0.0614(350) \\ A &= 21.49 \end{aligned}$$

37.
$$\begin{aligned} \frac{x}{100} &= \frac{2.05}{3.28} \\ x(3.28) &= 100(2.05) \\ 3.28x &= 205 \\ \frac{3.28x}{3.28} &= \frac{205}{3.28} \\ x &= 62.5 \Rightarrow 62.5\% \end{aligned}$$

39.
$$\begin{aligned} \frac{1}{9} &= \frac{x}{1845} \\ 1(1845) &= 9x \\ 1845 &= 9x \\ \frac{1845}{9} &= \frac{9x}{9} \\ 205 &= x \end{aligned}$$

41.
$$\begin{aligned} A &= R \times B \\ 41 &= R \times 85 \\ \frac{41}{85} &= \frac{R \times 85}{85} \\ 0.482 &\approx R \Rightarrow R \approx 48.2\% \end{aligned}$$

43.
$$\begin{aligned} A &= R \times B \\ 82 &= 0.248B \\ \frac{82}{0.248} &= \frac{0.248B}{0.248} \\ 330.65 &\approx B \end{aligned}$$

45.
$$\begin{aligned} \frac{327}{1000} &= \frac{x}{695} \\ 327(695) &= 1000x \\ 227{,}265 &= 1000x \\ \frac{227{,}265}{1000} &= \frac{1000x}{1000} \\ 227.265 &= x \end{aligned}$$

47.
$$\begin{aligned} \frac{x}{100} &= \frac{37}{156} \\ x(156) &= 100(37) \\ 156x &= 3700 \\ \frac{156x}{156} &= \frac{3700}{156} \\ x &\approx 23.7 \Rightarrow 23.7\% \end{aligned}$$

49.

$$\begin{aligned} \frac{5\frac{1}{8}}{100} &= \frac{x}{8\frac{1}{3}} \\ 5\frac{1}{8}\cdot 8\frac{1}{3} &= 100x \\ \frac{41}{8}\cdot\frac{25}{3} &= 100x \\ \frac{1025}{24} &= 100x \\ \frac{1025}{24(100)} &= \frac{100x}{100} \\ \frac{41}{96} &= x \end{aligned}$$

51.

$$\begin{aligned} \frac{8\frac{7}{15}}{100} &= \frac{x}{1350} \\ 8\frac{7}{15}\cdot 1350 &= 100x \\ \frac{127}{15}\cdot\frac{1350}{1} &= 100x \\ 11{,}430 &= 100x \\ \frac{11{,}430}{100} &= \frac{100x}{100} \\ \frac{1143}{10} &= x \\ 114\frac{3}{10} &= x \end{aligned}$$

53.

$$\begin{aligned} A &= R \times B \\ 45.87 &= R \times 34.76 \\ \frac{45.87}{34.76} &= \frac{R\times 34.76}{34.76} \\ 1.320 &\approx R \Rightarrow R \approx 132.0\% \end{aligned}$$

55. A 40% profit would have to be based on the cost of 30¢, or 40% of 30¢. Thus, a 40% profit would be $0.40(30) = 12$¢. However, the actual profit here is 40¢. A 40¢ profit is about 133% of the cost.

57.

$$\begin{aligned} \frac{\frac{1}{2}}{100} &= \frac{x}{45\frac{1}{3}} \\ \frac{1}{2}\cdot 45\frac{1}{3} &= 100x \\ \frac{1}{2}\cdot\frac{136}{3} &= 100x \\ \frac{68}{3} &= 100x \\ \frac{68}{3(100)} &= \frac{100x}{100} \\ \frac{68}{300} &= x \\ \frac{17}{75} &= x \end{aligned}$$

59.

$$\begin{aligned} \frac{\frac{3}{4}}{100} &= \frac{x}{13\frac{1}{3}} \\ \frac{3}{4}\cdot 13\frac{1}{3} &= 100x \\ \frac{3}{4}\cdot\frac{40}{3} &= 100x \\ 10 &= 100x \\ \frac{10}{100} &= \frac{100x}{100} \\ 0.1 &= x \end{aligned}$$

61.

$$\begin{aligned} \frac{18}{24} &= \frac{x}{60} \\ 18(60) &= 24x \\ 1080 &= 24x \\ \frac{1080}{24} &= \frac{24x}{24} \\ 45 &= x \end{aligned}$$

63.

$$\begin{aligned} \frac{a}{\frac{5}{8}} &= \frac{1\frac{1}{2}}{3\frac{3}{4}} \\ a\left(3\frac{3}{4}\right) &= \frac{5}{8}\cdot 1\frac{1}{2} \\ a\cdot\frac{15}{4} &= \frac{5}{8}\cdot\frac{3}{2} \\ a\cdot\frac{15}{4} &= \frac{15}{16} \\ a\cdot\frac{15}{4}\cdot\frac{4}{15} &= \frac{15}{16}\cdot\frac{4}{15} \\ a &= \frac{1}{4} \end{aligned}$$

65.

$$\begin{aligned} \frac{1.4}{0.21} &= \frac{w}{3.03} \\ 1.4(3.03) &= 0.21w \\ 4.242 &= 0.21w \\ \frac{4.242}{0.21} &= \frac{0.21w}{0.21} \\ 20.2 &= w \end{aligned}$$

67.
$$\frac{1\frac{3}{4}\text{ in.}}{70\text{ mi}} = \frac{3\frac{5}{8}\text{ in.}}{x\text{ mi}}$$
$$1\frac{3}{4}x = 70 \cdot 3\frac{5}{8}$$
$$\frac{7}{4}x = \frac{70}{1} \cdot \frac{29}{8}$$
$$\frac{7}{4}x = \frac{1015}{4}$$
$$4 \cdot \frac{7}{4}x = 4 \cdot \frac{1015}{4}$$
$$7x = 1015$$
$$\frac{7x}{7} = \frac{1015}{7}$$
$$x = 145$$
The distance is 145 miles.

69.
$$\frac{1\frac{3}{4}\text{ in.}}{70\text{ mi}} = \frac{x\text{ in.}}{105\text{ mi}}$$
$$1\frac{3}{4} \cdot 105 = 70x$$
$$\frac{7}{4} \cdot \frac{105}{1} = 70x$$
$$\frac{735}{4} = 70x$$
$$\frac{735}{4(70)} = \frac{70x}{70}$$
$$\frac{735}{280} = x$$
$$2\frac{5}{8} = x$$
The number of inches needed is $2\frac{5}{8}$.

Exercises 6.6 (page 450)

1.
$$A = R \times B$$
$$A = 0.16(5240)$$
$$A = 838.4$$
There are about 838 undergraduate students from low-income families.

3.
$$\frac{44.8}{100} = \frac{x}{10{,}395{,}000}$$
$$44.8(10{,}395{,}000) = 100x$$
$$465{,}696{,}000 = 100x$$
$$\frac{465{,}696{,}000}{100} = \frac{100x}{100}$$
$$4{,}656{,}960 = x$$
The Hispanic or Latino population was about 4,657,000.

5.
$$A = R \times B$$
$$27 = R \times 34$$
$$\frac{27}{34} = \frac{R \times 34}{34}$$
$$0.79 \approx R \Rightarrow R \approx 79\%$$
Her score was 79%.

7.
$$A = R \times B$$
$$7 = R \times 43.21$$
$$\frac{7}{43.21} = \frac{R \times 43.21}{43.21}$$
$$0.16 \approx R \Rightarrow R \approx 16\%$$
The tip was about 16% of the check.

9.
$$\text{Total} = 300 + 100 + 200 = 600$$
$$A = R \times B$$
$$200 = R \times 600$$
$$\frac{200}{600} = \frac{R \times 600}{600}$$
$$0.\overline{3} = R \Rightarrow R = 33\tfrac{1}{3}\%$$
Sand is $33\frac{1}{3}\%$ of the mixture.

11.
$$\frac{x}{100} = \frac{360}{1180}$$
$$x(1180) = 100(360)$$
$$1180x = 36{,}000$$
$$\frac{1180x}{1180} = \frac{36{,}000}{1180}$$
$$x \approx 31$$
About 31% of the acreage is soybeans.

13.

$$A = R \times B$$
$$360{,}372{,}000 = R \times 864{,}030{,}000$$
$$\frac{360{,}372{,}000}{864{,}030{,}000} = \frac{R \times 864{,}030{,}000}{864{,}030{,}000}$$
$$0.417 \approx R \Rightarrow R \approx 41.7\%$$

The General Fund Budget is about 41.7% of the total budget.

15.

$$\frac{x}{100} = \frac{2415}{8102}$$
$$x(8102) = 100(2415)$$
$$8102x = 241{,}500$$
$$\frac{8102x}{8102} = \frac{241{,}500}{8102}$$
$$x \approx 29.8$$

Mickey Mantle got a hit about 29.8% of the time.

17.

$$\frac{4}{100} = \frac{5}{x}$$
$$4x = 100(5)$$
$$4x = 500$$
$$\frac{4x}{4} = \frac{500}{4}$$
$$x = 125$$

The population of Japan is about 125 million.

19.

$$\frac{16\frac{2}{3}}{100} = \frac{4}{x}$$
$$16\frac{2}{3}x = 100(4)$$
$$\frac{50}{3}x = 400$$
$$3 \cdot \frac{50}{3}x = 3(400)$$
$$50x = 1200$$
$$\frac{50x}{50} = \frac{1200}{50}$$
$$x = 24$$

The manager had 24 boxes in stock.

21.

$$A = R \times B$$
$$641.20 = 0.14B$$
$$\frac{641.20}{0.14} = \frac{0.14B}{0.14}$$
$$4580 = B$$
$$4580 \div 1145 = 4$$

She must sell 4 refrigerators.

23.

$$146.5 + 25.6 = 172.1$$
$$A = R \times B$$
$$25.6 = R \times 172.1$$
$$\frac{25.6}{172.1} = \frac{R \times 172.1}{172.1}$$
$$0.149 \approx R \Rightarrow R \approx 14.9\%$$

She lost about 14.9% of her body weight.

25.

$$A = R \times B \qquad A = R \times B \qquad A = R \times B \qquad A = R \times B$$
$$2.5 = 0.04B \qquad 220 = 0.09B \qquad 180 = 0.05B \qquad 3 = 0.14B$$
$$\frac{2.5}{0.04} = \frac{0.04B}{0.04} \qquad \frac{220}{0.09} = \frac{220B}{0.09} \qquad \frac{180}{0.05} = \frac{0.05B}{0.05} \qquad \frac{3}{0.14} = \frac{0.14B}{0.14}$$
$$63 \approx B \qquad 2444 \approx B \qquad 3600 \approx B \qquad 21 \approx B$$

The recommendations are 63 g fat, 2444 mg sodium, 3600 mg potassium, and 21 g dietary fiber.

27.
$$\frac{13.6}{100} = \frac{10.3}{x}$$
$$13.6x = 100(10.3)$$
$$13.6x = 1030$$
$$\frac{13.6x}{13.6} = \frac{1030}{13.6}$$
$$x \approx 75.7$$
In 1900, the population was about 75.7 million.

29.
$$A = R \times B$$
$$A = 0.015(8.23)$$
$$A = 0.12345 \approx 0.12$$
$$8.23 + 0.12 = 8.35$$
Viewers will watch about 8.35 hours of TV each day.

31.
$$\text{Increase} = 415 - 345 = 70$$
$$\frac{x}{100} = \frac{70}{345}$$
$$x(345) = 100(70)$$
$$345x = 7000$$
$$\frac{345x}{345} = \frac{7000}{345}$$
$$x \approx 20.3 \Rightarrow 20.3\%$$
The percent increase is about 20.3%.

33.
$$\text{New Amount} = 764 + 124 = 888$$
$$A = R \times B$$
$$124 = R \times 764$$
$$\frac{124}{764} = \frac{R \times 764}{764}$$
$$0.162 \approx R \Rightarrow R \approx 16.2\%$$
The percent increase is about 16.2%.

35.
$$A = R \times B$$
$$A = 0.15(2900)$$
$$A = 435$$
The increase is $435.

New Amount $= 2900 + 435 = \$3335$

37.
$$\text{Decrease} = 17{,}364 - 17{,}223 = 141$$
$$\frac{x}{100} = \frac{141}{17{,}223}$$
$$x(17{,}223) = 100(141)$$
$$17{,}223x = 14{,}100$$
$$\frac{17{,}223x}{17{,}223} = \frac{14{,}100}{17{,}223}$$
$$x \approx 0.8 \Rightarrow 0.8\%$$
The percent decrease is about 0.8%.

39.
$$\text{Increase} = 1{,}107{,}126 - 1{,}052{,}415 = 54{,}710$$
$$\frac{x}{100} = \frac{54{,}710}{1{,}052{,}415}$$
$$x(1{,}052{,}415) = 100(54{,}710)$$
$$1{,}052{,}415x = 5{,}471{,}000$$
$$\frac{1{,}052{,}415x}{1{,}052{,}415} = \frac{5{,}471{,}000}{1{,}052{,}415}$$
$$x \approx 5.2 \Rightarrow 5.2\%$$
The percent increase is about 5.2%.

41.
$$\text{Increase} = 27{,}467 - 18{,}678 = 8789$$
$$\frac{x}{100} = \frac{8789}{18{,}678}$$
$$x(18{,}678) = 100(8789)$$
$$18{,}678x = 878{,}900$$
$$\frac{18{,}678x}{18{,}678} = \frac{878{,}900}{18{,}678}$$
$$x \approx 47.1 \Rightarrow 47.1\%$$
The percent increase is about 47.1%.

43.

$$\text{Increase} = 620 - 330 = 290$$
$$A = R \times B$$
$$290 = R \times 330$$
$$\frac{290}{330} = \frac{R \times 330}{330}$$
$$0.879 \approx R \Rightarrow R \approx 87.9\%$$

The percent increase is about 87.9%.

45.

$$\text{Decrease} = 2.34 - 2.2 = 0.14$$
$$\frac{x}{100} = \frac{0.14}{2.34}$$
$$x(2.34) = 100(0.14)$$
$$2.34x = 14$$
$$\frac{2.34x}{2.34} = \frac{14}{2.34}$$
$$x \approx 6.0 \Rightarrow 6.0\%$$

The percent decrease is about 6.0%.

47.

$$\text{Increase} = 29.6 - 27.5 = 2.1$$
$$\frac{x}{100} = \frac{2.1}{27.5}$$
$$x(27.5) = 100(2.1)$$
$$27.5x = 210$$
$$\frac{27.5x}{27.5} = \frac{210}{27.5}$$
$$x \approx 7.6 \Rightarrow 7.6\%$$

The percent increase is about 7.6%.

49.

$$\text{Increase} = 2224 - 1650 = 574$$
$$A = R \times B$$
$$574 = R \times 1650$$
$$\frac{574}{1650} = \frac{R \times 1650}{1650}$$
$$0.348 \approx R \Rightarrow R \approx 34.8\%$$

The size increased by about 34.8%.

51. Asians are the smallest identifiable population in TX.

53. The second largest ethnic population in TX is Hispanic.

55. Chrysler sold more cars than Nissan.

57. $\frac{1}{8} = 0.125 = 12.5\%$; $\frac{3}{8} = 0.375 = 37.5\%$

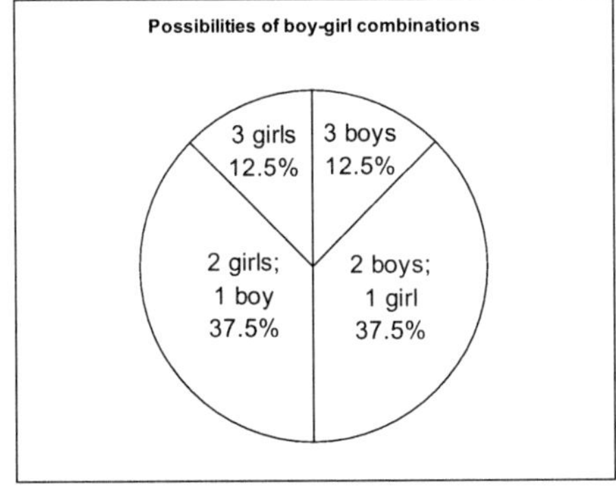

59.

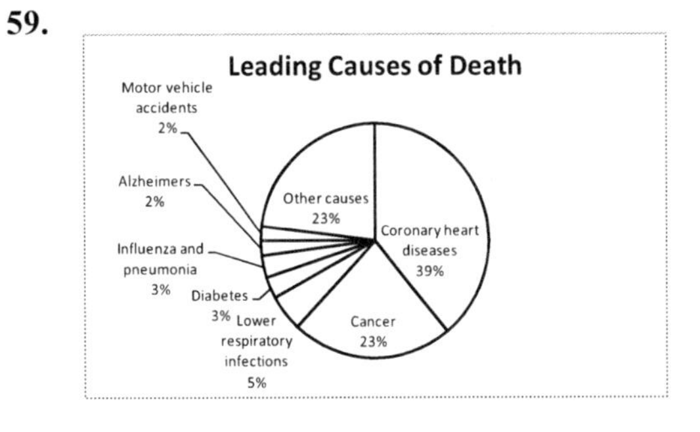

61. The percent of increase is 100%.

63. Amount of increase $= 24\frac{3}{8} - 7\frac{3}{4} = 24.375 - 7.75 = 16.625$

Percent increase:
$$A = R \times B$$
$$16.625 = R \times 7.75$$
$$\frac{16.625}{7.75} = \frac{R \times 7.75}{7.75}$$
$$2.15 \approx R \Rightarrow R \approx 215\%$$

The baby increased in weight by about 215%.

65. Total: $0.89 + 2.46 + 1.78 + 1.15 + 4.97 + 6.78 + 3.21 + 1.89 + 1.78 + 1.56 + 2.35 + 6.80 + 1.98 + 2.07 + 2.89 = 42.56$

Nonfood: $1.15 + 4.97 + 6.80 = 12.92$

Meat: $6.78 + 3.21 = 9.99$

$$A = R \times B$$
$$12.92 = R \times 42.56$$
$$\frac{12.92}{42.56} = \frac{R \times 42.56}{42.56}$$
$$0.304 \approx R$$
$$30.4\% \approx R$$

$$A = R \times B$$
$$9.99 = R \times 42.56$$
$$\frac{9.99}{42.56} = \frac{R \times 42.56}{42.56}$$
$$0.235 \approx R$$
$$23.5\% \approx R$$

Nonfood items accounted for about 30.4% of the total, while meat products accounted for about 23.5% of the total.

67.
$$\frac{6 \text{ Sally}}{10 \text{ total}} = \frac{x}{12{,}600}$$
$$6(12{,}600) = 10x$$
$$75{,}600 = 10x$$
$$\frac{75{,}600}{10} = \frac{10x}{10}$$
$$7560 = x$$
$$12{,}600 - 7560 = 5040$$

Sally should get \$7560, and Rita should get \$5040.

69. $321(48) = 15{,}408$

$2100 + 15{,}408 = 17{,}508$

Joe pays a total of \$17,508 for the car.

71.
$$\frac{400}{260} = \frac{175}{x}$$
$$400x = 260(175)$$
$$400x = 45{,}500$$
$$\frac{400x}{400} = \frac{45{,}500}{400}$$
$$x = 113.75$$

The engine has 113.75 horsepower.

73.
$$A = R \times B$$
$$18 = R \times 20$$
$$\frac{18}{20} = \frac{R \times 20}{20}$$
$$0.9 = R \Rightarrow 90\% = R$$

Peter attended 90% of the classes.

75.
$$A = R \times B$$
$$17 = R \times 20$$
$$\frac{17}{20} = \frac{R \times 20}{20}$$
$$0.85 \approx R$$

$85\% \approx R \Rightarrow$ The score is 85%.

Exercises 6.7 (page 462)

1.
$$\text{Tax} = R \times \text{Price}$$
$$\text{Tax} = 0.058(238)$$
$$\text{Tax} = \$13.80$$
$$\text{Cost} = \text{Price} + \text{Tax}$$
$$= 238 + 13.80 = \$251.80$$

3.
$$\text{Tax} = R \times \text{Price}$$
$$\text{Tax} = 0.038(90.10)$$
$$\text{Tax} = \$3.42$$
$$\text{Cost} = \text{Price} + \text{Tax}$$
$$= 90.10 + 3.42 = \$93.52$$

5.
$$\text{Cost} = \text{Price} + \text{Tax}$$
$$= 628 + 28.26 = \$656.26$$
$$\text{Tax} = R \times \text{Price}$$
$$28.26 = R \times 628$$
$$\frac{28.26}{628} = \frac{R \times 628}{628}$$
$$0.045 = R$$
The sales tax rate is 4.5%.

7.
$$\text{Tax} = \text{Cost} - \text{Price}$$
$$= 1552.96 - 1499 = \$53.96$$
$$\text{Tax} = R \times \text{Price}$$
$$53.96 = R \times 1499$$
$$\frac{53.96}{1499} = \frac{R \times 1499}{1499}$$
$$0.036 \approx R$$
The sales tax rate is about 3.6%.

9.
$$\text{Tax} = R \times \text{Price}$$
$$\text{Tax} = 0.068(895.89)$$
$$\text{Tax} = \$60.92$$

11.
$$\text{Tax} = R \times \text{Price}$$
$$\text{Tax} = 0.063(476.45)$$
$$\text{Tax} = \$30.02$$
$$\text{Cost} = \text{Price} + \text{Tax}$$
$$= 476.45 + 30.02 = \$506.47$$

13.
$$\text{Tax} = \text{Cost} - \text{Price}$$
$$= 1143.67 - 1075.89 = \$67.78$$
$$\text{Tax} = R \times \text{Price}$$
$$67.78 = R \times 1075.89$$
$$\frac{67.78}{1075.89} = \frac{R \times 1075.89}{1075.89}$$
$$0.063 = R$$
The sales tax rate is 6.3%.

15.
$$\text{Price} = 749.99 - 125 = 624.99$$
$$\text{Tax} = R \times \text{Price}$$
$$\text{Tax} = 0.048(624.99)$$
$$\text{Tax} = \$30.00$$
$$\text{Cost} = \text{Price} + \text{Tax}$$
$$= 624.99 + 30.00 = \$654.99$$

17.
$$\text{Price} = 189.90 - 22 = \$167.90$$
$$\text{Tax} = R \times \text{Price}$$
$$6.05 = R \times 167.90$$
$$\frac{6.05}{167.90} = \frac{R \times 167.90}{167.90}$$
$$0.036 \approx R$$
The sales tax rate is about 3.6%.

19.
$$\text{Tax} = \text{Cost} - \text{Price}$$
$$= 726.65 - 675.95 = \$50.70$$
$$\text{Tax} = R \times \text{Price}$$
$$50.70 = R \times 675.95$$
$$\frac{50.70}{675.95} = \frac{R \times 675.95}{675.95}$$
$$0.075 = R$$
The sales tax rate is 7.5%.

SECTION 6.7

21. Discount $= R \times$ Orig. Price
$= 0.18(75.82) = \$13.65$
Sale Price $=$ Orig. Price $-$ Discount
$= 75.82 - 13.65 = \$62.17$

23. Discount $= R \times$ Orig. Price
$= 0.30(320) = \$96$
Sale Price $=$ Orig. Price $-$ Discount
$= 320 - 96 = \$224$

25. Sale Price $=$ Orig. Price $-$ Discount
$= 15.95 - 0.49 = \$15.46$
Discount $= R \times$ Orig. Price
$0.49 = R \times 15.95$
$$\frac{0.49}{15.95} = \frac{R \times 15.95}{15.95}$$
$0.031 \approx R \Rightarrow R \approx 3.1\%$

27. Discount $=$ Orig. Price $-$ Sale Price
$= 1798 - 1706.30 = \$91.70$
Discount $= R \times$ Orig. Price
$91.70 = R \times 1798$
$$\frac{91.70}{1798} = \frac{R \times 1798}{1798}$$
$0.051 \approx R \Rightarrow R \approx 5.1\%$

29. Discount $= R \times$ Orig. Price
$= 0.15(189.95) = \$28.49$
Sale Price $=$ Orig. Price $-$ Discount
$= 189.95 - 28.49 = \$161.46$
Joan pays \$161.46 for the painting.

31. Discount $= R \times$ Orig. Price
$= 0.09(52{,}890) = \$4760.10$
Sale Price $=$ Orig. Price $-$ Discount
$= 52{,}890 - 4760.10$
$= \$48{,}129.90$
Larry pays \$48,129.90 for the SUV.

33. Discount $= R \times$ Orig. Price
$= 0.28(979.99) = \$274.40$
The discount is \$274.40.
Sale Price $=$ Orig. Price $-$ Discount
$= 979.99 - 274.40 = \$705.59$
Melvin pays \$705.59 for the generator.

35. Discount $= R \times$ Orig. Price
$= 0.44(598) = \$263.12$
Sale Price $=$ Orig. Price $-$ Discount
$= 598 - 263.12 = 334.88$
Tax $=$ Cost $-$ Price
$= 350.95 - 334.88 = \$16.07$
Tax $= R \times$ Price
$16.07 = R \times 334.88$
$$\frac{16.07}{334.88} = \frac{R \times 334.88}{334.88}$$
$0.048 \approx R$
The discount price is \$334.88, and the sales tax rate is about 4.8%.

37. Reduced Price $= 89.99 - 20 = 69.99$
Discount $= R \times$ Reduced Price
$= 0.10(69.99) = \$7.00$
Total Savings $= 20 + 7 = 27$
Heather will save \$27.

39. Discount $= R \times$ Orig. Price
$= 0.25(536.95) = \$134.24$
Sale Price $=$ Orig. Price $-$ Discount
$= 536.95 - 134.24 = \$402.71$
The sale price is \$402.71.

41. $\text{Rebate} = R \times \text{Price}$
$= 0.06(445.60) = \$26.74$
The rebate is $26.74.

43. $\text{Discount} = R \times \text{Orig. Price}$
$= 0.25(42.75) = \$10.69$
$\text{Sale Price} = \text{Orig. Price} - \text{Discount}$
$= 42.75 - 10.69 = \$32.06$
The sale price is $32.06.

45. $\text{1st Disc.} = R \times \text{Orig. Price}$
$= 0.45(249) = \$112.05$
$\text{Reduced Price} = \text{Orig. Price} - \text{1st Disc.}$
$= 249 - 112.05 = \$136.95$
$\text{2nd Disc.} = R \times \text{Reduced Price}$
$= 0.24(136.95) = \$32.87$
$\text{Final Price} = \text{Reduced Price} - \text{2nd Disc.}$
$= 136.95 - 32.87 = \$104.08$
$\text{Tot. Disc.} = \text{Orig. Price} - \text{Final Price}$
$= 249 - 104.08 = \$144.92$
$\text{Discount} = R \times \text{Orig. Price}$
$144.92 = R \times 249$
$$\frac{144.92}{249} = \frac{R \times 249}{249}$$
$0.582 \approx R \Rightarrow R \approx 58.2\%$
The final price is $104.08, which reflects a total discount of about 58.2%.

47. $\text{Discount} = R \times \text{Orig. Price}$
$= 0.50(42.98) = \$21.49$
$\text{Sale Price} = \text{Orig. Price} - \text{Discount}$
$= 42.98 - 21.49 = \$21.49$
$\text{Total Cost} = 55.99 + 21.49 = \$77.48$
$\text{Original Total} = 55.99 + 42.98$
$= 98.97$
$\text{Discount} = R \times \text{Original Total}$
$21.49 = R \times 98.97$
$$\frac{21.49}{98.97} = \frac{R \times 98.97}{98.97}$$
$0.217 \approx R \Rightarrow R = 21.7\%$
The final cost is $77.48, a savings of 21.7% over the original total cost.

49. $\text{1st Discount} = R \times \text{Orig. Price}$
$= 0.25(1375) = \$343.75$
$\text{Red. Pr} = \text{Orig. Price} - \text{1st Disc}$
$= 1375 - 343.75 = \$1031.25$
$\text{2nd Discount} = R \times \text{Red. Price}$
$= 0.10(1031.25) = \$103.13$
$\text{Final} = \text{Reduced} - \text{2nd Disc}$
$= 1031.25 - 103.13 = \$928.12$
The final price will be $928.12.

51. $\text{Commission} = R \times \text{Total Sales}$
$= 0.09(4890) = \$440.10$

53. $\text{Commission} = R \times \text{Total Sales}$
$= 0.085(67{,}320) = \$5722.20$

55. $\text{Commission} = R \times \text{Total Sales}$
$44 = R \times 1100$
$$\frac{44}{1100} = \frac{R \times 1100}{1100}$$
$0.04 = R \Rightarrow R = 4\%$

57. $\text{Commission} = R \times \text{Total Sales}$
$31{,}212 = 0.09 \times \text{Total Sales}$
$$\frac{31{,}212}{0.09} = \frac{0.09 \times \text{Total Sales}}{0.09}$$
$346{,}800 = \text{Total Sales}$

59. $\text{Commission} = R \times \text{Total Sales}$
$= 0.075(8340.90) = \$625.57$
The commission was $625.57.

61. $8(45.50) = \$364$
$\text{Commission} = R \times \text{Total Sales}$
$= 0.07(364) = \$25.48$
The commission is $25.48.

63. $4678.50 - 2000 = \$2678.50$

$$\begin{aligned}\text{Comm} &= R \times \text{Sales over } \$2000 \\ &= 0.09(2678.50) = \$241.07\end{aligned}$$

$500 + 241.07 = \$741.07$

He earned \$741.07.

65.

$$\begin{aligned}\text{Commission} &= R \times \text{Total Sales} \\ &= 0.07(6025) = \$421.75\end{aligned}$$

The commission was \$421.75.

67. $705 - 260 = \$445$ in commission.

$$\begin{aligned}\text{Commission} &= R \times \text{Total Sales} \\ 445 &= 0.075 \times \text{Total Sales} \\ \frac{445}{0.075} &= \frac{0.075 \times \text{Total Sales}}{0.075} \\ 5933.33 &= \text{Total Sales}\end{aligned}$$

His total sales were \$5933.33.

69. $0.04(2000) = 80$

$15{,}000 - 2000 = 13{,}000$

$0.05(13{,}000) = 650$

$26{,}725 - 15{,}000 = 11{,}725$

$0.06(11{,}725) = 703.50$

$80 + 650 + 703.50 = 1433.50$

She earned \$1433.50.

71. A sales tax is a percent increase in the price paid by the consumer because it added to the sales price.

73. $0.09(18{,}700) = 1683$

$1500 + 1683 = 3183$

She earned \$3183 in July.

$0.09(30{,}000) = 2700$

$49{,}500 - 30{,}000 = 19{,}500$

$0.05(19{,}500) = 975$

$1500 + 2700 + 975 = 5175$

She earned \$5175 in August.

75.

$$\begin{aligned}\frac{11}{24} + \frac{13}{36} &= \frac{11}{24} \cdot \frac{3}{3} + \frac{13}{36} \cdot \frac{2}{2} \\ &= \frac{33}{72} + \frac{26}{72} = \frac{59}{72}\end{aligned}$$

77.

$$\begin{aligned}\frac{9}{16} - \frac{1}{12} &= \frac{9}{16} \cdot \frac{3}{3} - \frac{1}{12} \cdot \frac{4}{4} \\ &= \frac{27}{48} - \frac{4}{48} = \frac{23}{48}\end{aligned}$$

79.

$$\frac{7}{16} \cdot \frac{8}{21} = \frac{\overset{1}{\cancel{7}}}{\underset{2}{\cancel{16}}} \cdot \frac{\overset{1}{\cancel{8}}}{\underset{3}{\cancel{21}}} = \frac{1 \cdot 1}{2 \cdot 3} = \frac{1}{6}$$

81.

$$\frac{11}{12} \div \frac{33}{52} = \frac{11}{12} \cdot \frac{52}{33} = \frac{\overset{1}{\cancel{11}}}{\underset{3}{\cancel{12}}} \cdot \frac{\overset{13}{\cancel{52}}}{\underset{3}{\cancel{33}}} = \frac{1 \cdot 13}{3 \cdot 3} = \frac{13}{9}, \text{ or } 1\frac{4}{9}$$

83.

$$\begin{aligned}6\frac{3}{5} + 2\frac{7}{8} + 6\frac{3}{5} + 2\frac{7}{8} &= 6 + 2 + 6 + 2 + \frac{3}{5} + \frac{7}{8} + \frac{3}{5} + \frac{7}{8} \\ &= 16 + \frac{3}{5} \cdot \frac{8}{8} + \frac{7}{8} \cdot \frac{5}{5} + \frac{3}{5} \cdot \frac{8}{8} + \frac{7}{8} \cdot \frac{5}{5} \\ &= 16 + \frac{24}{40} + \frac{35}{40} + \frac{24}{40} + \frac{35}{40} = 16 + \frac{118}{40} = 16 + 2\frac{38}{40} = 18\frac{19}{20}\end{aligned}$$

The perimeter is $18\frac{19}{20}$ inches.

Exercises 6.8 (page 472)

1. Interest = Principal × Rate × Time
$= 10{,}000(0.05)(1) = \$500$
Amount due $= 10{,}000 + 500 = \$10{,}500$

3. Interest = Principal × Rate × Time
$= 5962(0.08)(1) = \$476.96$
Amount due $= 5962 + 476.96 = \$6438.96$

5. Interest = Principal × Rate × Time
$= 32{,}000(0.08)(4) = \$10{,}240$
Amount due $= 32{,}000 + 10{,}240 = \$42{,}240$

7. Interest = Principal × Rate × Time
$= 4500(0.055)\left(\frac{9}{12}\right) = \$185.63$
Amount due $= 4500 + 185.63 = \$4685.63$

9. Interest = Principal × Rate × Time
$= 2400(0.075)(1) = \$180$
Maria earns \$180 in interest.

11. Interest = Principal × Rate × Time
$= 8500(0.055)(3) = \$1402.50$
Nancy earns \$1402.50 in interest.

13. Interest = Principal × Rate × Time
$= 5765(0.045)\left(\frac{8}{12}\right) = \$172.95$
Tyra earns \$172.95 in interest.

15. Interest = Principal × Rate × Time
$= 2300(0.065)\left(\frac{10}{12}\right) = \$124.58$
$2300 + 124.58 = \$2424.58$
Janna pays a total of \$2424.58.

17. **a.** Interest $= 50{,}838 - 45{,}800 = \$5038$
b. Interest = Principal × Rate × Time
$5038 = 45{,}800 \times R \times 1$
$$\frac{5038}{45{,}800} = \frac{45{,}800 \times R}{45{,}800}$$
$0.11 = R \Rightarrow R = 11\%$

19. End Balance = Principal × Comp. Factor
$= 21{,}000(1.4148)$
$= \$29{,}710.80$

21. End Balance = Principal × Comp. Factor
$= 60{,}000(3.3102)$
$= \$198{,}612$

23. End Balance = Principal × Comp. Factor
$= 9800(2.1072)$
$= \$20{,}650.56$
Interest $= 20{,}650.56 - 9800 = \$10{,}850.56$

25. End Balance = Principal × Comp. Factor
$= 95{,}000(2.4594)$
$= \$233{,}643$
Interest $= 233{,}643 - 95{,}000 = \$138{,}643$

27.

Jose
End Balance = Principal × Comp. Factor
$= 15{,}000(1.8167)$
$= \$27{,}250.50$
Interest $= 27{,}250.50 - 15{,}000 = \$12{,}250.50$

Juanita
End Balance = Principal × Comp. Factor
$= 15{,}000(1.8221)$
$= \$27{,}331.50$
Interest $= 27{,}331.50 - 15{,}000 = \$12{,}331.50$

$12{,}331.50 - 12{,}250.50 = 81$; Juanita earns \$81 more interest.

SECTION 6.8

29.

Lucy

Interest = Principal × Rate × Time
= 15,000(0.03)(5)
= $2250

Carl

End Balance = Principal × Comp. Factor
= 15,000(1.1618)
= $17427

Interest = 17,427 − 15,000 = $2427

2427 − 2250 = 177; Carl earns $177 more interest.

31. **a.** Interest = (Balance × Rate) ÷ 12 = (3967.10)(0.175) ÷ 12 = $57.85
b. Amount applied to balance = Payment − Interest = 200 − 57.85 = $142.15
c. New balance = Old balance − Amount applied = 3967.10 − 142.15 = $3824.95

33. **a.** Minimum payment = 0.04(4277) ≈ 171
b. Interest = (Balance × Rate) ÷ 12 = (4277)(0.1479) ÷ 12 = $52.71
c. Amount applied to balance = Payment − Interest = 171 − 52.71 = $118.29
d. New balance = Old balance − Amount applied = 4277 − 118.29 = $4158.71

35. **a.**

	Beginning balance	Minimum payment	Interest paid	Balance reduction	Ending balance
Month 1	10,000	400	108.33	291.67	9708.33
Month 2	9708.33	388	105.17	282.83	9425.50
Month 3	9425.50	377	102.11	274.89	9150.61

b. The amounts are decreasing because they are based on the balance, which is also decreasing.
c. Since the amount paid off is decreasing by less than $10 per month, he will pay off at least $200 each of the remaining 9 months, for $1800 more in balance reductions. The balance will probably be below $8000.

37. The minimum monthly payment is generally only a little bit larger than the interest payment so that making the minimum payment decreases the balance owed by only a small amount. Thus, additional charges or fees could easily result in a higher balance after the minimum payment has been made.

39.

Month	1	2	3	4	5	6	7
Starting balance	1235.60	1154.75	1072.65	989.28	904.61	818.63	731.32
Payment	100	100	100	100	100	100	100
Interest paid	19.15	17.90	16.63	15.33	14.02	12.69	11.34
App. to balance	80.85	82.10	83.37	84.67	85.98	87.31	88.66
Ending balance	1154.75	1072.65	989.28	904.61	818.63	731.32	642.66

Month	8	9	10	11	12	13	14
Starting balance	642.66	552.62	461.19	368.34	274.05	178.30	81.06
Payment	100	100	100	100	100	100	82.32
Interest paid	9.96	8.57	7.15	5.71	4.25	2.76	1.26
App. to balance	90.04	91.43	92.85	94.29	95.75	97.24	82.32
Ending balance	552.62	461.19	368.34	274.05	178.30	81.06	0

Linn will pay off the balance on the14th payment (which will be $82.32). She will pay a total of $146.72 in interest.

41.

$$\begin{array}{r} {\scriptstyle 1\,1\,2\;\;2} \\ 32.670 \\ 45.098 \\ 102.500 \\ +\ 134.760 \\ \hline 315.028 \end{array}$$

43.

$$\begin{array}{r} {\scriptstyle 12} \\ {\scriptstyle 0\,\not{2}\;14\quad 5\,10} \\ \not{1}\not{3}\not{4}.5\not{6}\not{0} \\ -\ 98.235 \\ \hline 36.325 \end{array}$$

45. Put four decimal places in the product.

$$\begin{array}{r} {\scriptstyle 1\;1} \\ {\scriptstyle \not{2}\;\not{2}} \\ 56.72 \\ \times\ 0.023 \\ \hline 17016 \\ 113440 \\ \hline 1.30456 \end{array}$$

47. $1.34\overline{)76.40}$

$$\begin{array}{r} 57.0149 \\ 134\overline{)7640.0000} \\ \underline{670} \\ 940 \\ \underline{938} \\ 20 \\ \underline{0} \\ 200 \\ \underline{134} \\ 660 \\ \underline{536} \\ 1240 \\ \underline{1206} \end{array}$$

$57.0149 \approx 57.015$

49. $0.05(1{,}456{,}000) = 72{,}800$;
$1{,}456{,}000 - 72{,}800 = 1{,}383{,}200$;
$1{,}383{,}200 \div 7 = 197{,}600$
Each received \$197,600.

Chapter 6 Review Exercises (page 480)

1. The whole = 20 squares. 15 squares are shaded. $\frac{15}{20}\cdot\frac{5}{5}=\frac{75}{100}=75\%$

2. The whole = 10 squares. 18 squares are shaded. $\frac{18}{10}\cdot\frac{10}{10}=\frac{180}{100}=180\%$

3. $\frac{39}{50}\cdot\frac{2}{2}=\frac{78}{100}=78\%$

4. $\frac{354}{120}=\frac{59}{20}\cdot\frac{5}{5}=\frac{295}{100}=295\%$

5.

$$\begin{aligned} \frac{44}{77} &= \frac{R}{100} \\ 44(100) &= 77R \\ 4400 &= 77R \\ \frac{4400}{77} &= R \\ 57\frac{1}{7} &= R \Rightarrow 57\frac{1}{7}\% \end{aligned}$$

6. The percent that used illegal drugs was 55%.

7. $0.652 = 65.2\%$

8. $0.508 = 50.8\%$

CHAPTER 6 REVIEW EXERCISES

9. $0.00017 = 0.017\%$

10. $73 = 7300\%$

11. $0.756 = 75.6\%$; They won 75.6% of their league games.

12. $0.70 = 70\%$; The sale price is 70% of the original. The store will advertise 30% off.

13. $48\% = 0.48$

14. $632\% = 6.32$

15. $\frac{1}{16}\% = 0.0625\% = 0.000625$

16. $81\frac{3}{4}\% = 81.75\% = 0.8175$

17. $18.45\% = 0.1845$; The decimal used to compute the interest is 0.1845.

18. $5.72\% = 0.0572$; The decimal used to compute the interest is 0.0572.

19. $\frac{11}{16} = 0.6875 = 68.75\%$

20. $7\frac{23}{64} = 7.359375 = 735.9375\%$

21. $\frac{13}{27} \approx 0.481 = 48.1\%$

22. $9\frac{33}{73} \approx 9.452 = 945.2\%$

23. $\frac{37}{46} \approx 0.8043 = 80.43\%$; The team won 80.43% of their games.

24. $\frac{37}{42} \approx 0.88 = 88\%$; Jose got about 88% of the problems correct.

25. $165\% = 165 \cdot \frac{1}{100} = \frac{165}{100} = \frac{33}{20} = 1\frac{13}{20}$

26. $6.4\% = \frac{64}{10} \cdot \frac{1}{100} = \frac{64}{1000} = \frac{8}{125}$

27. $32.5\% = \frac{325}{10} \cdot \frac{1}{100} = \frac{325}{1000} = \frac{13}{40}$

28. $382\% = \frac{382}{1} \cdot \frac{1}{100} = \frac{382}{100} = \frac{191}{50} = 3\frac{41}{50}$

29. $42\% = \frac{42}{1} \cdot \frac{1}{100} = \frac{42}{100} = \frac{21}{50}$; The defense played $\frac{29}{50}$ of the time.

30. $92.4\% = \frac{924}{10} \cdot \frac{1}{100} = \frac{924}{1000} = \frac{231}{250}$; The spraying eliminated $\frac{231}{250}$ of the moths.

31. $\frac{17}{25} = 17 \div 25 = 0.68$

32. $0.68 = 68\%$

33. $0.74 = \frac{74}{100} = \frac{37}{50}$

34. $0.74 = 74\%$

35. $1.5\% = \frac{15}{10} \cdot \frac{1}{100} = \frac{15}{1000} = \frac{3}{200}$

36. $1.5\% = 0.015$

37. $3\frac{11}{40} = 3 + 11 \div 40 = 3.275$

38. $3.275 = 327.5\%$

CHAPTER 6 REVIEW EXERCISES

39.
$$\frac{22}{100}=\frac{x}{455}$$
$$22(455)=100x$$
$$10{,}010=100x$$
$$\frac{10{,}010}{100}=\frac{100x}{100}$$
$$100.1=x$$

40.
$$\frac{45}{100}=\frac{36}{x}$$
$$45x=100(36)$$
$$45x=3600$$
$$\frac{45x}{45}=\frac{3600}{45}$$
$$x=80$$

41.
$$\frac{x}{100}=\frac{17}{80}$$
$$x(80)=100(17)$$
$$80x=1700$$
$$\frac{80x}{80}=\frac{1700}{80}$$
$$x=21.25$$
21.25%

42.
$$A=R\times B$$
$$37=R\times 125$$
$$\frac{37}{125}=\frac{R\times 125}{125}$$
$$0.296=R$$
$$29.6\%=R$$

43.
$$A=R\times B$$
$$A=0.034(370)$$
$$A=12.58$$

44.
$$A=R\times B$$
$$2385=0.53B$$
$$\frac{2385}{0.53}=\frac{0.53B}{0.53}$$
$$4500=B$$

45.
$$\frac{x}{100}=\frac{123}{677}$$
$$x(677)=100(123)$$
$$677x=12{,}300$$
$$\frac{677x}{677}=\frac{12{,}300}{677}$$
$$x\approx 18.2$$
18.2%

46.
$$\frac{154.8}{100}=\frac{254}{x}$$
$$154.8x=100(254)$$
$$154.8x=25{,}400$$
$$\frac{154.8x}{154.8}=\frac{25{,}400}{154.8}$$
$$x\approx 164.08$$

47.
$$A=R\times B$$
$$6345.24=0.264B$$
$$\frac{6345.24}{0.264}=\frac{0.264B}{0.264}$$
$$24{,}035=B$$
Melinda's salary was $24,035.

48. $5721-3564=2157$
$$A=R\times B$$
$$2157=R\times 3564$$
$$\frac{2157}{3564}=\frac{R\times 3564}{3564}$$
$$0.605\approx R\Rightarrow$$
$R\approx 60.5\%$;
The population increased by about 60.5%.

49.
$$A=R\times B$$
$$A=0.32(325)$$
$$A=104$$
$$325+104=429$$
There are now 429 employees.

50.
$$\text{Total}=13+7+5+11$$
$$=36$$
$$A=R\times B$$
$$7=R\times 36$$
$$\frac{7}{36}=\frac{R\times 36}{36}$$
$$0.194\approx R$$
The class is about 19.4% African American.

51. $57{,}480-50{,}650=6830$
$$A=R\times B$$
$$6830=R\times 57{,}480$$
$$\frac{6830}{57{,}480}=\frac{R\times 57{,}480}{57{,}480}$$
$0.12\approx R\Rightarrow R\approx 12\%$;
The value decreased by about 12%.

52.
$$A=R\times B$$
$$74=R\times 110$$
$$\frac{74}{110}=\frac{R\times 110}{110}$$
$0.67\approx R\Rightarrow R\approx 67\%$;
Toni does not qualify for an interview.

53. Most students received a grade of B.

54. Yes. More A and C grades were given out than B and D grades.

CHAPTER 6 REVIEW EXERCISES

55. Tax $= R \times$ Price
Tax $= 0.0635(465.75)$
Tax $= \$29.58$
Cost = Price + Tax
$= 465.75 + 29.58 = \$495.33$

56. Tax = Cost − Price
$= 3233.58 - 3025 = \$208.58$
Tax $= R \times$ Price
$208.58 = R \times 3025$
$\frac{208.58}{3025} = \frac{R \times 3025}{3025}$
$0.0690 = R$
The sales tax rate is 6.90%.

57. $A = R \times B$
$A = 0.25 \times 675.90$
$A = 168.98$
$675.90 - 168.98 = 506.92$ (price after 25% off)

$A = R \times B$
$A = 0.20 \times 506.92$
$A = 101.38$
$506.92 - 101.38 = 405.54$ (final price)

58. Amount over 9500 is $21{,}300 - 9500$
$= 11{,}800$
$S = 1500 + 0.115(11{,}800)$
$= 1500 + 1357 = 2857$
She earned \$2857.

59. Decrease $= 110.95 - 79.49 = 31.46$
$A = R \times B$
$31.46 = R \times 110.95$
$\frac{31.46}{110.95} = \frac{R \times 110.95}{110.95}$
$0.28 \approx R \Rightarrow R \approx 28\%$
They are sold at about 28% off.

60. $A = R \times B$
$A = 0.30 \times 235$
$A = 70.50$
$235 - 70.50 = 164.50$
(price after 30% off)

$A = R \times B$
$A = 0.15 \times 164.50$
$A = 24.68$
$164.50 - 24.68 = 139.83$
(after 15% off)

$A = R \times B$
$A = 0.0475 \times 139.83$
$A = 6.64$
$139.83 + 6.64 = 146.47$
(final price)

61. Interest = Principal × Rate × Time
$= 5500(0.065)(2) = \$715$
Amount due $= 5500 + 715 = \$6215$

62. 1st month Int = Principal × Rate × Time
$= 2000(.08)\left(\frac{1}{12}\right) = 13.33$
2nd month Int = Principal × Rate × Time
$= 2013.33(.08)\left(\frac{1}{12}\right) = 13.42$
Final Balance = \$2026.75

63. Interest = (Balance × Rate) ÷ 12 $= (1345.60)(0.186) \div 12 = \$20.86$
Amount applied to balance = Payment − Interest $= 55 - 20.86 = \$34.14$
New balance = Old balance − Amount applied $= 1345.60 - 34.14 = \$1311.46$

64. **a.** Interest = (Balance × Rate) ÷ 12 $= (4446.60)(0.198) \div 12 = \$73.37$
b. Amount applied to balance = Payment − Interest $= 178 - 73.37 = \$104.63$
c. New balance = Old balance − Amount applied $= 4446.60 - 104.63 = \$4341.97$

Chapter 6 True/False Concept Review (page 484)

1. true

2. true

3. false; Rewrite the fraction with a denominator of 100 and then write the numerator and the percent symbol.

4. false; Move the decimal point two positions to the right and write the percent symbol.

5. true

6. false; The base unit is always 100.

7. true

8. true

9. false; The proportion is $\frac{A}{B} = \frac{X}{100}$.

10. true

11. true

12. false; $4\frac{3}{4} = 475\%$

13. false; $0.009\% = 0.00009$

14. false; $B = 43{,}000$

15. true

16. true

17. false; It is a decrease of 27.75%.

18. False; It is 99% of Monday's value.

19. true

20. true

21. true

22. false; $\frac{1}{2}\% = 0.005$

23. true

24. false; The rate is 6.5%.

Chapter 6 Test (page 485)

1.

$$\begin{aligned} \frac{x}{100} &= \frac{415.50}{1495} \\ x(1495) &= 100(415.50) \\ 1495x &= 41{,}550 \\ \frac{1495x}{1495} &= \frac{41{,}550}{1495} \\ x &\approx 27.8 \end{aligned}$$

The discount is about 27.8%.

2. $0.03542 = 3.542\%$

CHAPTER 6 TEST

3. The population is 56% female.

4.
$$\frac{x}{100} = \frac{6\frac{1}{4}}{8\frac{3}{8}}$$
$$x\left(8\frac{3}{8}\right) = 100\left(6\frac{1}{4}\right)$$
$$x\left(\frac{67}{8}\right) = \frac{100}{1} \cdot \frac{25}{4}$$
$$\frac{67}{8}x = \frac{625}{1}$$
$$\frac{8}{67} \cdot \frac{67}{8}x = \frac{8}{67} \cdot \frac{625}{1}$$
$$x \approx 74.6 \Rightarrow 74.6\%$$

5. $\frac{27}{32} = 27 \div 32 = 0.84375 = 84.375\%$

6.
$$A = R \times B$$
$$113.85 = 2.53 \times B$$
$$\frac{113.85}{2.53} = \frac{2.53 \times B}{2.53}$$
$$45 = B$$

7. $16\frac{8}{13}\% = \frac{216}{13} \cdot \frac{1}{100} = \frac{216}{1300} = \frac{54}{325}$

8. $0.0078 = 0.78\%$

9.
$$\frac{15.6}{100} = \frac{x}{75}$$
$$15.6(75) = 100x$$
$$1170 = 100x$$
$$\frac{1170}{100} = \frac{100x}{100}$$
$$11.7 = x$$

10. $6\frac{9}{11} = \frac{75}{11} = 75 \div 11 = 6.818 = 681.8\%$

11. $272\% = \frac{272}{1} \cdot \frac{1}{100} = \frac{272}{100} = \frac{68}{25} = 2\frac{18}{25}$

12. $7.89\% = 0.0789$

13. $A = R \times B$

$A = 0.175(6400)$

$A = 1120$

They spend \$1120 on rent.

14. $\frac{13}{16} = 13 \div 16 = 0.8125$

15. $\frac{13}{16} = 13 \div 16 = 0.8125 = 81.25\%$

16. $0.624 = \frac{624}{1000} = \frac{78}{125}$

17. $0.624 = 62.4\%$

18. $18.5\% = \frac{185}{10} \cdot \frac{1}{100} = \frac{185}{1000} = \frac{37}{200}$

19. $18.5\% = 0.185$

20.
$$A = R \times B$$
$$87.6 = R \times 115.9$$
$$\frac{87.6}{115.9} = \frac{R \times 115.9}{115.9}$$
$$0.756 \approx R \Rightarrow R \approx 75.6\%$$

21. $4.765\% = 0.04765$

22.
$$\text{Tax} = R \times \text{Price}$$
$$\text{Tax} = 0.063(357.85)$$
$$\text{Tax} = \$22.54$$
$$\text{Cost} = \text{Price} + \text{Tax}$$
$$= 357.85 + 22.54 = \$380.39$$

23.
$$D = R \times B$$
$$D = 0.30 \times 212.95$$
$$D = 63.885$$
$$212.95 - 63.885 = 149.065$$
The sale price is \$149.07.

24.
$$2{,}410{,}758 - 1{,}998{,}257 = 412{,}501$$
$$A = R \times B$$
$$412{,}501 = R \times 1{,}998{,}257$$
$$\frac{412{,}501}{1{,}998{,}257} = \frac{R \times 1{,}998{,}257}{1{,}998{,}257}$$
$$0.206 \approx R \Rightarrow R \approx 20.6\%$$
The population increased by about 20.6%.

25.
$$\text{Total} = 10 + 18 + 2 + 5 + 42 = 77$$
$$\frac{x}{100} = \frac{18}{77}$$
$$x(77) = 100(18)$$
$$77x = 1800$$
$$\frac{77x}{77} = \frac{1800}{77}$$
$$x \approx 23.4$$
The percent of B grades is about 23.4%.

26.
$$\text{Commission} = R \times \text{Total Sales}$$
$$= 0.06(75{,}850) = \$4551$$
$$600 + 4551 = 5151$$
He earned \$5151.

27.
$$\text{Interest} = \text{Principal} \times \text{Rate} \times \text{Time}$$
$$= 6500(0.07)(1) = \$455$$
$$\text{Amount due} = 6500 + 455 = \$6955$$

28.
$$37(10) = 370 \text{ calories from fat}$$
$$A = R \times B$$
$$370 = R \times 780$$
$$\frac{370}{780} = \frac{R \times 780}{780}$$
$$0.474 = R \Rightarrow R = 47.4\%$$
The percent of calories from fat is 47.4%

29. $\text{End Balance} = \text{Principal} \times \text{Comp. Factor} = 9000(1.6470) = \$14{,}823$

30.
$$\text{Minimum payment} = 0.04(6732.50) = 269$$
$$\text{Interest} = (\text{Balance} \times \text{Rate}) \div 12 = (6732.50)(0.185) \div 12 = \$103.79$$
$$\text{Amount applied to balance} = \text{Payment} - \text{Interest} = 269 - 103.79 = \$165.21$$
$$\text{New} = \text{Old} - \text{Amt applied} + \text{Charges} = 6732.50 - 165.21 + 234.50 = \$6801.79$$

Exercises 7.1 (page 498)

1. feet or meters

3. inches or millimeters

5. inches or centimeters

7. feet or meters

9. feet or meters

11. inches or centimeters

13. feet or meters

15. yards, feet, or meters

17. $3 \text{ yd} = \frac{3 \text{ yd}}{1} \cdot \frac{3 \text{ ft}}{1 \text{ yd}} \cdot \frac{12 \text{ in.}}{1 \text{ ft}} = 108 \text{ in.}$

19. $125 \text{ m} = \frac{125 \text{ m}}{1} \cdot \frac{100 \text{ cm}}{1 \text{ m}} = 12{,}500 \text{ cm}$

21. $12 \text{ yd} = \frac{12 \text{ yd}}{1} \cdot \frac{3 \text{ ft}}{1 \text{ yd}} = 36 \text{ ft}$

23. $968.5 \text{ m} = \frac{968.5 \text{ m}}{1} \cdot \frac{1 \text{ km}}{1000 \text{ m}} = 0.9685 \text{ km}$

25. $6 \text{ km} = \frac{6 \text{ km}}{1} \cdot \frac{1000 \text{ m}}{1 \text{ km}} \cdot \frac{100 \text{ cm}}{1 \text{ m}}$
$= 600{,}000 \text{ cm}$

27. $4573 \text{ yd} = \frac{4573 \text{ yd}}{1} \cdot \frac{3 \text{ ft}}{1 \text{ yd}} \cdot \frac{1 \text{ mi}}{5280 \text{ ft}}$
$\approx 2.60 \text{ mi}$

29. $27 \text{ in.} = \frac{27 \text{ in.}}{1} \cdot \frac{1 \text{ ft}}{12 \text{ in.}} = 2.25 \text{ ft}$

31. $245 \text{ in.} = \frac{245 \text{ in.}}{1} \cdot \frac{2.54 \text{ cm}}{1 \text{ in.}} \cdot \frac{1 \text{ m}}{100 \text{ cm}}$
$= 6.22 \text{ m}$

33. $120 \text{ km} = \frac{120 \text{ km}}{1} \cdot \frac{0.6214 \text{ mi}}{1 \text{ km}} \approx 74.57 \text{ mi}$

35. $17.5 \text{ m} = \frac{17.5 \text{ m}}{1} \cdot \frac{3.2808 \text{ ft}}{1 \text{ m}} \approx 57.41 \text{ ft}$

37. $11(13 \text{ yd}) = 143 \text{ yd}$

39. $24.7 \text{ ft} \div 5 = 4.94 \text{ ft}$

41. $25(8 \text{ mm}) = 200 \text{ mm}$

43. $27 \text{ in.} + 45 \text{ in.} + 13 \text{ in.} + 31 \text{ in.} = 116 \text{ in.}$

45. $100(26 \text{ mi}) = 2600 \text{ mi}$

47.
$$\begin{array}{rr} & 3 \text{ yd } 2 \text{ ft} \\ + & 5 \text{ ft} \\ \hline & 3 \text{ yd } 7 \text{ ft} \end{array} = 5 \text{ yd } 1 \text{ ft}$$

49. $256 \text{ mi} \div 13 \approx 19.69 \text{ mi}$

51. $75(321 \text{ mm}) = 24{,}075 \text{ mm}$

53.
$$\begin{array}{rr} & 14 \text{ ft } 7 \text{ in.} \\ - & 8 \text{ ft } 10 \text{ in.} \\ \hline \end{array} \Rightarrow \begin{array}{rr} & 13 \text{ ft } 19 \text{ in.} \\ - & 8 \text{ ft } 10 \text{ in.} \\ \hline & 5 \text{ ft } 9 \text{ in.} \end{array}$$

55. $(107 \text{ km})23 = 2461 \text{ km}$

57.
$$\begin{array}{rr} & 7 \text{ ft } 3 \text{ in.} \\ & 2 \text{ ft } 9 \text{ in.} \\ + & 9 \text{ ft } 7 \text{ in.} \\ \hline & 18 \text{ ft } 19 \text{ in.} \end{array} \Rightarrow 19 \text{ ft } 7 \text{ in.}$$

59. $(2 \text{ yd } 2 \text{ ft } 1 \text{ in.}) \cdot 3 = 6 \text{ yd } 6 \text{ ft } 3 \text{ in.}$
$= 8 \text{ yd } 3 \text{ in.}$

61. 1 door: $2(7 \text{ ft}) + 40 \text{ in.} = 14 \text{ ft} + 40 \text{ in.}$
2 doors: $2(14 \text{ ft} + 40 \text{ in.}) = 28 \text{ ft} + 80 \text{ in.}$
$= 34 \text{ ft} + 8 \text{ in.}$
She should buy 34 ft, 8 in. of garland.

63. $\frac{137 \text{ cm} + 142 \text{ cm} + 123 \text{ cm}}{3} = \frac{402 \text{ cm}}{3}$
$= 134 \text{ cm}$
$= 1.34 \text{ m}$
Their average height is 1.34 meters.

65. $343 \text{ m} = \frac{343 \text{ m}}{1} \cdot \frac{3.2808 \text{ ft}}{1 \text{ m}} \approx 1125 \text{ ft}$
The Millan Viaduct is about 1125 ft tall.

67. Nine lanes will require 8 dividers.
$8(50 \text{ m}) = 400 \text{ m}$; 400 m will be needed.

69. The Amazon and Congo lengths end in multiple zeros, so these may be estimates.

71. The Amazon River is about twice as long as the Sao Francisco River.

73. The yard is 20 ft wide and 28 ft long.

75. The walkways are 2 ft wide and 4 ft wide.

77. **Answers may vary.**

79. $34.76(100{,}000) = 3{,}476{,}000$

81. $\frac{3}{8} + \frac{5}{24} = \frac{9}{24} + \frac{5}{24} = \frac{14}{24} = \frac{7}{12}$

83. $\begin{array}{r} 5\frac{1}{3} \\ -2\frac{5}{6} \\ \hline \end{array} \Rightarrow \begin{array}{r} 5\frac{2}{6} \\ -2\frac{5}{6} \\ \hline \end{array} \Rightarrow \begin{array}{r} 4\frac{8}{6} \\ -2\frac{5}{6} \\ \hline 2\frac{3}{6} = 2\frac{1}{2} \end{array}$

85. $(2.13)^2 = (2.13)(2.13) = 4.5369$

87. $65 + 82 + 92 + 106 + 77 = 422$;
$422 \div 5 = 84.4$; The mean is 84.4.

Exercises 7.2 (page 507)

1. $35 \text{ oz} + 72 \text{ oz} = 107 \text{ oz}$

3. $(400 \text{ mL}) \div 16 = 25 \text{ mL}$

5. $(80 \text{ gal}) \div 20 = 4 \text{ gal}$

7. $34 \text{ c} = \frac{34 \text{ c}}{1} \cdot \frac{1 \text{ pt}}{2 \text{ c}} \cdot \frac{1 \text{ qt}}{2 \text{ pt}} = 8.5 \text{ qt}$

9. $21 \text{ gal} = \frac{21 \text{ gal}}{1} \cdot \frac{4 \text{ qt}}{1 \text{ gal}} = 84 \text{ qt}$

11. $42(3 \text{ lb}) = 126 \text{ lb}$

13. $33 \text{ oz} + 49 \text{ oz} = 82 \text{ oz}$

15. $63 \text{ gal} - 29 \text{ gal} - 4 \text{ gal} = 30 \text{ gal}$

17. $5.5 \text{ gal} = \frac{5.5 \text{ gal}}{1} \cdot \frac{4 \text{ qt}}{1 \text{ gal}} \cdot \frac{2 \text{ pt}}{1 \text{ qt}} \cdot \frac{2 \text{ c}}{1 \text{ pt}} = 88 \text{ c}$

19. $48.4 \text{ L} = \frac{48.4 \text{ L}}{1} \cdot \frac{0.2642 \text{ gal}}{1 \text{ L}} \approx 12.79 \text{ gal}$

21. $212 \text{ kg} - 157 \text{ kg} = 55 \text{ kg}$

23. $37 \text{ lb} + 43 \text{ lb} = 80 \text{ lb}$

25. $(62 \text{ lb})9 = 558 \text{ lb}$

27. $8 \text{ kg} = \frac{8 \text{ kg}}{1} \cdot \frac{1000 \text{ g}}{1 \text{ kg}} = 8000 \text{ g}$

SECTION 7.2

29. $8000 \text{ mg} = \frac{8000 \text{ mg}}{1} \cdot \frac{1 \text{ g}}{1000 \text{ mg}} = 8 \text{ g}$

31. $(1360 \text{ oz}) \div 16 = 85 \text{ oz}$

33. $214 \text{ lb} + 406 \text{ lb} = 620 \text{ lb}$

35. $$\begin{aligned} 234 \text{ g} + 147 \text{ g} - 158 \text{ g} &= 381 \text{ g} - 158 \text{ g} \\ &= 223 \text{ g} \end{aligned}$$

37. $72 \text{ g} = \frac{72 \text{ g}}{1} \cdot \frac{1000 \text{ mg}}{1 \text{ g}} = 72{,}000 \text{ mg}$

39. $13 \text{ oz} = \frac{13 \text{ oz}}{1} \cdot \frac{28.3495 \text{ g}}{1 \text{ oz}} = 368.54 \text{ g}$

41. $$\begin{aligned} C &= \frac{5}{9}(F - 32) \\ &= \frac{5}{9}(32 - 32) = \frac{5}{9}(0) = 0° \text{ C} \end{aligned}$$

43. $$\begin{aligned} C &= \frac{5}{9}(F - 32) \\ &= \frac{5}{9}(104 - 32) = \frac{5}{9}(72) = 40° \text{ C} \end{aligned}$$

45. $C = \frac{5}{9}(F - 32) = \frac{5}{9}(68 - 32) = \frac{5}{9}(36) = 20° \text{ C}$

47. $8 \text{ min} = \frac{8 \text{ min}}{1} \cdot \frac{60 \text{ sec}}{1 \text{ min}} = 480 \text{ sec}$

49. $19 \text{ weeks} = \frac{19 \text{ wk}}{1} \cdot \frac{7 \text{ days}}{1 \text{ wk}} = 133 \text{ days}$

51. $$\begin{aligned} C &= \frac{5}{9}(F - 32) \\ &= \frac{5}{9}(152 - 32) = \frac{5}{9}(120) \approx 66.7° \text{ C} \end{aligned}$$

53. $$\begin{aligned} C &= \frac{5}{9}(F - 32) \\ &= \frac{5}{9}(48 - 32) = \frac{5}{9}(16) \approx 8.9° \text{ C} \end{aligned}$$

55. $$\begin{aligned} F &= \frac{9}{5} \cdot C + 32 \\ &= \frac{9}{5} \cdot 71 + 32 = 127.8 + 32 = 159.8° \text{ F} \end{aligned}$$

57. $$\begin{aligned} F &= \frac{9}{5} \cdot C + 32 \\ &= \frac{9}{5} \cdot 14 + 32 = 25.2 + 32 = 57.2° \text{ F} \end{aligned}$$

59. $$\begin{aligned} 2.6 \text{ yr} &= \frac{2.6 \text{ yr}}{1} \cdot \frac{365 \text{ days}}{1 \text{ yr}} \cdot \frac{24 \text{ hr}}{1 \text{ day}} \\ &= 22{,}776 \text{ hr} \end{aligned}$$

61. $$\begin{aligned} 250 \text{ min} &= \frac{250 \text{ min}}{1} \cdot \frac{1 \text{ hr}}{60 \text{ min}} \cdot \frac{1 \text{ day}}{24 \text{ hr}} \\ &\approx 0.2 \text{ day} \end{aligned}$$

63. $$\begin{array}{rr} 12 \text{ lb} & 2 \text{ oz} \\ 2 \text{ lb} & 15 \text{ oz} \\ + \; 11 \text{ lb} & 5 \text{ oz} \\ \hline 25 \text{ lb} & 22 \text{ oz} \end{array} \Rightarrow 26 \text{ lb } 6 \text{ oz}$$

65. $$\begin{array}{rrr} 2 \text{ gal} & 3 \text{ qt} & 1 \text{ pt} \\ + \; 4 \text{ gal} & 2 \text{ qt} & 1 \text{ pt} \\ \hline 6 \text{ gal} & 5 \text{ qt} & 2 \text{ pt} \\ 6 \text{ gal} & 6 \text{ qt} & 0 \text{ pt} \\ 7 \text{ gal} & 2 \text{ qt} & 0 \text{ pt} \end{array}$$

67. $$\begin{array}{rr} 20 \text{ lb} & 6 \text{ oz} \\ 13 \text{ lb} & 8 \text{ oz} \\ + \; 21 \text{ lb} & 9 \text{ oz} \\ \hline 54 \text{ lb} & 23 \text{ oz} \\ 55 \text{ lb} & 7 \text{ oz} \end{array}$$
The grocery sold 55 lb, 7 oz of hamburger.

69. $(300 \text{ ml}) \div 24 = 12.5 \text{ ml}$; Each student will receive 12.5 ml of acid.

71. $C = \frac{5}{9}(F - 32) = \frac{5}{9}(84 - 32) = \frac{5}{9}(52) \approx 28.9° \text{ C}$; The average temperature is 28.9° C.

73. $\frac{1}{2}$ gal $\div 10 = 0.5$ gal $\div 10 = 0.05$ gal; $0.05 \text{ gal} = \frac{0.05 \text{ gal}}{1} \cdot \frac{4 \text{ qt}}{1 \text{ gal}} \cdot \frac{2 \text{ pt}}{1 \text{ qt}} \cdot \frac{2 \text{ c}}{1 \text{ pt}} \cdot \frac{8 \text{ oz}}{1 \text{ c}} = 6.4 \text{ oz}$
Each serving is 0.05 gal, or 6.4 oz.

75. $397 \text{ g} \cdot 4 = 1588$ g; They contain 1588 grams of chips.

77. Take # kg $\times$ 2.2046 lb/kg;

kg	54	58	63	69	76	85	97	130
lb	119	128	139	152	168	187	214	287

79. Use the answer to #78.
$$1{,}483{,}820{,}800 \text{ lb} = \frac{1{,}483{,}820{,}800 \text{ lb}}{1} \cdot \frac{0.4536 \text{ kg}}{1 \text{ lb}} \approx 673{,}000{,}000 \text{ kg}$$
It weighs about 673 million kg fully loaded.

81. $3 \text{ g} + 4.5 \text{ g} = 7.5$ g per serving
$78 \text{ g} \div 7.5 \text{ g} = 10.4$ servings
He needs to eat 11 servings.

83. The units are different types, so they cannot be added.

85. $$1 \text{ oz t} = \frac{1 \text{ oz t}}{1} \cdot \frac{20 \text{ dwt}}{1 \text{ oz t}} \cdot \frac{24 \text{ grains}}{1 \text{ dwt}} = 480 \text{ grains}$$
$$1 \text{ lb t} = \frac{1 \text{ lb t}}{1} \cdot \frac{12 \text{ oz t}}{1 \text{ lb t}} \cdot \frac{20 \text{ dwt}}{1 \text{ oz t}} \cdot \frac{24 \text{ grains}}{1 \text{ dwt}} = 5760 \text{ grains}$$

87. 2 lb t 8 oz t 17 dwt = 32 oz t 17 dwt = 657 dwt; 657 dwt $\div 6 = 109.5$ dwt per child
109.5 dwt = 5 oz t 9 dwt 12 grains per child

89. $451 + 88 + 309 = 848$

91. $34.7 + 6.8 + 0.44 = 41.94$

93. $$23\frac{7}{9} + 19\frac{5}{6} \Rightarrow 23\frac{14}{18} + 19\frac{15}{18} = 42\frac{29}{18} \Rightarrow 43\frac{11}{18}$$

95. $5.2 + 148 + \frac{1}{4} = 5.2 + 148 + 0.25 = 153.45$

97. $$56.84 \text{ m} = \frac{56.84 \text{ m}}{1} \cdot \frac{1 \text{ km}}{1000 \text{ m}} = 0.05684 \text{ km}$$

Exercises 7.3 (page 516)

1. $2(7 \text{ ft}) + 2(17 \text{ ft}) = 14 \text{ ft} + 34 \text{ ft} = 48 \text{ ft}$

3. 39 in. + 42 in. + 49 in. = 130 in.

5. $2(14 \text{ mm}) + 2(23 \text{ mm}) = 28 \text{ mm} + 46 \text{ mm} = 74 \text{ mm}$

7. $2(6 \text{ m}) + 2(24 \text{ m}) = 12 \text{ m} + 48 \text{ m} = 60 \text{ m}$

SECTION 7.3

9. $16 \text{ mm} + 27 \text{ mm} + 40 \text{ mm} = 83 \text{ mm}$

11. $2(275 \text{ ft}) + 2(125 \text{ ft}) = 550 \text{ ft} + 250 \text{ ft}$
$= 800 \text{ ft}$

13. $2(30 \text{ ft}) + 2(15 \text{ ft}) = 60 \text{ ft} + 30 \text{ ft} = 90 \text{ ft}$

15. $2(84 \text{ cm}) + 2(35 \text{ cm}) + 2(10 \text{ cm}) = 168 \text{ cm} + 70 \text{ cm} + 20 \text{ cm} = 258 \text{ cm}$

17. $5(8 \text{ yd}) = 40 \text{ yd}$

19. $19 \text{ m} + 16 \text{ m} + 30 \text{ m} + 19 \text{ m} + 49 \text{ m} + 35 \text{ m} = 168 \text{ m}$

21. $2(58 \text{ ft}) + 2(27 \text{ ft}) = 116 \text{ ft} + 54 \text{ ft} = 170 \text{ ft}$

23. $C = \pi d = (3.14)7 \text{ in.} = 21.98 \text{ in.}$

25. $C = \pi d = (3.14)12 \text{ mm} = 37.68 \text{ mm}$

27. $C = 2\pi r = 2(3.14)8 \text{ km} = 50.24 \text{ km}$

29. Circle: $\pi d = \pi(8 \text{ cm}) \approx 25.13 \text{ cm}$
$12 \text{ cm} + 12 \text{ cm} + 25.13 \text{ cm} = 49.13 \text{ cm}$

31. $\frac{1}{2}$ circle: $\frac{1}{2}\pi d = \frac{1}{2}\pi(29 \text{ yd}) = 45.55 \text{ yd}$;
$45.55 \text{ yd} + 29 \text{ yd} = 74.55 \text{ yd}$

33. $\frac{1}{2}$ circle: $\frac{1}{2}\pi d = \frac{1}{2}\pi(18 \text{ mm}) = 28.27 \text{ mm}$; $32 \text{ mm} + 32 \text{ mm} + 18 \text{ mm} + 28.27 \text{ mm} = 110.27 \text{ mm}$

35. $2(42 \text{ in.}) + 2(65 \text{ in.}) = 84 \text{ in.} + 130 \text{ in.} = 214 \text{ in.}$

37. Circle: $C = \pi d = \pi(12 \text{ in.}) \approx 37.70 \text{ in.}$; $7 \text{ in.} + 7 \text{ in.} + 24 \text{ in.} + 37.70 \text{ in.} = 75.70 \text{ in.}$

39.

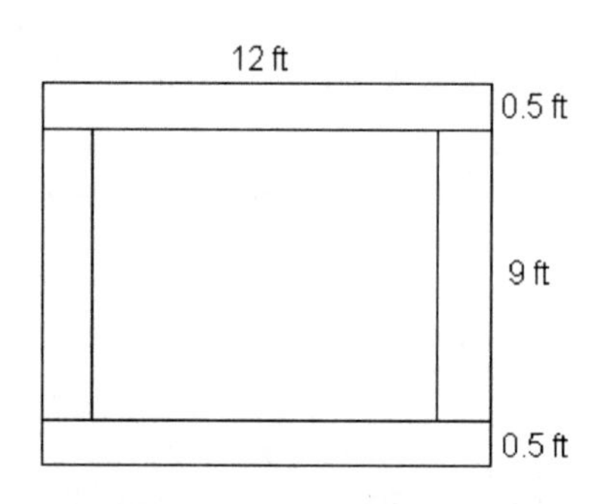

One bed: $2(12 \text{ ft}) + 2(9 \text{ ft}) = 24 \text{ ft} + 18 \text{ ft} = 42 \text{ ft}$
$4(42 \text{ ft}) = 168 \text{ ft}$
June will need 168 ft of railroad ties.

41. Circle: $C = \pi d = \pi(60 \text{ yd}) \approx 188.50 \text{ yd}$; $125.75 \text{ yd} + 125.75 \text{ yd} + 188.50 \text{ yd} = 440 \text{ yd}$
The track is about 440 yards long.

43. $20 + 3 + 3 = 26$; $14 + 3 + 3 = 20$; The dimensions are 26 inches by 20 inches.

45. $12 \text{ ft} + 12 \text{ ft} + 24 \text{ ft} + 24 \text{ ft} + 45 \text{ ft} = 117 \text{ ft}$; They need 117 ft of fence.

47. $2+3+3+5+4+6+8+8+7+10=56$ laps
Each lap: $2(120\text{ yd})+2(53\text{ yd})=240\text{ yd}+106\text{ yd}=346\text{ yd}$
$\#\text{ yd}=56(346\text{ yd})=19{,}376\text{ yd}; \frac{19{,}376\text{ yd}}{1}\cdot\frac{3\text{ ft}}{1\text{ yd}}\cdot\frac{1\text{ mi}}{5280\text{ ft}}\approx 11\text{ mi}$
He ran 19,376 yards, which is about 11 miles.

49. $3\cdot 2(4\text{ ft}+8\text{ ft})=6(12\text{ ft})=72\text{ ft}$
She needs 72 feet of lumber.

51. $72\text{ ft}\div 8\text{ ft}=9$ boards; $9(15.49)=139.41$
The cost of the lumber is \$139.41.

53. Diameter with 1st row $=50+4=54$ in. $C=\pi d=\pi(54\text{ in.})\approx 170$ in.
She will need about 170 inches.

55. $8+7+8+7+8+7+8+7=60$ ft;
The perimeter is 60 feet.

57. The center line divides the yard into identical but mirror-imaged halves.

59. **Answers may vary.**

61. **Answers may vary.**

63. **Answers may vary.**

65. 850,085

67. $32{,}571{,}600\approx 32{,}570{,}000$

69. $673(412)=277{,}276$

71.
$$\begin{array}{r} 2\text{ lb }\ \ 3\text{ oz} \\ +\ 3\text{ lb }14\text{ oz} \\ \hline 5\text{ lb }17\text{ oz} \\ 6\text{ lb }\ \ 1\text{ oz} \end{array}$$

73. $85+78+91+86+95$
$=435$
$435\div 5=87$
The average score is 87.

Exercises 7.4 (page 529)

1. $A=s^2=(11\text{ ft})^2=121\text{ ft}^2$

3. $A=\frac{1}{2}bh=\frac{1}{2}(5\text{ m})(7\text{ m})=\frac{1}{2}(35\text{ m}^2)$
$=17.5\text{ m}^2$

5. $A=bh=(9\text{ ft})(6\text{ ft})=54\text{ ft}^2$

7. $A=\frac{1}{2}(b_1+b_2)h=\frac{1}{2}(9\text{ yd}+15\text{ yd})(5\text{ yd})=\frac{1}{2}(24\text{ yd})(5\text{ yd})=\frac{1}{2}(120\text{ yd}^2)=60\text{ yd}^2$

9. $A=\pi r^2=\frac{22}{7}\left(\frac{21}{2}\text{ cm}\right)^2=\frac{22}{7}\cdot\frac{441}{4}\text{ cm}^2$
$=\frac{693}{2}\text{ cm}^2$
$=346\frac{1}{2}\text{ cm}^2$

11. $A=lw=(17\text{ m})(15\text{ m})=255\text{ m}$

13. $A=\frac{1}{2}bh=\frac{1}{2}(20\text{ ft})(17\text{ ft})$
$=\frac{1}{2}(340\text{ ft}^2)=170\text{ ft}^2$

15. $A=bh=(54\text{ in.})(30\text{ in.})=1620\text{ in.}^2$

SECTION 7.4

17. $A = \frac{1}{2}bh = \frac{1}{2}(36 \text{ cm})(22 \text{ cm})$
$= \frac{1}{2}(792 \text{ cm}^2) = 396 \text{ cm}^2$

19. $A = bh = (9 \text{ in.})(15 \text{ in.}) = 135 \text{ in.}^2$

21. $A = \frac{1}{2}(b_1 + b_2)h = \frac{1}{2}(48 \text{ in.} + 32 \text{ in.})(17 \text{ in.}) = \frac{1}{2}(80 \text{ in.})(17 \text{ in.}) = \frac{1}{2}(1360 \text{ in.}^2) = 680 \text{ in.}^2$

23. $A = \pi r^2 = 3.14(9.6 \text{ cm})^2 = 3.14(92.16 \text{ cm}^2) = 289.3824 \text{ cm}^2$

25. $6780 \text{ in.}^2 = 6780(\text{in.})(\text{in.}) \cdot \frac{1 \text{ ft}}{12 \text{ in.}} \cdot \frac{1 \text{ ft}}{12 \text{ in.}} = \frac{6780}{(12)(12)} \text{ ft}^2 \approx 47.08 \text{ ft}^2$

27. $12 \text{ yd}^2 = 12(\text{yd})(\text{yd}) \cdot \frac{3 \text{ ft}}{1 \text{ yd}} \cdot \frac{3 \text{ ft}}{1 \text{ yd}} \cdot \frac{12 \text{ in.}}{1 \text{ ft}} \cdot \frac{12 \text{ in.}}{1 \text{ ft}} = 12(3)(3)(12)(12) \text{ in.}^2 = 15{,}552 \text{ in.}^2$

29. $1 \text{ mi}^2 = 1(\text{mi})(\text{mi}) \cdot \frac{5280 \text{ ft}}{1 \text{ mi}} \cdot \frac{5280 \text{ ft}}{1 \text{ mi}} = 1(5280)(5280) \text{ ft}^2 = 27{,}878{,}400 \text{ ft}^2$

31. $4056 \text{ in.}^2 = 1296(\text{in.})(\text{in.}) \cdot \frac{1 \text{ ft}}{12 \text{ in.}} \cdot \frac{1 \text{ ft}}{12 \text{ in.}} \cdot \frac{1 \text{ yd}}{3 \text{ ft}} \cdot \frac{1 \text{ yd}}{3 \text{ ft}} = \frac{4056}{(12)(12)(3)(3)} \text{ yd}^2 \approx 3.13 \text{ yd}^2$

33. $345 \text{ ft}^2 = 345(\text{ft})(\text{ft}) \cdot \frac{1 \text{ yd}}{3 \text{ ft}} \cdot \frac{1 \text{ yd}}{3 \text{ ft}} = \frac{345}{(3)(3)} \text{ yd}^2 \approx 38.33 \text{ yd}^2$

35. $44 \text{ yd}^2 = 44(\text{yd})(\text{yd}) \cdot \frac{3 \text{ ft}}{1 \text{ yd}} \cdot \frac{3 \text{ ft}}{1 \text{ yd}} = 44(3)(3) \text{ ft}^2 = 396 \text{ ft}^2$

37. $15{,}000{,}000 \text{ ft}^2 = 15{,}000{,}000(\text{ft})(\text{ft}) \cdot \frac{1 \text{ mi}}{5280 \text{ ft}} \cdot \frac{1 \text{ mi}}{5280 \text{ ft}} = \frac{15{,}000{,}000}{(5280)(5280)} \text{ mi}^2 \approx 0.54 \text{ mi}^2$

39. $1600 \text{ cm}^2 = 1600(\text{cm})(\text{cm}) \cdot \frac{1 \text{ in.}}{2.54 \text{ cm}} \cdot \frac{1 \text{ in.}}{2.54 \text{ cm}} = \frac{1600}{(2.54)(2.54)} \text{ in.}^2 \approx 248.00 \text{ in.}^2$

41. $12 \text{ m}^2 = 12(\text{m})(\text{m}) \cdot \frac{3.2808 \text{ ft}}{1 \text{ m}} \cdot \frac{3.2808 \text{ ft}}{1 \text{ m}} = 12(3.2808)(3.2808) \text{ ft}^2 \approx 129.16 \text{ ft}^2$

43. $15 \text{ ft}^2 = 15(\text{ft})(\text{ft}) \cdot \frac{12 \text{ in.}}{1 \text{ ft}} \cdot \frac{12 \text{ in.}}{1 \text{ ft}} \cdot \frac{2.54 \text{ cm}}{1 \text{ in.}} \cdot \frac{2.54 \text{ cm}}{1 \text{ in.}} = 15(12)(12)(2.54)(2.54) \text{ cm}^2$
$\approx 13{,}935.46 \text{ cm}^2$

45. $A = (17 \text{ ft})(11 \text{ ft}) + (10 \text{ ft})(5 \text{ ft}) = 187 \text{ ft}^2 + 50 \text{ ft}^2 = 237 \text{ ft}^2$

47. $A = (15 \text{ m})(14 \text{ m}) + (22 \text{ m})(5 \text{ m}) = 210 \text{ m}^2 + 110 \text{ m}^2 = 320 \text{ m}^2$

49. $A = (40 \text{ yd})(12 \text{ yd}) + \frac{1}{2}(29 \text{ yd})(14 \text{ yd}) = 480 \text{ yd}^2 + \frac{1}{2}(406 \text{ yd}^2) = 480 \text{ yd}^2 + 203 \text{ yd}^2 = 683 \text{ yd}^2$

51. $A = \pi(12.5 \text{ cm})^2 - (12 \text{ cm})^2 = 156.25\pi \text{ cm}^2 - 144 \text{ cm}^2 \approx 346.87 \text{ cm}^2$

53. $A = (35 \text{ ft})(24 \text{ ft}) = 840 \text{ ft}^2$; $840 \text{ ft}^2 \div 250 \text{ ft}^2 = 3.36$ gallons; He will need about 3.4 gallons.

55. $A = \pi r^2 = \pi(6 \text{ in.})^2 = 36(3.14) \text{ in.}^2 \approx 113.04 \text{ in}^2$
$A = \pi r^2 = \pi(5 \text{ in.})^2 = 25(3.14) \text{ in.}^2 \approx 78.5 \text{ in}^2$
$113.04 \text{ in}^2 - 78.5 \text{ in}^2 = 34.54 \text{ in.}^2 \approx 35 \text{ in.}^2$

57. $(30 \text{ m})(15 \text{ m}) = 450 \text{ m}^2$; $450 \text{ m}^2 \cdot 1.25 = 562.5$ oz; Debbie needs 562.5 oz.

59. $2(3 \text{ ft})(6 \text{ ft}) = 2(18 \text{ ft}^2) = 36 \text{ ft}^2$; The doors require 36 ft^2 of glass.

61. $\frac{1}{2}(36 \text{ ft})(9 \text{ ft}) = \frac{1}{2}(324 \text{ ft}^2) = 162 \text{ ft}^2$; The gable requires 162 ft^2 of sheathing.

63. $2 - \frac{1}{4} - \frac{1}{4} = \frac{8}{4} - \frac{2}{4} = \frac{6}{4} = \frac{3}{2}$; $48 \div \frac{3}{2} = \frac{48}{1} \cdot \frac{2}{3} = 32$ ribbons needed
$32(2 \text{ in.})(60 \text{ in.}) = 3840 \text{ in.}^2$; She needs 32 ribbons, for a total of 3840 in.^{2} of fabric.

65. Bottom: $A = \pi r^2 = \pi(4 \text{ ft})^2 = 16\pi \text{ ft}^2 \approx 50 \text{ ft}^2$
Circumference of bottom $= \pi d = \pi(8 \text{ ft}) = 8\pi$ ft
Sides $= C \cdot \text{depth} = (8\pi \text{ ft})(4 \text{ ft}) = 32\pi \text{ ft}^2 \approx 101 \text{ ft}^2$; $50 \text{ ft}^2 + 101 \text{ ft}^2 = 151 \text{ ft}^2$ total area

67. Left trapezoid: $A = \frac{1}{2}(b_1 + b_2)h = \frac{1}{2}(10 \text{ ft} + 16 \text{ ft})(10 \text{ ft}) = \frac{1}{2}(26 \text{ ft})(10 \text{ ft}) = \frac{1}{2}(260 \text{ ft}^2)$
$= 130 \text{ ft}^2$

Right trapezoid: $A = \frac{1}{2}(b_1 + b_2)h = \frac{1}{2}(30 \text{ ft} + 20 \text{ ft})(16 \text{ ft}) = \frac{1}{2}(50 \text{ ft})(16 \text{ ft}) = \frac{1}{2}(800 \text{ ft}^2)$
$= 400 \text{ ft}^2$

Unshaded: $A = (15 \text{ ft})(8 \text{ ft}) = 120 \text{ ft}^2$; $130 + 400 - 120 = 410 \text{ ft}^2$
$410 \div 70 \approx 5.86$; About 6 bags of seed will be needed.

69. $(14 \text{ ft})(8 \text{ ft}) + (10 \text{ ft})(8 \text{ ft}) = 112 \text{ ft}^2 + 80 \text{ ft}^2 = 192 \text{ ft}^2$; $(8 \text{ ft})(4 \text{ ft}) = 32 \text{ ft}^2$;
$192 \text{ ft}^2 \div 32 \text{ ft}^2 = 6$; Six sheets will be needed.

71. Patio: $2 \cdot \frac{1}{2}(14 \text{ ft} + 8 \text{ ft})(6 \text{ ft}) + (14 \text{ ft})(8 \text{ ft}) = (14 \text{ ft} + 8 \text{ ft})(6 \text{ ft}) + (14 \text{ ft})(8 \text{ ft})$
$= (22 \text{ ft})(6 \text{ ft}) + (14 \text{ ft})(8 \text{ ft})$
$= 132 \text{ ft}^2 + 112 \text{ ft}^2 = 244 \text{ ft}^2$

Walkways: $2 \cdot \frac{1}{2}(6 \text{ ft} + 9 \text{ ft})(2 \text{ ft}) + 2 \cdot \frac{1}{2}(4 \text{ ft} + 6 \text{ ft})(2 \text{ ft}) + (8 \text{ ft})(4 \text{ ft})$
$= (6 \text{ ft} + 9 \text{ ft})(2 \text{ ft}) + (4 \text{ ft} + 6 \text{ ft})(2 \text{ ft}) + (8 \text{ ft})(4 \text{ ft})$
$= (15 \text{ ft})(2 \text{ ft}) + (10 \text{ ft})(2 \text{ ft}) + (8 \text{ ft})(4 \text{ ft})$
$= 30 \text{ ft}^2 + 20 \text{ ft}^2 + 32 \text{ ft}^2 = 82 \text{ ft}^2$

$244 \text{ ft}^2 + 82 \text{ ft}^2 = 326 \text{ ft}^2$; The area is 326 ft^2.

73. **Answers may vary.**

75. **Answers may vary.**

77. Inside dimensions: 190 ft by 110 ft $\Rightarrow$ Inside area $= (190 \text{ ft})(110 \text{ ft}) = 20{,}900 \text{ ft}^2$
Outside area $= (200 \text{ ft})(120 \text{ ft}) = 24{,}000 \text{ ft}^2$; $24{,}000 - 20{,}900 = 3100$;
The walkway will be 3100 ft^2, while the lawn will be $20{,}900 \text{ ft}^2$.

79. $6.708 \times 100{,}000 = 670{,}800$

81. $2(46 - 28) + 50 \div 5 = 2(18) + 10$
$= 36 + 10 = 46$

83. $3^2 + 2^3 = 3 \cdot 3 + 2 \cdot 2 \cdot 2 = 9 + 8 = 17$

85. Average $= (16 + 17 + 18 + 19 + 16 + 214) \div 6 = 300 \div 6 = 50$;
In order: 16, 16, 17, 18, 19, 214; median $= (17 + 18)/2 = 17.5$
mode $= 16$

87.

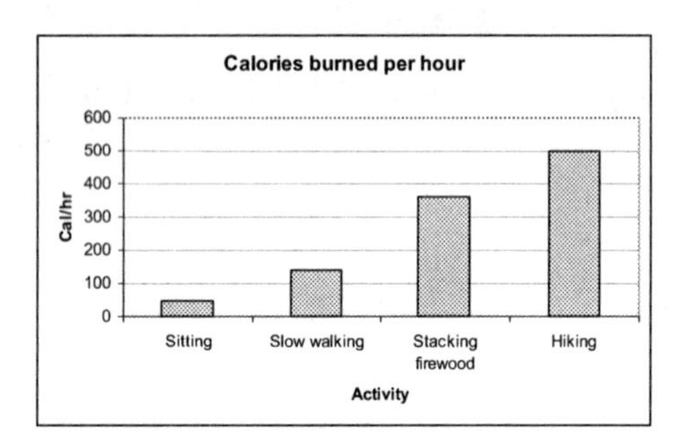

Exercises 7.5 (page 542)

1. $V = lwh = (10 \text{ ft})(4 \text{ ft})(3 \text{ ft}) = 120 \text{ ft}^3$

3. $V = lwh = (33 \text{ cm})(13 \text{ cm})(11 \text{ cm})$
$= 4719 \text{ cm}^3$

5. $V = Bh = (14 \text{ cm}^2)(12 \text{ cm}) = 168 \text{ cm}^3$

7. $V = \pi r^2 h = 3.14(2 \text{ ft})^2(4 \text{ ft})$
$= 3.14(4 \text{ ft}^2)(4 \text{ ft})$
$= 50.24 \text{ ft}^3$

9. $V = lwh = (50 \text{ cm})(30 \text{ cm})(18 \text{ cm})$
$= 27{,}000 \text{ cm}^3 = 27{,}000 \text{ mL}$

11. $V = \pi r^2 h = 3.14(9 \text{ in.})^2(24 \text{ in.})$
$= 3.14(81 \text{ in.}^2)(24 \text{ in.})$
$= 6104.16 \text{ in.}^3$

13. $V = \frac{4}{3}\pi r^3 = \frac{4}{3}(3.14)(1 \text{ ft})^3 = \frac{4}{3} \cdot \frac{3.14}{1} \cdot \frac{1}{1} \text{ ft}^3 = \frac{12.56}{3} \text{ ft}^3 = 4.19 \text{ ft}^3$

15. $V = lwh = (45 \text{ in.})(15 \text{ in.})(7 \text{ in.})$
$= 4725 \text{ in.}^3$

17. $V = Bh = (425 \text{ cm}^2)(90 \text{ cm})$
$= 38{,}250 \text{ cm}^3$

SECTION 7.5

19. $2592 \text{ in.}^3 = 2592(\text{in.})(\text{in.})(\text{in.}) \cdot \frac{1 \text{ ft}}{12 \text{ in.}} \cdot \frac{1 \text{ ft}}{12 \text{ in.}} \cdot \frac{1 \text{ ft}}{12 \text{ in.}} = \frac{2592}{(12)(12)(12)} \text{ ft}^3 = 1.5 \text{ ft}^3$

21. $1 \text{ m}^3 = 1(\text{m})(\text{m})(\text{m}) \cdot \frac{100 \text{ cm}}{1 \text{ m}} \cdot \frac{100 \text{ cm}}{1 \text{ m}} \cdot \frac{100 \text{ cm}}{1 \text{ m}} = 1{,}000{,}000 \text{ cm}^3$

23. $1 \text{ yd}^3 = 1(\text{yd})(\text{yd})(\text{yd}) \cdot \frac{3 \text{ ft}}{1 \text{ yd}} \cdot \frac{3 \text{ ft}}{1 \text{ yd}} \cdot \frac{3 \text{ ft}}{1 \text{ yd}} \cdot \frac{12 \text{ in.}}{1 \text{ ft}} \cdot \frac{12 \text{ in.}}{1 \text{ ft}} \cdot \frac{12 \text{ in.}}{1 \text{ ft}}$
$= 1(3)(3)(3)(12)(12)(12) \text{ in.}^3 = 46{,}656 \text{ in.}^3$

25. $13{,}824 \text{ in.}^3 = 13{,}824(\text{in.})(\text{in.})(\text{in.}) \cdot \frac{1 \text{ ft}}{12 \text{ in.}} \cdot \frac{1 \text{ ft}}{12 \text{ in.}} \cdot \frac{1 \text{ ft}}{12 \text{ in.}} = \frac{13{,}824}{(12)(12)(12)} \text{ ft}^3 = 8 \text{ ft}^3$

27. $10{,}000 \text{ mm}^3 = 10{,}000(\text{mm})(\text{mm})(\text{mm}) \cdot \frac{1 \text{ cm}}{10 \text{ mm}} \cdot \frac{1 \text{ cm}}{10 \text{ mm}} \cdot \frac{1 \text{ cm}}{10 \text{ mm}} = \frac{10{,}000}{(10)(10)(10)} \text{ cm}^3 = 10 \text{ cm}^3$

29. $13{,}560 \text{ in.}^3 = 13{,}560(\text{in.})(\text{in.})(\text{in.}) \cdot \frac{1 \text{ ft}}{12 \text{ in.}} \cdot \frac{1 \text{ ft}}{12 \text{ in.}} \cdot \frac{1 \text{ ft}}{12 \text{ in.}} = \frac{13{,}560}{(12)(12)(12)} \text{ ft}^3 \approx 7.85 \text{ ft}^3$

31. $350{,}000 \text{ cm}^3 = 350{,}000(\text{cm})(\text{cm})(\text{cm}) \cdot \frac{1 \text{ m}}{100 \text{ cm}} \cdot \frac{1 \text{ m}}{100 \text{ cm}} \cdot \frac{1 \text{ m}}{100 \text{ cm}} = \frac{350{,}000}{(100)(100)(100)} \text{ m}^3$
$= 0.35 \text{ m}^3$

33. $8.93 \text{ km}^3 = 8.93(\text{km})(\text{km})(\text{km}) \cdot \frac{0.6214 \text{ mi}}{1 \text{ km}} \cdot \frac{0.6214 \text{ mi}}{1 \text{ km}} \cdot \frac{0.6214 \text{ mi}}{1 \text{ km}} \approx 2.14 \text{ mi}^3$

35. $2826 \text{ ft}^2 = \frac{2826 \text{ ft}^2}{1} \cdot \frac{1 \text{ yd}}{3 \text{ ft}} \cdot \frac{1 \text{ yd}}{3 \text{ ft}} = \frac{2826 \text{ yd}^2}{(3)(3)} = 314 \text{ yd}^2; V = Bh = (314 \text{ yd}^2)(14 \text{ yd}) = 4396 \text{ yd}^3$

37. $B = \frac{1}{2}bh = \frac{1}{2}(24 \text{ cm})(10 \text{ cm}) = 120 \text{ cm}^2; V = Bh = (120 \text{ cm}^2)(46 \text{ cm}) = 5520 \text{ cm}^3$

39. $V = Bh = \frac{1}{2}(12 \text{ ft} + 9 \text{ ft})(6 \text{ ft}) \cdot 5 \text{ ft} = \frac{1}{2}(21 \text{ ft})(6 \text{ ft})(5 \text{ ft}) = \frac{1}{2}(630 \text{ ft}^3) = 315 \text{ ft}^3$

41. $V = lwh + lwh = (72 \text{ cm})(15 \text{ cm})(54 \text{ cm}) + (110 \text{ cm})(72 \text{ cm})(15 \text{ cm})$
$= 58{,}320 \text{ cm}^3 + 118{,}800 \text{ cm}^3 = 177{,}120 \text{ cm}^3$

43. $V = lwh = \left(4\frac{3}{4} \text{ in.}\right)\left(7\frac{1}{4} \text{ in.}\right)\left(2\frac{3}{4} \text{ in.}\right) = \left(\frac{19}{4} \text{ in.}\right)\left(\frac{29}{4} \text{ in.}\right)\left(\frac{11}{4} \text{ in.}\right)$
$= \frac{6061}{64} \text{ in.}^3 = 94\frac{45}{64} \text{ in.}^3$

45. From #72 in 7.4, 2282 bricks are needed. $2282 \div 1000 = 2.282$
$2.282(9) = 20.538 \approx 21$; The bricklayer needs about 21 ft^3 of mortar.

47. If the answer were rounded down, there would not be enough cement.

49. Total area $= (20 \text{ ft})(28 \text{ ft}) = 560 \text{ ft}^2$; From #71 in 7.4, the area of the patio and walkways is 326 ft^2.
$560 - 326 = 234 \text{ ft}^2$; $234(\text{ft})(\text{ft}) \cdot \frac{12 \text{ in.}}{1 \text{ ft}} \cdot \frac{12 \text{ in.}}{1 \text{ ft}} = 33{,}696 \text{ in.}^2$
$V = 33{,}696 \text{ in.}^2(8 \text{ in.}) = 269{,}568 \text{ in.}^3 \approx 6 \text{ yd}^3$; They need about 6 cubic yards of top soil.

51. $36 \cdot 231 = 8316 \text{ in.}^3$; $B = \pi r^2 = \pi(12 \text{ in.})^2 \approx 452.39 \text{ in.}^3$; $V = Bh \Rightarrow h = \frac{V}{B} = \frac{8316}{452.39} \approx 18.4$
The can is about 18.4 inches tall.

53. $V = lwh = (48 \text{ in.})(15 \text{ in.})(24 \text{ in.}) = 17{,}280 \text{ in.}^3$; $\frac{17280}{231} \approx 74.8$ gallon capacity

55. $6 \cdot 3 = 18$ gallons of water per angel fish; $\frac{62}{18} \approx 3.4$; The aquarium will support 3 angel fish.

57. **Answers may vary.**

59. $V = lwh + Bh = (30 \text{ ft})(15 \text{ ft})(3 \text{ ft}) + \frac{1}{2}(20 \text{ ft})(7 \text{ ft}) \cdot 15 \text{ ft}$
$= 1350 \text{ ft}^3 + 1050 \text{ ft}^3 = 2400 \text{ ft}^3$; The pool holds 2400 ft^3 of water.

61. $9^3 = 9 \cdot 9 \cdot 9 = 729$

63. $2^3 + 4^3 + 5^3 = 8 + 64 + 125 = 197$

65. $(6^2 + 7^2)^2 = (36 + 49)^2 = 85^2 = 7225$

67. $(10^2)(2^2)(3^2) = (100)(4)(9) = 3600$

69. $A = (3 \text{ in.})(2 \text{ ft}) = (3 \text{ in.})(24 \text{ in.}) = 72 \text{ in.}^2$; $A = (3 \text{ in.})(2 \text{ ft}) = \left(\frac{1}{4} \text{ ft}\right)(2 \text{ ft}) = \frac{1}{2} \text{ ft}^2$

Exercises 7.6 (page 551)

1. $\sqrt{49} = 7$

3. $\sqrt{36} = 6$

5. $\sqrt{100} = 10$

7. $\sqrt{900} = 30$

9. $\sqrt{\frac{49}{100}} = \frac{7}{10}$

11. $\sqrt{0.81} = 0.9$

13. $\sqrt{175} \approx 13.23$

15. $\sqrt{821} \approx 28.65$

17. $\sqrt{13.46} \approx 3.67$

19. $\sqrt{\frac{14}{23}} \approx 0.78$

21. $\sqrt{11{,}500} \approx 107.24$

23. $c = \sqrt{6^2 + 8^2} = \sqrt{100} = 10$

25. $c = \sqrt{21^2 + 28^2} = \sqrt{1225} = 35$

27. $a = \sqrt{65^2 - 60^2} = \sqrt{625} = 25$

29. $b = \sqrt{15^2 - 9^2} = \sqrt{144} = 12$

31. $b = \sqrt{85^2 - 35^2} = \sqrt{6000} \approx 77.46$

33. $c = \sqrt{7^2 + 28^2} = \sqrt{833} \approx 28.86$

35. $b = \sqrt{60^2 - 15^2} = \sqrt{3375} \approx 58.09$

37. $a = \sqrt{58^2 - 40^2} = \sqrt{1764} = 42$

39. $c = \sqrt{21^2 + 72^2} = \sqrt{5625} = 75$
The hypotenuse is 75 cm.

41. $S = \sqrt{A} = \sqrt{144 \text{ ft}^2} = 12 \text{ ft}$
The side of the square is 12 ft.

43. $c = \sqrt{42^2 + 24^2} = \sqrt{2340} \approx 48.37$
The TV has a 48 in. screen.

45. $b = \sqrt{12^2 - 3^2} = \sqrt{135} \approx 11.62$
The top is about 11.62 ft high.

47. $c = \sqrt{35^2 + 16^2} = \sqrt{1481} \approx 38.48$
Lynzie is about 38.48 mi from the start.

49. $c = \sqrt{90^2 + 90^2} = \sqrt{16{,}200} \approx 127.28$
It is about 127.28 ft from 1st to 3rd.

51. $c = \sqrt{3^2 + 4^2} = \sqrt{25} = 5;\ 5 + 1 = 6;$
$2(6) = 12$; 12 ft of wire is needed.

53. **Answers may vary.**

55. The rod will fit in the 2nd, but not the 1st.

57. $48.3 + 562 + 0.931 = 611.231$

59. $80 = 16 \cdot 5 = 2^4 \cdot 5$

61. $9.03 \div 0.005 = 1806$

63. $72.5 \text{ m}^2 = 72.5(\text{m})(\text{m}) \cdot \dfrac{3.2808 \text{ ft}}{1 \text{ m}} \cdot \dfrac{3.2808 \text{ ft}}{1 \text{ m}} = 72.5(3.2808)(3.2808) \text{ ft}^2 \approx 780.36 \text{ ft}^2$

65. $V = \pi r^2 h = \pi(4 \text{ in.})^2(10 \text{ in.}) = \pi(16 \text{ in.}^2)(10 \text{ in.}) = 160\pi \text{ in.}^3 \approx 502.65 \text{ in.}^3$

Chapter 7 Review Exercises (page 559)

1. $23 \cdot 2 \text{ mm} = 46 \text{ mm}$

2. $330 \text{ m} \div 15 = 22 \text{ m}$

3. $\begin{array}{r} 5 \text{ yd} \\ -\quad 12 \text{ ft} \\ \hline \end{array} \Rightarrow \begin{array}{r} 1 \text{ yd } 12 \text{ ft} \\ -\quad 12 \text{ ft} \\ \hline 1 \text{ yd} \end{array}$

4. $5 \text{ ft} + 29 \text{ ft} + 8 \text{ ft} = 42 \text{ ft}$

5. $\begin{array}{r} 6 \text{ ft } 2 \text{ in.} \\ -\ 4 \text{ ft } 10 \text{ in.} \\ \hline \end{array} \Rightarrow \begin{array}{r} 5 \text{ ft } 14 \text{ in.} \\ -\ 4 \text{ ft } 10 \text{ in.} \\ \hline 1 \text{ ft } 4 \text{ in.} \end{array}$

6. $245 \text{ cm} = \dfrac{245 \text{ cm}}{1} \cdot \dfrac{1 \text{ m}}{100 \text{ cm}} = 2.45 \text{ m}$

7. $4 \text{ yd} = \dfrac{4 \text{ yd}}{1} \cdot \dfrac{36 \text{ in.}}{1 \text{ yd}} = 144 \text{ in.}$

8. $5 \text{ km} = \dfrac{5 \text{ km}}{1} \cdot \dfrac{1 \text{ mi}}{1.6093 \text{ km}} \approx 3.11 \text{ mi}$

9. $15 \text{ in.} = \dfrac{15 \text{ in.}}{1} \cdot \dfrac{2.54 \text{ cm}}{1 \text{ in.}} \cdot \dfrac{10 \text{ mm}}{1 \text{ cm}} = 381 \text{ mm}$

10. $2(3 \cdot 8 \text{ ft} + 6 \cdot 2 \text{ ft}) = 2(24 \text{ ft} + 12 \text{ ft}) = 2(36 \text{ ft}) = 72 \text{ ft}$; Ted needs 72 ft of tubing.

11. $6 \text{ qt} \cdot 6 = 36 \text{ qt}$

12. $(322 \text{ hr}) \div 14 = 23 \text{ hr}$

CHAPTER 7 REVIEW EXERCISES

13. 34 gal + 52 gal = 86 gal

14. 34 L − 8 L = 26 L

15.
$$\begin{array}{r} 6 \text{ hr } 42 \text{ min} \\ + \ 3 \text{ hr } 38 \text{ min} \\ \hline 9 \text{ hr } 80 \text{ min} \end{array} = 10 \text{ hr } 20 \text{ min}$$

16. $6 \text{ g} = \dfrac{6 \text{ g}}{1} \cdot \dfrac{1000 \text{ mg}}{1 \text{ g}} = 6000 \text{ mg}$

17. $14 \text{ lb} = \dfrac{14 \text{ lb}}{1} \cdot \dfrac{16 \text{ oz}}{1 \text{ lb}} = 224 \text{ oz}$

18. $4 \text{ L} = \dfrac{4 \text{ L}}{1} \cdot \dfrac{1.057 \text{ qt}}{1 \text{ L}} \approx 4.23 \text{ qt}$

19.
$$\begin{aligned} C &= \frac{5}{9}(F - 32) \\ &= \frac{5}{9}(95 - 32) = \frac{5}{9}(63) = 35° \text{ C} \end{aligned}$$

20. $5 \cdot 45 \text{ min} = 225 \text{ min}$
$4 \text{ hr} = 4 \cdot 60 \text{ min} = 240 \text{ min}$
She needs 15 more minutes.

21.
$$\begin{aligned} 2(34 \text{ cm}) + 2(15 \text{ cm}) &= 68 \text{ cm} + 30 \text{ cm} \\ &= 98 \text{ cm} \end{aligned}$$

22. 5 ft + 12 ft + 13 ft = 30 ft

23. 10 ft + 12 ft + 21 ft + 26 ft = 69 ft

24. 2(12 m) + 2(27 m) = 24 m + 54 m = 78 m

25. $\pi d = 3.14(13 \text{ yd}) = 40.82 \text{ yd}$

26. 6 mm + 18 mm + 3 mm + 14 mm + 12 mm + 14 mm + 3 mm + 18 mm = 88 mm

27. 2(12 m) + 2(15 m) = 24 m + 30 m = 54 m; The perimeter is 54 m.

28. $C = \pi d = 3.14(8 \text{ ft}) = 25.12 \text{ ft}$; The distance around is 25.12 ft.

29. 2(230 ft) + 2(275 ft) = 460 ft + 550 ft = 1010 ft; The city needs 1010 ft of fencing.

30. 2(9 ft) + 2(12 ft) − 4 ft = 18 ft + 24 ft − 4 ft = 38 ft; Larry needs 38 ft of baseboard.

31. $A = lw = (83 \text{ mm})(25 \text{ mm}) = 2075 \text{ mm}^2$

32.
$$\begin{aligned} A = \frac{1}{2}bh &= \frac{1}{2}(24 \text{ ft})(7 \text{ ft}) \\ &= \frac{1}{2}\left(168 \text{ ft}^2\right) = 84 \text{ ft}^2 \end{aligned}$$

33. $A = \frac{1}{2}(b_1 + b_2)h = \frac{1}{2}(24 \text{ cm} + 18 \text{ cm})(11 \text{ cm}) = \frac{1}{2}(42 \text{ cm})(11 \text{ cm}) = \frac{1}{2}(462 \text{ cm}^2) = 231 \text{ cm}^2$

34. $A = \frac{1}{2}(b_1 + b_2)h = \frac{1}{2}(30 \text{ in.} + 16 \text{ in.})(8 \text{ in.}) = \frac{1}{2}(46 \text{ in.})(8 \text{ in.}) = \frac{1}{2}(368 \text{ in.}^2) = 184 \text{ in.}^2$

35. $A = (8 \text{ yd})(6 \text{ yd}) + \dfrac{1}{2}\pi(3 \text{ yd})^2 = 48 \text{ yd}^2 + \dfrac{1}{2}(3.14)(9 \text{ yd})^2 = 48 \text{ yd}^2 + 14.13 \text{ yd}^2 = 62.13 \text{ yd}^2$

36. $A = (10 \text{ m})(30 \text{ m}) + (20 \text{ m})(25 \text{ m}) = 300 \text{ m}^2 + 500 \text{ m}^2 = 800 \text{ m}^2$

37. $A = S^2 = (600 \text{ mm})^2 = 360{,}000 \text{ mm}^2$
The area is 360,000 mm^2.

38. $A = (35 \text{ ft})(22 \text{ ft}) = 770 \text{ ft}^2$
The area is 770 ft^2.

CHAPTER 7 REVIEW EXERCISES

39. $A = 2(6 \text{ ft})(8 \text{ ft}) + 2(8 \text{ ft})(8 \text{ ft}) - (3 \text{ ft})(6 \text{ ft}) = 96 \text{ ft}^2 + 128 \text{ ft}^2 - 18 \text{ ft}^2 = 206 \text{ ft}^2$
2 coats would require 412 ft^2 of paint. One can is not enough.

40. Bottom: $A = (50 \text{ m})(30 \text{ m}) = 1500 \text{ m}^2$
Sides: $A = 2(50 \text{ m})(3 \text{ m}) + 2(30 \text{ m})(3 \text{ m}) = 300 \text{ m}^2 + 180 \text{ m}^2 = 480 \text{ m}^2$
Total: $A = 1500 \text{ m}^2 + 480 \text{ m}^2 = 1980 \text{ m}^2$

41. $V = lwh = (16 \text{ cm})(18 \text{ cm})(16 \text{ cm}) = 4608 \text{ cm}^3$

42. $V = Bh = (45 \text{ in.}^2)(3 \text{ in.}) = 135 \text{ in.}^3$

43. $V = Bh = (25 \text{ in.}^2)(4 \text{ in.}) = 100 \text{ in.}^3$

44. $V = lwh = (5 \text{ ft})(5 \text{ ft})(5 \text{ ft}) = 125 \text{ ft}^3$

45. $V = lwh = (8 \text{ m})(21 \text{ m})(2 \text{ m}) = 336 \text{ m}^3$

46. $V = (15 \text{ yd})(8 \text{ yd})(7 \text{ yd}) + (7 \text{ yd})(8 \text{ yd})(5 \text{ yd}) = 840 \text{ yd}^3 + 280 \text{ yd}^3 = 1120 \text{ yd}^3$

47. $V = \pi r^2 h = 3.14(22.5 \text{ ft})^2(30 \text{ ft}) = 3.14(506.25 \text{ ft}^2)(30 \text{ ft}) \approx 47{,}689 \text{ ft}^3$

48. $V = lwh = (10 \text{ ft})(8 \text{ ft})(5 \text{ ft}) = 400 \text{ ft}^3$; The volume is 400 ft^3.

49. $1 \text{ yd}^3 = 1 \text{ (yd)(yd)(yd)} \cdot \dfrac{36 \text{ in.}}{1 \text{ yd}} \cdot \dfrac{36 \text{ in.}}{1 \text{ yd}} \cdot \dfrac{36 \text{ in.}}{1 \text{ yd}} = 46{,}656 \text{ in.}^3$

50. $(20 \text{ ft})(6 \text{ ft})(3 \text{ in.}) = (240 \text{ in.})(72 \text{ in.})(3 \text{ in.}) = 51{,}840 \text{ in.}^3$
A cubic yard is 46,656 cubic inches, which will not be enough.

51. $\sqrt{529} = 23$

52. $\sqrt{60} \approx 7.75$

53. $\sqrt{\dfrac{81}{49}} = \dfrac{9}{7}$, or 1.29

54. $\sqrt{62.37} = 7.90$

55. $b = \sqrt{30^2 - 24^2} = \sqrt{324} = 18$

56. $c = \sqrt{10^2 + 24^2} = \sqrt{676} = 26$

57. $b = \sqrt{17^2 - 9^2} = \sqrt{208} \approx 14.42$

58. $c = \sqrt{38^2 + 42^2} = \sqrt{3208} \approx 56.64$

59. $b = \sqrt{18^2 - 3^2} = \sqrt{315} \approx 17.75$
The ladder is about 17.75 ft high.

60. $c = \sqrt{3^2 + 6^2} = \sqrt{45} \approx 6.71$
Each diagonal is about 6.71 ft long, so Mark will need 13.42 ft of stripping.

Chapter 7 True/False Concept Review (page 562)

1. true

2. false; Equivalent measurements have different units of measurement, such as 2 yd and 6 ft.

CHAPTER 7 TRUE/FALSE CONCEPT REVIEW

3. true

4. false; Perimeter is measured in units of length, such as meters or inches.

5. true

6. false; The area of a square is $A = s^2$.

7. false; The area of a circle is $A = \pi r^2$.

8. true

9. true

10. true

11. false; $3^2 = 9$, or $\sqrt{9} = 3$

12. true

13. true

14. false; 1 milliliter is equivalent to 1 cubic centimeter.

15. false; Measurements can be added or subtracted only if they are expressed with the same unit of measurement.

16. true

17. false; Volume can also be measured in cubic units of length, such as ft^3.

18. false; The sum of the squares of the lengths of the legs of a right triangle is equal to the square of the length of the hypotenuse.

19. false; $1 \text{ yd}^2 = 9 \text{ ft}^2$

20. true

Chapter 7 Test (page 563)

1. $8 \text{ m} + 329 \text{ mm} = 8000 \text{ mm} + 329 \text{ mm} = 8329 \text{ mm}$

2. $2(24 \text{ in.}) + 2(35 \text{ in.}) = 118 \text{ in.}$

3. $V = lwh = (6 \text{ in.})(20 \text{ in.})(12 \text{ in.}) = 1440 \text{ in.}^3$

4. $A = (17 \text{ ft})(8 \text{ ft}) + (6 \text{ ft})(8 \text{ ft}) = 184 \text{ ft}^2$; The room needs 184 ft^2 of tile.

5. $\sqrt{310} \approx 17.607$

6. $6 \text{ cm} + 8 \text{ cm} + 4 \text{ cm} + 3 \text{ cm} + 8 \text{ cm} = 29 \text{ cm}$

7. $A = \frac{1}{2}bh = \frac{1}{2}(16 \text{ in.})(5 \text{ in.}) = \frac{1}{2}(80 \text{ in.}^2) = 40 \text{ in.}^2$

8.
$$\begin{array}{r} 6 \text{ yd } 2 \text{ ft} \\ -\ 3 \text{ yd } 2 \text{ ft } 5 \text{ in.} \\ \hline \end{array} \Rightarrow \begin{array}{r} 6 \text{ yd } 1 \text{ ft } 12 \text{ in.} \\ -\ 3 \text{ yd } 2 \text{ ft } \ 5 \text{ in.} \\ \hline \end{array} \Rightarrow \begin{array}{r} 5 \text{ yd } 4 \text{ ft } 12 \text{ in.} \\ -\ 3 \text{ yd } 2 \text{ ft } \ 5 \text{ in.} \\ \hline 2 \text{ yd } 2 \text{ ft } \ 7 \text{ in.} \end{array}$$

9. In the English system, there are pounds, ounces, and tons. In the metric system, there are grams, milligrams, and kilograms.

10. $6 \text{ gal} = 6 \text{ gal} \cdot \frac{4 \text{ qt}}{1 \text{ gal}} \cdot \frac{2 \text{ pt}}{1 \text{ qt}} = 48 \text{ pt}$

11. Add 4 inches to each dimension, and convert to inches from feet:
$2(124) + 2(100) = 448$ in. (picture window); $2(52) + 2(100) = 304$ in. (side window);
$2(304 \text{ in.}) = 608$ in.; 448 in. + 608 in. = 1056 in. = 88 ft; The windows need 88 ft of molding.

12. $V = (12 \text{ in.})(5 \text{ in.})(4 \text{ in.}) + \frac{1}{2}(2 \text{ in.})(5 \text{ in.})(4 \text{ in.}) = 240 \text{ in.}^3 + 20 \text{ in.}^3 = 260 \text{ in.}^3$

13. $A = bh = (12 \text{ cm})(4 \text{ cm}) = 48 \text{ cm}^2$

14. $V = Bh = \left(4 \text{ ft}^2\right)(6 \text{ ft}) = 24 \text{ ft}^3$
The tank has a volume of 24 ft^3.

15. $192 \text{ oz} \div 24 = 8$ oz; Each student will receive 8 oz of juice.

16. $6(9 \text{ ft}) + 4(25 \text{ ft}) = 154$ ft;
$1.2(154) = 184.8$; It will cost \$184.80.

17. $b = \sqrt{10^2 - 8^2} = \sqrt{36} = 6$ cm

18. $A = (7 \text{ km})(11 \text{ km}) + \frac{1}{2}\pi(3.5 \text{ km})^2 = 77 \text{ km}^2 + \frac{1}{2}(3.14)(12.25 \text{ km})^2 \approx 96.2 \text{ km}^2$

19. $4 \cdot 4 = 16$ laps; $16(2 \cdot 100 + 2 \cdot 60) = 16(200 + 120) = 16(320) = 5120$; Each runs 5120 yards.

20. Area is a two-dimensional measure, while volume is a three-dimensional measure.

21. $340 + 495 + 432 + 510 + 670 = 2447$; $2447 \div 5 = 489.4$; The average weight is 489.4 pounds.

22. $C = \pi d = 3.14(90 \text{ in.}) = 282.6$ in.; 282.6 in. $\div$ 36 in. ≈ 7.85; Li needs 282.6 inches of lace, so she will need to buy 8 yards.

23. bag: $V = lwh = (52 \text{ in.})(16 \text{ in.})(5 \text{ in.}) = 4160 \text{ in.}^3$
box: $V = lwh = (24 \text{ in.})(18 \text{ in.})(12 \text{ in.}) = 5184 \text{ in.}^3$; All the food will fit in the box.

24. $c = \sqrt{3.5^2 + 1.2^2} = \sqrt{13.69} = 3.7$ mi; She is 3.7 miles from her starting point.

25. $2(12 \text{ in.})(14 \text{ in.}) + 2(12 \text{ in.})(4 \text{ in.}) + 2(14 \text{ in.})(4 \text{ in.}) = 336 \text{ in.}^2 + 96 \text{ in.}^2 + 112 \text{ in.}^2 = 544 \text{ in.}^2$
She needs at least 544 in.^2 of wrapping paper.

Exercises 8.1 (page 574)

1. opposite of $-4 = 4$

3. opposite of $5 = -5$

5. opposite of $-2.3 = 2.3$

7. opposite of $\frac{3}{5} = -\frac{3}{5}$

9. opposite of $-\frac{6}{7} = \frac{6}{7}$

11. -61

13. opposite of $-42 = 42$

15. opposite of $-3.78 = 3.78$

17. opposite of $\frac{17}{5} = -\frac{17}{5}$

19. opposite of $0.55 = -0.55$

21. opposite of $113.8 = -113.8$

SECTION 8.1

23. opposite of $-0.0123 = 0.0123$

25. $|-4| = 4$

27. $|24| = 24$

29. $|-3.17| = 3.17$

31. $\left|\frac{5}{11}\right| = \frac{5}{11}$

33. $\left|-\frac{3}{11}\right| = \frac{3}{11}$

35. 32 or -32

37. $|0.0065| = 0.0065$

39. $|-355| = 355$

41. $\left|-\frac{11}{3}\right| = \frac{11}{3}$

43. $\left|-\frac{33}{7}\right| = \frac{33}{7}$

45. $|0| = 0$

47. $|-0.341| = 0.341$

49. opposite of $\left|-\frac{5}{17}\right|$ = opposite of $\frac{5}{17} = -\frac{5}{17}$

51. opposite of $|78|$ = opposite of $78 = -78$

53. opposite of $1.54 = -1.54$
The opposite of a gain of 1.54 is -1.54.

55. opposite of $14°C = -14°C$
The opposite of a reading of 14°C is $-14°C$.

57. opposite of $-1915 = 1915$
The opposite of 1915 B.C. is A.D. 1915.

59. The distance up the mountain is $+1415$, while the distance down the mountain is -1143.

61. opposite of $+80 = -80$; Eighty miles south is represented by -80 miles.

63.

Ocean	Deepest Part	(ft)	Average Depth (ft)
Pacific	Mariana Trench	−35,840	−12,925
Atlantic	Puerto Rico Trench	−28,232	−11,730
Indian	Java Trench	−23,376	−12,598
Arctic	Eurasia Basin	−17,881	−3407
Mediterranean	Ionian Basin	−16,896	−4926

65. $-(-24) = 24$

67. $-(-(-14)) = -(-14) = -14$

69. On the first play, the team made -8 yd. The opposite of the loss is $+8$ yd.

71. The distances driven can be represented as $+137$ mi, $+212$ mi, -98 mi, and -38 mi.

73. **Answers may vary.**

75. $|16 - 10| - |14 - 9| + 6 = |6| - |5| + 6$
$= 6 - 5 + 6 = 7$

77. If n is negative, then $-n$ is positive.

79. $7 + 11 + 32 + 9 = 59$

81. $17 - 12 = 5$

83.
$$\begin{array}{r} 6 \text{ m } 250 \text{ cm} \\ + \quad 7 \text{ m } 460 \text{ cm} \\ \hline 13 \text{ m } 710 \text{ cm} \approx 20.1 \text{ m} \end{array}$$

85.
$$\begin{array}{r} 3 \text{ yd } 2 \text{ ft } \ \ 8 \text{ in.} \\ - \ 1 \text{ yd } 2 \text{ ft } 11 \text{ in.} \\ \hline \end{array} \Rightarrow \begin{array}{r} 3 \text{ yd } 1 \text{ ft } 20 \text{ in.} \\ - \ 1 \text{ yd } 2 \text{ ft } 11 \text{ in.} \\ \hline \end{array} \Rightarrow \begin{array}{r} 2 \text{ yd } 4 \text{ ft } 20 \text{ in.} \\ - \ 1 \text{ yd } 2 \text{ ft } 11 \text{ in.} \\ \hline 1 \text{ yd } 2 \text{ ft } \ \ 9 \text{ in.} = 69 \text{ in.} \end{array}$$

87. $12 \text{ m} \div 8 = 1.5 \text{ m}$; Each part contains 1.5 m of wire.

Exercises 8.2 (page 582)

1. $-7 + 9 = +(9 - 7) = 2$

3. $8 + (-6) = +(8 - 6) = 2$

5. $-9 + (-6) = -(9 + 6) = -15$

7. $-11 + (-3) = -(11 + 3) = -14$

9. $-11 + 11 = 0$

11. $0 + (-12) = -12$

13. $-23 + (-23) = -(23 + 23) = -46$

15. $3 + (-17) = -(17 - 3) = -14$

17. $-19 + (-7) = -(19 + 7) = -26$

19. $24 + (-17) = +(24 - 17) = 7$

21. $-5 + (-7) + 6 = -(5 + 7) + 6 = -12 + 6 = -(12 - 6) = -6$

23. $-65 + (-43) = -(65 + 43) = -108$

25. $-98 + 98 = 0$

27. $-45 + 72 = +(72 - 45) = 27$

29. $-36 + 43 + (-17) = +(43 - 36) + (-17) = 7 + (-17) = -(17 - 7) = -10$

31. $62 + (-56) + (-13) = +(62 - 56) + (-13) = 6 + (-13) = -(13 - 6) = -7$

33. $-4.6 + (-3.7) = -(4.6 + 3.7) = -8.3$

35. $10.6 + (-7.8) = +(10.6 - 7.8) = 2.8$

37. $-\frac{5}{6} + \frac{1}{4} = -\frac{10}{12} + \frac{3}{12} = -\left(\frac{10}{12} - \frac{3}{12}\right) = -\frac{7}{12}$

39. $-\frac{7}{10} + \left(-\frac{7}{15}\right) = -\frac{21}{30} + \left(-\frac{14}{30}\right) = -\left(\frac{21}{30} + \frac{14}{30}\right) = -\frac{35}{30} = -\frac{7}{6}$

41. $135 + (-256) = -(256 - 135) = -121$

43. $-81 + (-32) + (-76) = -(81 + 32) + (-76) = -113 + (-76) = -(113 + 76) = -189$

45.
$$\begin{aligned} -31 + 28 + (-63) + 36 = -(31 - 28) + (-63) + 36 &= -3 + (-63) + 36 \\ &= -(3 + 63) + 36 \\ &= -66 + 36 = -(66 - 36) = -30 \end{aligned}$$

47.
$$\begin{aligned} 49 + (-67) + 27 + 72 = -(67 - 49) + 27 + 72 = -18 + 27 + 72 &= +(27 - 18) + 72 \\ &= 9 + 72 = 81 \end{aligned}$$

SECTION 8.2

49. $356 + (-762) + (-892) + 541 = -(762 - 356) + (-892) + 541 = -406 + (-892) + 541$
$= -(406 + 892) + 541$
$= -1298 + 541$
$= -(1298 - 541) = -757$

51. $542 + (-481) + (-175) = +(542 - 481) + (-175) = 61 + (-175) = -(175 - 61) = -114$

53. $-57 + (-49) + (-86) + (-93) + (-64) = -(57 + 49) + (-86) + (-93) + (-64)$
$= -106 + (-86) + (-93) + (-64)$
$= -(106 + 86) + (-93) + (-64)$
$= -192 + (-93) + (-64)$
$= -(192 + 93) + (-64)$
$= -285 + (-64) = -(285 + 64) = -349$

The sum of the temperatures at 9:00 a.m. is $-349°$ C.

55. $-107 + (-105) + (-105) + (-98) + (-90) = -(107 + 105) + (-105) + (-98) + (-90)$
$= -212 + (-105) + (-98) + (-90)$
$= -(212 + 105) + (-98) + (-90)$
$= -317 + (-98) + (-90)$
$= -(317 + 98) + (-90)$
$= -415 + (-90) = -(415 + 90) = -505$

The sum of the temperatures at 11:00 p.m. is $-505°$ C.

57. $-963 + 855 = -(963 - 855) = -108$; The net change is -108 lb.

59. $30{,}000 + 220 + (-200) + 55 + (-110) + (-55) + (-40)$
$= 30{,}220 + (-200) + 55 + (-110) + (-55) + (-40)$
$= +(30{,}220 - 200) + 55 + (-110) + (-55) + (-40)$
$= 30{,}020 + 55 + (-110) + (-55) + (-40)$
$= 30{,}075 + (-110) + (-55) + (-40)$
$= +(30{,}075 - 110) + (-55) + (-40)$
$= 29{,}965 + (-55) + (-40)$
$= +(29{,}965 - 55) + (-40) = 29{,}910 + (-40) = +(29{,}910 - 40) = 29{,}870$

The altitude at 4:00 p.m. is 29,870 ft.

61. $34{,}945 + 3456 + (-2024) + (-3854) + 612 + (-2765)$
$= 38{,}401 + (-2024) + (-3854) + 612 + (-2765)$
$= +(38{,}401 - 2024) + (-3854) + 612 + (-2765)$
$= 36{,}377 + (-3854) + 612 + (-2765)$
$= +(36{,}377 - 3854) + 612 + (-2765)$
$= 32{,}523 + 612 + (-2765) = 33{,}135 + (-2765) = +(33{,}135 - 2765) = 30{,}370$

The inventory is 30,370 books.

63. $-8 + 9 + 7 = +(9 - 8) + 7 = 1 + 7 = 8$; The Patriots were two yards short of a first down.

65. **a.**
$$\begin{aligned}0.62+(-0.44)+(-1.12)+0.49+(-0.56) &= +(0.62-0.44)+(-1.12)+0.49+(-0.56)\\ &= 0.18+(-1.12)+0.49+(-0.56)\\ &= -(1.12-0.18)+0.49+(-0.56)\\ &= -0.94+0.49+(-0.56)\\ &= -(0.94-0.49)+(-0.56)\\ &= -0.45+(-0.56) = -(0.45+0.56) = -1.01\end{aligned}$$

The net change was -1.01, or down 1.01.

b. $34.02+(-1.01) = +(34.02-1.01) = 33.01$; The closing price was \$33.01.

67.
$$\begin{aligned}-2376+5230+(-1055)+3278 &= +(5230-2376)+(-1055)+3278\\ &= 2854+(-1055)+3278\\ &= +(2854-1055)+3278 = 1799+3278 = 5077\end{aligned}$$

The net gain is \$5077.

69. **Answers may vary.**

71. $[(-73)+(-32)]+(-19) = [-(73+32)]+(-19) = -105+(-19) = -(105+19) = -124$

73. 32

75. $145-67=78$

77. $178-95=83$

79.
$$\begin{aligned}2(42\text{ m})+2(29\text{ m}) &= 84\text{ m}+58\text{ m}\\ &= 142\text{ m}\end{aligned}$$

The perimeter is 142 m.

81. $4(31\text{ in.}) = 124\text{ in.}$
The perimeter is 124 in.

83.
$$\begin{aligned}\frac{3\text{ lb}}{\$2.16} &= \frac{1\text{ lb}}{x}\\ 3x &= 2.16(1)\\ 3x &= 2.16\\ \frac{3x}{3} &= \frac{2.16}{3}\\ x &= 0.72\end{aligned}$$
$\Rightarrow$ One pound of bananas costs \$0.72.

Exercises 8.3 (page 588)

1. $7-3=7+(-3)=4$

3. $-7-5=-7+(-5)=-12$

5. $-8-(-3)=-8+3=-5$

7. $11-8=11+(-8)=3$

9. $-12-8=-12+(-8)=-20$

11. $-23-(-14)=-23+14=-9$

13. $19-(-13)=19+13=32$

15. $-23-11=-23+(-11)=-34$

17. $-15-13=-15+(-13)=-28$

19. $-13-(-14)=-13+14=1$

21. $-16-(-16)=-16+16=0$

23. $-9-9=-9+(-9)=-18$

SECTION 8.3

25. $-40 - 40 = -40 + (-40) = -80$

27. $72 - (-46) = 72 + 46 = 118$

29. $-57 - 62 = -57 + (-62) = -119$

31. $-48 - (-59) = -48 + 59 = 11$

33. $-91 - 91 = -91 + (-91) = -182$

35. $102 - (-102) = 102 + 102 = 204$

37. $-69 - (-69) = -69 + 69 = 0$

39. $134 - (-10) = 134 + 10 = 144$

41. $132 - (-41) = 132 + 41 = 173$

43. $-6.74 - 3.24 = -6.74 + (-3.24) = -9.98$

45. $-4.65 - (-3.21) = -4.65 + 3.21 = -1.44$

47. $-23.43 - 32.71 = -23.43 + (-32.71) = -56.14$

49. $43 - (-73) = 43 + 73 = 116$

51. $-359 - 328 = -359 + (-328) = -687$

53. $-24 - (-109) = -24 + 109 = 85$
The temperature difference was 85°C.

55. $782.45 - (-13.87) = 782.45 + 13.87 = 796.32$
Joe wrote checks totaling $796.32.

57.

Africa	$19{,}340 - (-512) = 19{,}340 + 512 = 19{,}852$
Antarctica	$16{,}864 - (-8327) = 16{,}864 + 8327 = 25{,}191$
Asia	$29{,}028 - (-1312) = 29{,}028 + 1312 = 30{,}340$
Australia	$7310 - (-52) = 7310 + 52 = 7362$
Europe	$18{,}510 - (-92) = 18{,}510 + 92 = 18{,}602$
North America	$20{,}320 - (-282) = 20{,}320 + 282 = 20{,}602$
South America	$22{,}834 - (-131) = 22{,}834 + 131 = 22{,}965$

The smallest range of elevation occurs in Australia, with 7362 ft. This indicates that Australia is more flat than the other continents.

59. $13{,}796 - 33{,}476 = 13{,}796 + (-33{,}476) = -19{,}680$; The depth is 19,680 ft.

61. $-105 - (-90) = -105 + 90 = -15$
The difference is $-15°$ C.

63. $-10 - (-115) = -10 + 115 = 105$
The greatest difference is 105° C.

65. $210.34 - 216.75 = 210.34 + (-216.75) = -6.41$
His balance is –$6.41 (overdrawn).

67. $467.82 - (-9.32) = 467.82 + 9.32 = 477.14$
Janna wrote checks totaling $477.14.

69. $-11.91 - (-5.67) = -11.91 + 5.67 = -6.24$
The US owed $6.24 trillion more, so the debt increased by $6.24 trillion.

71. $-8 - (-2) = -8 + 2 = -6$; The difference in scores is -6.

73. **Answers may vary.**

75. **Answers may vary.**

77. $-9.46 - [-(-3.22)] = -9.46 - [+3.22] = -9.46 - 3.22 = -9.46 + (-3.22) = -12.68$

79. $57 - |-67 - (-51)| - 82 = 57 - |-67 + 51| - 82 = 57 - |-16| - 82 = 57 - 16 - 82 = -41$

81.
$$\begin{array}{cccc} & & {\scriptstyle 4} & \\ & & 1 & 6 \\ \times & & & 7 \\ \hline & 1 & 1 & 2 \end{array}$$

83.
$$\begin{array}{ccccc} & & {\scriptstyle 1} & {\scriptstyle 1} & \\ & & {\scriptstyle \not{3}} & {\scriptstyle \not{4}} & \\ & & 1 & 5 & 6 \\ \times & & & 3 & 7 \\ \hline & 1 & 0 & 9 & 2 \\ & 4 & 6 & 8 & 0 \\ \hline & 5 & 7 & 7 & 2 \end{array}$$

85. $A = \pi r^2 = (3.14)(14 \text{ in.})^2$
$= 3.14(196 \text{ in.}^2) = 615.44 \text{ in.}^2$
The area is 615.44 in.^{2}

87. $A = lw = (13.6 \text{ cm})(9.4 \text{ cm})$
$= 127.84 \text{ cm}^2$
The area is 127.84 cm^2.

89. $A = (22.5 \text{ ft})(30.5 \text{ ft}) = 686.25 \text{ ft}^2$; $686.25 \div 9 = 76.25$; It requires 76.25 yd^2.

Exercises 8.4 (page 595)

1. $-2(4) = -8$

3. $-5(2) = -10$

5. $10(-8) = -80$

7. $-11(7) = -77$

9. 11

11. $-9(34) = -306$

13. $-17(15) = -255$

15. $2.5(-3.6) = -9$

17. $-0.35(1000) = -350$

19. $-\frac{2}{3} \cdot \frac{3}{8} = -\frac{1}{4}$

21. $(-1)(-3) = 3$

23. $-7(-4) = 28$

25. $-11(-9) = 99$

27. $-12(-5) = 60$

29. -11

31. $-14(-15) = 210$

33. $-1.2(-4.5) = 5.4$

35. $(-5.5)(-4.4) = 24.2$

37. $\left(-\frac{3}{14}\right)\left(-\frac{7}{9}\right) = \frac{1}{6}$

39. $(-0.35)(-4.7) = 1.645$

41. $(-56)(45) = -2520$

43. $(15)(31) = 465$

45. $(-1.4)(-5.1) = 7.14$

47. $\left(-\frac{9}{16}\right)\left(\frac{8}{15}\right) = -\frac{3}{10}$

49. $(-4.01)(3.5) = -14.035$

51. $-3.19(-1.7)(0.1) = 5.423(0.1) = 0.5423$

53. $-2(4)(-1)(0)(-5) = 0$

55. $-0.07(0.3)(-10)(100) = -0.021(-10)(100) = 0.21(100) = 21$

57. $-2(-5)(-6)(-4)(-1) = 10(-6)(-4)(-1) = -60(-4)(-1) = 240(-1) = -240$

SECTION 8.4

59. $\left(-\frac{2}{3}\right)\left(-\frac{3}{4}\right)\left(-\frac{4}{5}\right)\left(-\frac{5}{6}\right) = \frac{2}{4}\left(-\frac{4}{5}\right)\left(-\frac{5}{6}\right) = -\frac{2}{5}\left(-\frac{5}{6}\right) = \frac{2}{6} = \frac{1}{3}$

61. $C = \frac{5}{9}(F - 32) = \frac{5}{9}(5 - 32) = \frac{5}{9}(-27) = -15°$ C; The Celsius temperature is $-15°$C.

63. $8(-2.6) = -20.8$; Ms. Riles' loss is -20.8 lb.

65. $12(-1.74) = -20.88$; The Dow Jones loss is -20.88 points.

67. $(15 - 8)(5 - 12) = 7(-7) = -49$

69. $(15 - 21)(13 - 6) = -6(7) = -42$

71. $(-12 + 30)(-4 - 10) = 18(-14) = -252$

73. $386(-17) = -6562$; Safeway's loss is -6562¢, or $-\$65.62$.

75. $523(-6) = -3138$; Winn Dixie's loss is -3138¢, or $-\$31.38$.

77. $C = \frac{5}{9}(F - 32) = \frac{5}{9}(-128.6 - 32) = \frac{5}{9}(-160.6) = \frac{5(-160.6)}{9} \approx -89°$ C

79. $670(9.34) = 6257.8$; $670(6.45) = 4321.5$; $4321.5 - 6257.8 = -1936.3$; The trader paid \$6257.80 for the stock and sold the stock for \$4321.50. The loss was $-\$1936.30$.

81. Asia

83. **Answers may vary.**

85. $|-(-5)|(-9 - [-(-5)]) = |5|(-9 - [5]) = 5(-14) = -70$

87. opposite of 7 $\Rightarrow -7$; -8(opposite of 7) $= -8(-7) = 56$

89.

$$\begin{array}{r} 6 \\ 11\overline{)66} \\ \underline{66} \\ 0 \end{array}$$

91.

$$\begin{array}{r} 67 \\ 15\overline{)1005} \\ \underline{90} \\ 105 \\ \underline{105} \\ 0 \end{array}$$

93.

$$\begin{array}{l} 143.411 \approx 143.41 \\ 34\overline{)4876.000} \\ \underline{34} \\ 147 \\ \underline{136} \\ 116 \\ \underline{102} \\ 140 \\ \underline{136} \\ 40 \\ \underline{34} \\ 60 \\ \underline{34} \end{array}$$

95. $(24 \text{ ft})(27 \text{ ft}) = 648 \text{ ft}^2$; $648 \div 9 = 72 \text{ yd}^2$
$72(34.75) = 2502$; It will cost \$2502.

97. Bottom: $(5.5)(4.5) = 24.75 \text{ ft}^2$
2 Sides: $2(5.5)(0.75) = 8.25 \text{ ft}^2$
2 Sides: $2(4.5)(0.75) = 6.75 \text{ ft}^2$
Total: $24.75 + 8.25 + 6.75 = 39.75$
It takes 39.75 ft^2 of sheet metal.

Exercises 8.5 (page 600)

1. $-10 \div 5 = -2$

3. $-16 \div 4 = -4$

5. $18 \div (-6) = -3$

7. $24 \div (-3) = -8$

9. 8

11. $72 \div (-12) = -6$

13. $6.06 \div (-3) = -2.02$

15. $-210 \div 6 = -35$

17. $\left(-\frac{6}{7}\right) \div \frac{2}{7} = -\frac{6}{7} \cdot \frac{7}{2}$
$= -3$

19. $0.75 \div (-0.625) = -1.2$

21. $-10 \div (-5) = 2$

23. $-12 \div (-4) = 3$

25. $-28 \div (-4) = 7$

27. $-54 \div (-9) = 6$

29. -5

31. $-98 \div (-14) = 7$

33. $-96 \div (-12) = 8$

35. $-12.12 \div (-3) = 4.04$

37. $\left(-\frac{3}{8}\right) \div \left(-\frac{3}{4}\right) = -\frac{3}{8}\left(-\frac{4}{3}\right) = \frac{1}{2}$

39. $-0.65 \div (-0.13) = 5$

41. $-540 \div 12 = -45$

43. $-3364 \div (-29) = 116$

45. $3.735 \div (-0.83) = -4.5$

47. $0 \div (-35) = 0$

49. $-0.26 \div 100 = -0.0026$

51. $\frac{-16{,}272}{36} = -452$

53. $-384 \div (-24) = 16$

55. $-753.90 \div 6 = -125.65$; Each member's share of the loss is $-\$125.65$.

57. $-115 \div 25 = -4.6$; The average weight loss is -4.6 lb per week.

59. $-45.84 \div 12 = -3.82$; The average daily loss is -3.82 points.

61. **Answers may vary.**

63. Australia

65. $-814.29 \div 16 \approx 50.893$; The monthly loss was about $-\$50.89$.

67. **Answers may vary.**

69. $[-|-10|(6-11)] \div [(8-13)(11-10)] = [-(10)(-5)] \div [(-5)(1)] = 50 \div (-5) = -10$

71. $\left(-\frac{5}{6}-\frac{1}{2}\right)\left(-\frac{2}{3}+\frac{1}{6}\right)\div\left(\frac{1}{3}-\frac{3}{4}\right)=\left(-\frac{5}{6}-\frac{3}{6}\right)\left(-\frac{4}{6}+\frac{1}{6}\right)\div\left(\frac{4}{12}-\frac{9}{12}\right)$

$=\left(-\frac{8}{6}\right)\left(-\frac{3}{6}\right)\div\left(-\frac{5}{12}\right)$

$=\frac{24}{36}\div\left(-\frac{5}{12}\right)=\frac{2}{3}\cdot\left(-\frac{12}{5}\right)=-\frac{8}{5}$, or $-1\frac{3}{5}$

73. $(-0.82-1.28)(1.84-2.12)\div[3.14+(-3.56)]=(-2.1)(-0.28)\div[-0.42]$
$=0.588\div(-0.42)=-1.4$

75. $75\div 3\cdot 5+3-(5-2)=25\cdot 5+3-3=125+3-3=125$

77. $(17-3\cdot 5)^3+[16-(19-2\cdot 8)]=(17-15)^3+[16-(19-16)]$
$=2^3+[16-3]=8+13=21$

79. $V=\frac{1}{3}\pi r^2h=\frac{1}{3}(3.14)(12\text{ in.})^2(9\text{ in.})=1356.48\text{ in.}^3$; The volume is 1356.48 in.^3

81. $V=(27\text{ ft})(16\text{ ft})(6\text{ ft})=2592\text{ ft}^3$; $2592\div 27=96$; The amount of dirt removed is 96 yd^3.

83. $(88.75\text{ ft})(180\text{ ft})=15{,}975\text{ ft}^2$; $15{,}975\cdot 2=31{,}950$; $31{,}950(0.08)=2556$; The broker made $2556.

Exercises 8.6 (page 605)

1. $2(-7)-10=-14-10=-24$

3. $(-2)(-4)+11=8+11=19$

5. $3(-6)+12=-18+12=-6$

7. $-7+3(-3)=-7+(-9)=-16$

9. $-3(8)\div 4=-24\div 4=-6$

11. $(-4)8\div(-4)=-32\div(-4)=8$

13. $(-8)\div 4(-2)=(-2)(-2)=4$

15. $(-3)^2+(-2)^2=9+4=13$

17. $(11-3)+(9-6)=8+3=11$

19. $(3-5)(6-10)=(-2)(-4)=8$

21. $(-3)^2+4(-2)=9+(-8)=1$

23. $-5+(6-8)-5(3)=-5+(-2)-15$
$=-22$

25. $(-13)(-2)+(-16)2=26+(-32)=-6$

27. $(9-7)(-2-5)+(15-9)(2+7)=(2)(-7)+(6)(9)=-14+54=40$

29. $7(-11+5)-44\div(-11)=7(-6)-(-4)=-42+4=-38$

31. $18(-2)\div(-6)-14=-36\div(-6)-14=6-14=-8$

33. $-120 \div (-20) - (9 - 11) = 6 - (-2) = 8$

35. $-2^3 - (-2)^3 = -1 \cdot 2^3 - (-2)^3 = -1 \cdot 8 - (-8) = -8 + 8 = 0$

37. $-35 \div 7(-5) - 7^2 = (-5)(-5) - 49$
$= 25 - 49 = -24$

39. $2^2(5 - 4)(7 - 3)^2 = 4(1)(4)^2$
$= 4(1)(16) = 64$

41. $(9 - 13) + (-5)(-2) - (-2)5 - 3^3 = (-4) + (10) - (-10) - 27 = -4 + 10 + 10 - 27 = -11$

43. $(-3)(-2)(-3) - (-4)(-3) - (3)(-5) = (-18) - (12) - (-15) = -18 - 12 + 15 = -15$

45. $(-1)(-6)^2(-2) - (-3)^2(-2)^3 = (-1)(36)(-2) - (9)(-8) = (72) - (-72) = 144$

47. $12(-4) + (-3)(-12) = -48 + 36 = -12$

49. $-74 + (-64) + (-10) + (-42) + (-90) = -280$; $-280 \div 5 = -56$; The average was -56°C.

51. $-45 + (-33) + (-46) + (-36) + (-10) = -170$; $-170 \div 5 = -34$; The average high was -34°C.

53. $[-3 + (-6)]^2 - [-8 - 2(-3)]^2 = [-9]^2 - [-8 + 6]^2 = 81 - [-2]^2 = 81 - 4 = 77$

55. $\left[46 - 3(-4)^2\right]^3 - \left[-7(1)^3 + (-5)(-8)\right] = [46 - 3(16)]^3 - [-7(1) + 40]$
$= [46 - 48]^3 - [-7 + 40]$
$= [-2]^3 - [33] = -8 - 33 = -41$

57. $-15 - \dfrac{8^2 - (-4)}{3^2 + 3} = -15 - \dfrac{64 + 4}{9 + 3} = -15 - \dfrac{68}{12} = -\dfrac{45}{3} - \dfrac{17}{3} = -\dfrac{62}{3}$

59. $\dfrac{12(8 - 24)}{5^2 - 3^2} \div (-12) = \dfrac{12(-16)}{25 - 9} \div (-12) = \dfrac{-192}{16} \div (-12) = -12 \div (-12) = 1$

61. $-8|125 - 321| - 21^2 + 8(-7) = -8|-196| - 441 - 56 = -8(196) - 441 - 56$
$= -1568 - 441 - 56 = -2065$

63. $-6\left(8^2 - 9^2\right)^2 - (-7)20 = -6(64 - 81)^2 - (-140) = -6(-17)^2 + 140$
$= -6(289) + 140 = -1734 + 140 = -1594$

65. $[28 \div (-7)] - [-4(-3)] = -4 - 12 = -16$

67. $102(175) - 27{,}438 = 17{,}850 - 27{,}438 = -9588$; $-9588 \div 102 = -94$; His total profit was $-\$9588$, for a loss of $-\$94$ per chair.

69. $25(-3.50) + 9(7.25) = -87.50 + 65.25 = -22.25$; Kmart had a total loss of $-\$22.25$.

71.

Africa	$19{,}340 + (-512) = 18{,}828$; $18{,}828 \div 2 = 9414$
Antarctica	$16{,}864 + (-8327) = 8537$; $8537 \div 2 = 4268.5$
Asia	$29{,}028 + (-1312) = 27{,}716$; $27{,}716 \div 2 = 13{,}858$
Australia	$7310 + (-52) = 7258$; $7258 \div 2 = 3629$
Europe	$18{,}510 + (-92) = 18{,}418$; $18{,}418 \div 2 = 9209$
North America	$20{,}320 + (-282) = 20{,}038$; $20{,}038 \div 2 = 10{,}019$
South America	$22{,}834 + (-131) = 22{,}703$; $22{,}703 \div 2 = 11{,}351.5$

The largest average is for Asia. The smallest average is for Australia.

73. Three cities are Athens, Bangkok, and New Delhi.

75. $4756 + 345 - 212 - 1218 - 15 + 98 - 450 - 78 = 3226$; The balance is \$3226.

77. $-14 + (-19) + (-12) + (-20) + (-13) + (-12) + (-5) = -95$; $-95 \div 7 \approx -13.57$
Tiger Woods' average winning score was about -13.6.

79. **Answers may vary.**

81. **Answers may vary.**

83. $$\frac{(5-9)^2 + (-6+8)^2 - (14-6)^2}{[3-4(7)+3^3]^2} = \frac{(-4)^2 + (2)^2 - (8)^2}{[3-28+27]^2} = \frac{16+4-64}{[2]^2} = \frac{-44}{4} = -11$$

85. $(-17.2) + (-18.6) + (-2.7) + 9.1 = -29.4$

87. $48 - (-136) = 48 + 136 = 184$

89. $(-36)(84)(-21) = (-3024)(-21) = 63{,}504$

91. $(-800) \div (-32) = 25$

93. $-5832 \div 4 = -1458$; Each loses \$1458.

Exercises 8.7 (page 611)

1.
$$\begin{aligned} -3x + 25 &= 4 \\ -3x + 25 - 25 &= 4 - 25 \\ -3x &= -21 \\ \frac{-3x}{-3} &= \frac{-21}{-3} \\ x &= 7 \end{aligned}$$

3.
$$\begin{aligned} -6 + 3x &= 9 \\ -6 + 6 + 3x &= 9 + 6 \\ 3x &= 15 \\ \frac{3x}{3} &= \frac{15}{3} \\ x &= 5 \end{aligned}$$

5.
$$\begin{aligned} 4y - 9 &= -29 \\ 4y - 9 + 9 &= -29 + 9 \\ 4y &= -20 \\ \frac{4y}{4} &= \frac{-20}{4} \\ y &= -5 \end{aligned}$$

7.
$$\begin{aligned} 2a - 11 &= 3 \\ 2a - 11 + 11 &= 3 + 11 \\ 2a &= 14 \\ \frac{2a}{2} &= \frac{14}{2} \\ a &= 7 \end{aligned}$$

SECTION 8.7

9.
$$\begin{aligned} -5x + 12 &= -23 \\ -5x + 12 - 12 &= -23 - 12 \\ -5x &= -35 \\ \frac{-5x}{-5} &= \frac{-35}{-5} \\ x &= 7 \end{aligned}$$

11.
$$\begin{aligned} 4x - 12 &= 28 \\ 4x - 12 + 12 &= 28 + 12 \\ 4x &= 40 \\ \frac{4x}{4} &= \frac{40}{4} \\ x &= 10 \end{aligned}$$

13.
$$\begin{aligned} -14 &= 2x - 8 \\ -14 + 8 &= 2x - 8 + 8 \\ -6 &= 2x \\ \frac{-6}{2} &= \frac{2x}{2} \\ -3 &= x \end{aligned}$$

15.
$$\begin{aligned} -40 &= 5x - 10 \\ -40 + 10 &= 5x - 10 + 10 \\ -30 &= 5x \\ \frac{-30}{5} &= \frac{5x}{5} \\ -6 &= x \end{aligned}$$

17.
$$\begin{aligned} -6 &= -8x - 6 \\ -6 + 6 &= -8x - 6 + 6 \\ 0 &= -8x \\ \frac{0}{-8} &= \frac{-8x}{-8} \\ 0 &= x \end{aligned}$$

19.
$$\begin{aligned} -10 &= -4x + 2 \\ -10 - 2 &= -4x + 2 - 2 \\ -12 &= -4x \\ \frac{-12}{-4} &= \frac{-4x}{-4} \\ 3 &= x \end{aligned}$$

21.
$$\begin{aligned} -14y - 1 &= -99 \\ -14y - 1 + 1 &= -99 + 1 \\ -14y &= -98 \\ \frac{-14y}{-14} &= \frac{-98}{-14} \\ y &= 7 \end{aligned}$$

23.
$$\begin{aligned} -3 &= -8a - 3 \\ -3 + 3 &= -8a - 3 + 3 \\ 0 &= -8a \\ \frac{0}{-8} &= \frac{-8a}{-8} \\ 0 &= a \end{aligned}$$

25.
$$\begin{aligned} -0.6x - 0.15 &= 0.15 \\ -0.6x - 0.15 + 0.15 &= 0.15 + 0.15 \\ -0.6x &= 0.3 \\ \frac{-0.6x}{-0.6} &= \frac{0.3}{-0.6} \\ x &= -0.5 \end{aligned}$$

27.
$$\begin{aligned} 0.03x + 2.3 &= 1.55 \\ 0.03x + 2.3 - 2.3 &= 1.55 - 2.3 \\ 0.03x &= -0.75 \\ \frac{0.03x}{0.03} &= \frac{-0.75}{0.03} \\ x &= -25 \end{aligned}$$

29.
$$\begin{aligned} -135x - 674 &= 1486 \\ -135x - 674 + 674 &= 1486 + 674 \\ -135x &= 2160 \\ \frac{-135x}{-135} &= \frac{2160}{-135} \\ x &= -16 \end{aligned}$$

31.
$$\begin{aligned} -102y + 6 &= 414 \\ -102y + 6 - 6 &= 414 - 6 \\ -102y &= 408 \\ \frac{-102y}{-102} &= \frac{408}{-102} \\ y &= -4 \end{aligned}$$

SECTION 8.7

33.
$$\begin{aligned} \frac{1}{2}a - \frac{3}{8} &= \frac{1}{40} \\ \frac{1}{2}a - \frac{3}{8} + \frac{3}{8} &= \frac{1}{40} + \frac{3}{8} \\ \frac{1}{2}a &= \frac{1}{40} + \frac{15}{40} \\ \frac{1}{2}a &= \frac{16}{40} \\ \frac{1}{2}a &= \frac{2}{5} \\ 2 \cdot \frac{1}{2}a &= 2 \cdot \frac{2}{5} \\ a &= \frac{4}{5} \end{aligned}$$

35. Let $x =$ the number.
$$\begin{aligned} 98 + 6x &= 266 \\ 98 - 98 + 6x &= 266 - 98 \\ 6x &= 168 \\ \frac{6x}{6} &= \frac{168}{6} \\ x &= 28 \end{aligned}$$
The number is 28.

37. Let $x =$ the number.
$$\begin{aligned} 15x - 181 &= -61 \\ 15x - 181 + 181 &= -61 + 181 \\ 15x &= 120 \\ \frac{15x}{15} &= \frac{120}{15} \\ x &= 8 \end{aligned}$$
The number is 8.

39.
$$\begin{aligned} 2d &= t^2a + 2v \\ 2(244) &= 4^2a + 2(-20) \\ 488 &= 16a - 40 \\ 488 + 40 &= 16a - 40 + 40 \\ 528 &= 16a \\ \frac{528}{16} &= \frac{16a}{16} \\ 33 &= a \end{aligned}$$
The acceleration is 33.

41.
$$\begin{aligned} D &= B - NP \\ 575 &= 925 - N(25) \\ 575 &= 925 - 25N \\ 575 - 925 &= 925 - 925 - 25N \\ -350 &= -25N \\ \frac{-350}{-25} &= \frac{-25N}{-25} \\ 14 &= N \end{aligned}$$
There will be 14 payments.

43.
$$\begin{aligned} G &= 78 - 6t \\ 30 &= 78 - 6t \\ 30 - 78 &= 78 - 78 - 6t \\ -48 &= -6t \\ \frac{-48}{-6} &= \frac{-6t}{-6} \\ 8 &= t \end{aligned}$$
It will take 8 minutes.

45. Let $L =$ the lowest elevation.
$$\begin{aligned} 7310 &= 4L + 12{,}558 \\ 7310 - 12{,}558 &= 4L + 12{,}558 - 12{,}558 \\ -5248 &= 4L \\ \frac{-5248}{4} &= \frac{4L}{4} \\ -1312 &= L \end{aligned}$$
Asia is the continent.

47. Let $D =$ the depth.
$$\begin{aligned} -35{,}840 &= 2D - 2000 \\ -35{,}840 + 2000 &= 2D - 2000 + 2000 \\ -33{,}840 &= 2D \\ \frac{-33{,}840}{2} &= \frac{2D}{2} \\ -16920 &= D \end{aligned}$$
The Mediterranean is the ocean, with its deepest part being the Ionian Basin.

49.
$$\begin{aligned} 8z - 12 + (-6) &= 38 \\ 8z - 18 &= 38 \\ 8z - 18 + 18 &= 38 + 18 \\ 8z &= 56 \\ \frac{8z}{8} &= \frac{56}{8} \\ z &= 7 \end{aligned}$$

51.
$$\begin{aligned} -5z - 15 + 6 &= -21 - 18 \\ -5z - 9 &= -39 \\ -5z - 9 + 9 &= -39 + 9 \\ -5z &= -30 \\ \frac{-5z}{-5} &= \frac{-30}{-5} \\ z &= 6 \end{aligned}$$

53. **Answers may vary.**

55.
$$\begin{aligned} 7x + 14 &= 3x - 2 \\ 7x - 3x + 14 &= 3x - 3x - 2 \\ 4x + 14 &= -2 \\ 4x + 14 - 14 &= -2 - 14 \\ 4x &= -16 \\ \frac{4x}{4} &= \frac{-16}{4} \\ x &= -4 \end{aligned}$$

57.
$$\begin{aligned} 10x + 16 &= 5x + 6 \\ 10x - 5x + 16 &= 5x - 5x + 6 \\ 5x + 16 &= 6 \\ 5x + 16 - 16 &= 6 - 16 \\ 5x &= -10 \\ \frac{5x}{5} &= \frac{-10}{5} \\ x &= -2 \end{aligned}$$

Chapter 8 Review Exercises (page 615)

1. opposite of $-39 = 39$

2. opposite of $57 = -57$

3. opposite of $-0.91 = 0.91$

4. opposite of $-0.134 = 0.134$

5. $|-16.5| = 16.5$

6. $|-386| = 386$

7. $|71| = 71$

8. $|3.03| = 3.03$

9. opposite of $|-6.4|$ = opposite of 6.4
$= -6.4$

10. Ninety-three miles south would be represented by -93 miles.

11. $-75 + (-23) = -(75 + 23) = -98$

12. $-75 + 23 = -(75 - 23) = -52$

13. $75 + (-23) = +(75 - 23) = 52$

14. $65 + (-45) + (-82) = +(65 - 45) + (-82) = 20 + (-82) = -(82 - 20) = -62$

15. $-7.8 + (-5.3) + 9.9 = -(7.8 + 5.3) + 9.9 = -13.1 + 9.9 = -(13.1 - 9.9) = -3.2$

16.
$$\begin{aligned} -24 + 65 + (-17) + 31 &= +(65 - 24) + (-17) + 31 = 41 + (-17) + 31 \\ &= +(41 - 17) + 31 = 24 + 31 = 55 \end{aligned}$$

CHAPTER 8 REVIEW EXERCISES

17. $-6.8+(-4.3)+7.12+3.45=-(6.8+4.3)+7.12+3.45=-11.1+7.12+3.45$
$=-(11.1-7.12)+3.45$
$=-3.98+3.45$
$=-(3.98-3.45)=-0.53$

18. $$-\frac{7}{12}+\frac{7}{15}+\left(-\frac{7}{10}\right)=-\frac{35}{60}+\frac{28}{60}+\left(-\frac{42}{60}\right)=-\left(\frac{35}{60}-\frac{28}{60}\right)+\left(-\frac{42}{60}\right)$$
$$=-\frac{7}{60}+\left(-\frac{42}{60}\right)=-\left(\frac{7}{60}+\frac{42}{60}\right)=-\frac{49}{60}$$

19. $9+(-5)+(-6)+12=+(9-5)+(-6)+12=4+(-6)+12$
$=-(6-4)+12=-2+12=+(12-2)=10$

The team gained 10 yd and reached a first down.

20. $0.78+(-1.34)+(-2.78)+3.12+(-0.15)=-(1.34-0.78)+(-2.78)+3.12+(-0.15)$
$=-0.56+(-2.78)+3.12+(-0.15)$
$=-(0.56+2.78)+3.12+(-0.15)$
$=-3.34+3.12+(-0.15)$
$=-(3.34-3.12)+(-0.15)$
$=-0.22+(-0.15)=-(0.22+0.15)=-0.37$

The net change was down 0.37 points.

21. $19-(-3)=19+3=22$

22. $-45-81=-45+(-81)=-126$

23. $16-(-75)=16+75=91$

24. $-134-(-134)=-134+134=0$

25. $-4.56-3.25=-4.56+(-3.25)$
$=-7.81$

26. $-4.56-(-3.25)=-4.56+3.25$
$=-1.31$

27. $-127-(-156)=-127+156=29$

28. $-45-(-56)=-45+56=11$

29. $562.75-(-123.15)=562.75+123.15$
$=685.9$

The checks totaled $685.90.

30. $24.82-25.62=24.82+(-25.62)=-0.8$

There was a loss of -0.8 points.

31. $-6(11)=-66$

32. $5(-28)=-140$

33. $-1.2(3.4)=-4.08$

34. $7.4(-5.1)=-37.74$

35. $-3(-17)=51$

36. $-7(-21)=147$

37. $-4.03(-2.1)=8.463$

38. $(-1)(-4)(-6)(5)(-2)=4(-6)(5)(-2)=-24(5)(-2)=-120(-2)=240$

39. $632(-45)=-28{,}440$

Kroger's loss is $-28{,}440$¢, or $-\$284.40$.

40. $723.5(-0.32)=-231.52$

Pedro's loss is $-\$231.52$.

CHAPTER 8 REVIEW EXERCISES

41. $-18 \div 6 = -3$

42. $153 \div (-3) = -51$

43. $-45 \div (-9) = 5$

44. $-4.14 \div (-1.2) = 3.45$

45. $-2448 \div 153 = -16$

46. $-8342 \div (-97) = 86$

47. $-84.3 \div 1.5 = -56.2$

48. $-712 \div (-32) = 22.25$

49. $\dfrac{-15.5}{4} = -3.875$; The average loss is -3.875 points per day.

50. $\dfrac{-240.50}{650} = -0.37$; The average loss is $-\$0.37$, or -37¢, per box.

51. $5(-9) - 11 = -45 - 11 = -56$

52. $-3(17) + 45 = -51 + 45 = -6$

53. $-18 + (-4)(5) - 12 = -18 + (-20) - 12 = -50$

54. $-84 \div 4(7) = -21(7) = -147$

55. $-72 \div (-12)(3) + 6(-7) = 6(3) + 6(-7) = 18 + (-42) = -24$

56. $(7-4)(4)(-3) + 6(7-9) = 3(4)(-3) + 6(-2) = 12(-3) + (-12) = -36 + (-12) = -48$

57.
$$\begin{aligned}(-4)^2(-1)(-1) + (-6)(-3) - 17 &= (16)(-1)(-1) + 18 - 17 = (-16)(-1) + 18 - 17 \\ &= 16 + 18 - 17 = 17\end{aligned}$$

58.
$$\begin{aligned}(-1)(3^2)(-4) + 4(-5) - (-18-5) &= (-1)(9)(-4) + 4(-5) - (-23) \\ &= -9(-4) + (-20) + 23 \\ &= 36 + (-20) + 23 = 16 + 23 = 39\end{aligned}$$

59. $[71 \div (-2.5)] - [(3.2)(-2.4)] = [-28.4] - [-7.68] = -20.72$

60. $65(324) + 81(211) = 21{,}060 + 17{,}091 = 38{,}151$; $65 + 81 = 146$; $146(256) = 37{,}376$; $38{,}151 - 37{,}376 = 775$; The airline made a profit of \$775.

61.
$$\begin{aligned}7x + 25 &= -10 \\ 7x + 25 - 25 &= -10 - 25 \\ 7x &= -35 \\ \frac{7x}{7} &= \frac{-35}{7} \\ x &= -5\end{aligned}$$

62.
$$\begin{aligned}-6x + 21 &= -21 \\ -6x + 21 - 21 &= -21 - 21 \\ -6x &= -42 \\ \frac{-6x}{-6} &= \frac{-42}{-6} \\ x &= 7\end{aligned}$$

63.

$$\begin{aligned} 71 - 5x &= -54 \\ 71 - 71 - 5x &= -54 - 71 \\ -5x &= -125 \\ \frac{-5x}{-5} &= \frac{-125}{-5} \\ x &= 25 \end{aligned}$$

64.

$$\begin{aligned} -55 &= 3x + 41 \\ -55 - 41 &= 3x + 41 - 41 \\ -96 &= 3x \\ \frac{-96}{3} &= \frac{3x}{3} \\ -32 &= x \end{aligned}$$

65.

$$\begin{aligned} -43 &= 7x - 43 \\ -43 + 43 &= 7x - 43 + 43 \\ 0 &= 7x \\ \frac{0}{7} &= \frac{7x}{7} \\ 0 &= x \end{aligned}$$

66.

$$\begin{aligned} 12y - 9 &= -45 \\ 12y - 9 + 9 &= -45 + 9 \\ 12y &= -36 \\ \frac{12y}{12} &= \frac{-36}{12} \\ y &= -3 \end{aligned}$$

67.

$$\begin{aligned} 78a + 124 &= -890 \\ 78a + 124 - 124 &= -890 - 124 \\ 78a &= -1014 \\ \frac{78a}{78} &= \frac{-1014}{78} \\ a &= -13 \end{aligned}$$

68.

$$\begin{aligned} -55b + 241 &= -144 \\ -55b + 241 - 241 &= -144 - 241 \\ -55b &= -385 \\ \frac{-55b}{-55} &= \frac{-385}{-55} \\ b &= 7 \end{aligned}$$

69.

$$\begin{aligned} 9C &= 5F - 160 \\ 9(-22) &= 5F - 160 \\ -198 &= 5F - 160 \\ -198 + 160 &= 5F - 160 + 160 \\ -38 &= 5F \\ \frac{-38}{5} &= \frac{5F}{5} \\ -7.6 &= F \end{aligned}$$

The Fahrenheit temperature is $-7.6°$ F.

70.

$$\begin{aligned} 9C &= 5F - 160 \\ 9(-8) &= 5F - 160 \\ -72 &= 5F - 160 \\ -72 + 160 &= 5F - 160 + 160 \\ 88 &= 5F \\ \frac{88}{5} &= \frac{5F}{5} \\ 17.6 &= F \end{aligned}$$

The Fahrenheit temperature is $17.6°$ F.

Chapter 8 True/False Concept Review (page 617)

1. true

2. false; The opposite of a positive number is negative.

3. false; The absolute value of a nonzero number is positive.

4. true

5. false; The sum of two signed numbers can be positive, negative, or zero.

6. false; The sum of a positive signed number and a negative signed number can be positive, negative, or zero.

CHAPTER 8 TRUE/FALSE CONCEPT REVIEW

7. true

8. false; To subtract two signed numbers, add the opposite of the number to be subtracted.

9. true

10. true

11. false; The product of a positive number and a negative number is always negative.

12. true

13. true

14. true

Chapter 8 Test (page 618)

1. $-32 + (-19) + 39 + (-21) = -51 + 39 + (-21) = -12 + (-21) = -33$

2. $(45 - 52)(-16 + 21) = (-7)(5) = -35$

3. $\left(-\frac{3}{8}\right) \div \left(\frac{3}{10}\right) = \left(-\frac{3}{8}\right)\left(\frac{10}{3}\right) = -\frac{5}{4}$

4. $\left(-\frac{7}{15}\right) - \left(-\frac{3}{5}\right) = \left(-\frac{7}{15}\right) - \left(-\frac{9}{15}\right) = -\frac{7}{15} + \frac{9}{15} = \frac{2}{15}$

5. $(-11 - 5) - (5 - 22) + (-6) = (-16) - (-17) + (-6) = -16 + 17 - 6 = -5$

6. $-5.78 + 6.93 = 1.15$

7. **a.** $-(-17) = 17$
b. $|-33| = 33$

8. $-65 - (-32) = -65 + 32 = -33$

9. $(-18 + 6) \div 3 \cdot 4 - (-7)(-2)(-1) = (-12) \div 3 \cdot 4 - (-7)(-2)(-1)$
$= (-4) \cdot 4 - (14)(-1) = -16 - (-14) = -16 + 14 = -2$

10. $-110 \div (-55) = 2$

11. $(-6)(-8)(2) = (48)(2) = 96$

12. $(|-7|)(-3)(-1)(-1) = (7)(-3)(-1)(-1) = (-21)(-1)(-1) = (21)(-1) = -21$

13. $-63.2 - 45.7 = -63.2 + (-45.7) = -108.9$

14. $(-2)^2(-2)^2 + 4^2 \div (2)(3) = (4)(4) + 16 \div (2)(3) = 16 + 8(3) = 16 + 24 = 40$

15. $21.84 \div (-0.7) = -31.2$

16. $\left(-\frac{1}{3}\right) + \frac{5}{6} + \left(-\frac{1}{2}\right) + \left(-\frac{1}{6}\right) = \left(-\frac{2}{6}\right) + \frac{5}{6} + \left(-\frac{3}{6}\right) + \left(-\frac{1}{6}\right) = \frac{3}{6} + \left(-\frac{3}{6}\right) + \left(-\frac{1}{6}\right)$
$= 0 + \left(-\frac{1}{6}\right) = -\frac{1}{6}$

CHAPTER 8 TEST

17. $-112 \div (-8) = 14$

18. $-56 - 24 = -56 + (-24) = -80$

19. $(-7)(4-13)(-2) - 5(-2-6) = -7(-9)(-2) - 5(-8) = 63(-2) - (-40)$
$= -126 + 40 = -86$

20. $\left(-\frac{3}{8}\right)\left(\frac{12}{15}\right) = -\frac{36}{120} = -\frac{3}{10}$

21. $-45 + (-23) = -68$

22. $6(-7) + 37 = -42 + 37 = -5$

23.
$$\begin{aligned} -16 &= 5x + 14 \\ -16 - 14 &= 5x + 14 - 14 \\ -30 &= 5x \\ \frac{-30}{5} &= \frac{5x}{5} \\ -6 &= x \end{aligned}$$

24.
$$\begin{aligned} 7x - 32 &= 17 \\ 7x - 32 + 32 &= 17 + 32 \\ 7x &= 49 \\ \frac{7x}{7} &= \frac{49}{7} \\ x &= 7 \end{aligned}$$

25.
$$\begin{aligned} 24 - 6a &= 45 \\ 24 - 24 - 6a &= 45 - 24 \\ -6a &= 21 \\ \frac{-6a}{-6} &= \frac{21}{-6} \\ x &= -3.5 \end{aligned}$$

26. $(-1.05)(16) = -16.8$
The total loss is -16.8 lb.

27. $-9 - 12 = -21$; The drop is $-21°$ F.

28. $17.65 + 0.37 - 0.67 + 1.23 - 0.87 + 0.26 = 17.97$; The closing value is 17.97.

29. $F = \frac{9}{5}C + 32 = \frac{9}{5}(-10) + 32 = -18 + 32 = 14$; The Fahrenheit temperature is $14°$ F.

30. $-11 + (-15) + 23 + (-19) + 10 + (-12) = -24$; $-24 \div 6 = -4$; The average is -4.

Midterm Examination (page 623)

1. thousand

2.

$$\begin{array}{rrrrrr} & {\scriptstyle 1} & {\scriptstyle 2} & {\scriptstyle 2} & {\scriptstyle 2} & \\ & & & 2 & 8 & 9 \\ & & 4 & 6 & 7 & 5 \\ & & & & 5 & 2 \\ & 7 & 8 & 6 & 1 & 2 \\ + & & & 5 & 5 & 5 \\ \hline & 8 & 4 & 1 & 8 & 3 \end{array}$$

3. sixty-seven thousand, five hundred nine

4.

$$\begin{array}{rrrrrr} & & {\scriptstyle 17} & {\scriptstyle 12} & & \\ & \not{6} & \not{7} & \not{2} & \not{12} & \\ & \not{7} & \not{8} & \not{3} & \not{2} & 9 \\ - & 6 & 9 & 5 & 4 & 3 \\ \hline & & 8 & 7 & 8 & 6 \end{array}$$

5.

$$\begin{array}{rrrrrr} & & {\scriptstyle 1} & {\scriptstyle 2} & {\scriptstyle 1} & \\ & & & 7 & 0 & 3 \\ & 2 & 5 & 7 & 7 & 2 \\ & & 1 & 0 & 9 & 8 \\ + & & & & 3 & 2 \\ \hline & 2 & 7 & 6 & 0 & 5 \end{array}$$

6.

$$\begin{array}{rrrrrr} & & & {\scriptstyle 6} & {\scriptstyle 6} & \\ & & & \not{3} & \not{3} & \\ & & & 3 & 6 & 7 \\ \times & & & & 9 & 5 \\ \hline & & 1 & 8 & 3 & 5 \\ & 3 & 3 & 0 & 3 & 0 \\ \hline & 3 & 4 & 8 & 6 & 5 \end{array}$$

7.

$$\begin{array}{rrrrrrr} & & & & & {\scriptstyle 2} & \\ & & & & & \not{1} & \\ & & & & 8 & 0 & 3 \\ \times & & & & 9 & 0 & 6 \\ \hline & & & 4 & 8 & 1 & 8 \\ & & & 0 & 0 & 0 & 0 \\ & 7 & 2 & 2 & 7 & 0 & 0 \\ \hline & 7 & 2 & 7 & 5 & 1 & 8 \end{array} \Rightarrow \begin{array}{rrrrrrr} & & & & 8 & 0 & 0 \\ \times & & & & 9 & 0 & 0 \\ \hline & & & & 0 & 0 & 0 \\ & & & 0 & 0 & 0 & 0 \\ & 7 & 2 & 0 & 0 & 0 & 0 \\ \hline & 7 & 2 & 0 & 0 & 0 & 0 \end{array}$$

8.

$$\begin{array}{r} 213 \\ 347\overline{)73911} \\ \underline{694} \\ 451 \\ \underline{347} \\ 1041 \\ \underline{1041} \\ 0 \end{array}$$

9.

$$\begin{array}{r} 729 \text{ R } 54 \\ 63\overline{)45981}\phantom{\text{ R }54} \\ \underline{441}\phantom{\text{ R }54} \\ 188\phantom{\text{ R }54} \\ \underline{126}\phantom{\text{ R }54} \\ 621\phantom{\text{ R }54} \\ \underline{567}\phantom{\text{ R }54} \\ 54\phantom{\text{ R }54} \end{array}$$

10. $19 - 3 \cdot 4 + 18 \div 3 = 19 - 12 + 6 = 13$

11. $(72 \div 9) + (33 \cdot 2) = 8 + 66 = 74$

12. $245 + 175 + 893 + 660 + 452 = 2425$; $2425 \div 5 = 485$

13. Arrange in order: 52, 64, 64, 64, 82, 97, 97, 128
average $= (52 + 64 + 64 + 64 + 82 + 97 + 97 + 128) \div 8 = 648 \div 8 = 81$
median $= (64 + 82) \div 2 = 146 \div 2 = 73$; mode $= 64$

MIDTERM EXAMINATION

14. **a.** Civic has the highest sales.
b. 5 more Accords are sold than Passports.

15. $30 = 2 \times 3 \times 5$
$24 = 2^3 \times 3$
$40 = 2^3 \times 5$
$\text{LCM} = 2^3 \times 3 \times 5 = 120$

16. 1785 is divisible by 3 and 5.

17. $107 \Rightarrow$ prime

18. $7 + 2 + 6 + 3 = 18$; 18 is a multiple of 9, so 7263 is a multiple of 9.

19. $32 \cdot 1 = 32$; $32 \cdot 2 = 64$; $32 \cdot 3 = 96$; $32 \cdot 4 = 128$; $32 \cdot 5 = 160$

20. 304: 1, 2, 4, 8, 16, 19, 38, 76, 152, 304

21. $540 = 27 \cdot 20 = 3^3 \cdot 2^2 \cdot 5 = 2^2 \cdot 3^3 \cdot 5$

22.
$$\begin{array}{r} 11 \\ 7\overline{)83} \\ \underline{7} \\ 13 \\ \underline{7} \\ 6 \end{array} \Rightarrow 11\frac{6}{7}$$

23. $9\frac{3}{7} = \frac{7(9) + 3}{7} = \frac{63 + 3}{7} = \frac{66}{7}$

24. proper: $\frac{3}{4}, \frac{7}{8}$

25.
$\frac{5}{9} = \frac{5}{9} \cdot \frac{8}{8} = \frac{40}{72}$
$\frac{5}{8} = \frac{5}{8} \cdot \frac{9}{9} = \frac{45}{72}$
$\frac{7}{12} = \frac{7}{12} \cdot \frac{6}{6} = \frac{42}{72}$
$\frac{2}{3} = \frac{2}{3} \cdot \frac{24}{24} = \frac{48}{72}$
$\frac{5}{9} < \frac{7}{12} < \frac{5}{8} < \frac{2}{3}$

26. $\frac{114}{150} = \frac{19 \cdot \cancel{6}}{25 \cdot \cancel{6}} = \frac{19}{25}$

27. $\frac{3}{16} \cdot \frac{4}{9} \cdot \frac{5}{6} = \frac{\overset{1}{\cancel{3}}}{\underset{4}{\cancel{16}}} \cdot \frac{\overset{1}{\cancel{4}}}{\underset{3}{\cancel{9}}} \cdot \frac{5}{6} = \frac{5}{72}$

28. $\left(3\frac{3}{5}\right)\left(12\frac{6}{7}\right) = \frac{18}{5} \cdot \frac{90}{7} = \frac{324}{7} = 46\frac{2}{7}$

29. $\frac{45}{70} \div \frac{9}{14} = \frac{45}{70} \cdot \frac{14}{9} = \frac{\overset{5}{\cancel{45}}}{\underset{5}{\cancel{70}}} \cdot \frac{\overset{1}{\cancel{14}}}{\underset{1}{\cancel{9}}} = \frac{5}{5} = 1$

30. reciprocal of $3\frac{4}{9} = \frac{31}{9} \Rightarrow \frac{9}{31}$

31. $\frac{7}{15} + \frac{11}{18} = \frac{7}{15} \cdot \frac{6}{6} + \frac{11}{18} \cdot \frac{5}{5}$
$= \frac{42}{90} + \frac{55}{90} = \frac{97}{90}$, or $1\frac{7}{90}$

32. $6\frac{2}{3}+11\frac{7}{8}=6+11+\frac{2}{3}+\frac{7}{8}=17+\frac{2}{3}\cdot\frac{8}{8}+\frac{7}{8}\cdot\frac{3}{3}=17+\frac{16}{24}+\frac{21}{24}=17+\frac{37}{24}=17+1\frac{13}{24}$
$=18\frac{13}{24}$

33.
$$\begin{array}{r} 25 \\ -\,11\frac{4}{7} \\ \hline \end{array} \Rightarrow \begin{array}{r} 24\frac{7}{7} \\ -\,11\frac{4}{7} \\ \hline 13\frac{3}{7} \end{array}$$

34.
$$\begin{array}{r} 54\frac{1}{9} \\ -\,36\frac{4}{7} \\ \hline \end{array} \Rightarrow \begin{array}{r} 54\frac{7}{63} \\ -\,36\frac{36}{63} \\ \hline \end{array} \Rightarrow \begin{array}{r} 53\frac{70}{63} \\ -\,36\frac{36}{63} \\ \hline 17\frac{34}{63} \end{array}$$

35. $4\frac{5}{6}+6\frac{2}{3}+11\frac{3}{4}=4+6+11+\frac{5}{6}+\frac{2}{3}+\frac{3}{4}=21+\frac{20}{24}+\frac{16}{24}+\frac{18}{24}=21+\frac{54}{24}=21+\frac{9}{4}$
$=21+2\frac{1}{4}$
$=23\frac{1}{4}$

$23\frac{1}{4}\div 3=\frac{93}{4}\cdot\frac{1}{3}=\frac{93}{12}=\frac{31}{4}=7\frac{3}{4}$

36. $\frac{3}{5}-\frac{1}{3}\cdot\frac{5}{6}\div\frac{5}{6}=\frac{3}{5}-\frac{5}{18}\cdot\frac{6}{5}=\frac{3}{5}-\frac{1}{3}=\frac{9}{15}-\frac{5}{15}=\frac{4}{15}$

37. 80,240.122

38. $7.95863 \approx 8.0$; $7.95863 \approx 7.96$; $7.95863 \approx 7.959$

39. $0.26=\frac{26}{100}=\frac{13}{50}$

40. 1.02, 1.034, 1.044, 1.07, 1.094, 1.109

41.
$$\begin{array}{r} {\scriptstyle 2\,2\;\;2\,1\;\;\;\;\;} \\ 134.760 \\ 7.113 \\ 0.094 \\ 5.923 \\ +\;\;25.870 \\ \hline 173.760 \end{array}$$

42.
$$\begin{array}{r} {\scriptstyle 12\,9\;\;} \\ {\scriptstyle 7\;\;\not{2}\;\;\not{1}\not{0}\,10} \\ \not{8}.\not{3}\;\not{0}\;\not{0} \\ -\;5.7\;6\;3 \\ \hline 2.5\;3\;7 \end{array}$$

43. Put five decimal places in the product.
$$\begin{array}{r} {\scriptstyle 2\,6\;\;5\;\;4} \\ {\scriptstyle \not{2}\;\;\not{1}\;\;\not{1}} \\ {\scriptstyle \not{1}\not{3}\;\;\not{2}\;\;\not{2}} \\ 42.7\;65 \\ \times\qquad\;\; 8.34 \\ \hline 1\;71\;0\;60 \\ 12\;82\;9\;50 \\ 342\;12\;0\;00 \\ \hline 356.66\;0\;10 \end{array}$$

44. $0.046 \times 100{,}000 = 4600$

45. $0.902 \div 1{,}000 = 0.000902$

46. $0.00058 = 5.8 \times 10^{-4}$

47. $7.85\overline{)445.88}$ $\Downarrow$

$$\begin{array}{r} 56.8 \\ 785\overline{)44588.0} \\ \underline{3925} \\ 5338 \\ \underline{4710} \\ 6280 \\ \underline{6280} \\ 0 \end{array}$$

48. $\frac{15}{23} = 15 \div 23 = 0.652$

$$\begin{array}{r} 0.6521 \\ 23\overline{)15.0000} \\ \underline{138} \\ 120 \\ \underline{115} \\ 50 \\ \underline{46} \\ 40 \\ \underline{23} \end{array}$$

49. **Average**
$4.5 + 6.9 + 8.3 + 9.5 + 3.1 + 10.3 + 2.2 = 44.8$
$44.8 \div 7 = 6.4$

Median
In order: 2.2, 3.1, 4.5, 6.9, 8.3, 9.5, 10.3
6.9

50. $(5.5)^2 - 2.3(4.1) + 11.8 - 9.3 \div 0.3 = 30.25 - 9.43 + 11.8 - 31 = 1.62$

51. $7 + 18 + 10 + 25 = 60$ cm

52. $\left(7\frac{3}{8}\right)\left(10\frac{2}{5}\right) = \frac{59}{8} \cdot \frac{52}{5} = \frac{3068}{40} = 76\frac{28}{40}\text{ ft}^2 = 76\frac{7}{10}\text{ ft}^2$

53. $1577 + 1589 + 1854 + 1361 + 295 + 274 = 6950$; 6950 salmon were counted.

54. $765.72(14) = 10{,}720.08$; Maria earned \$107.20 as a dividend.

55. $\frac{\$5.79}{6.5\text{ lb}} = \$0.89$; The cost is \$0.89 per pound.

Final Examination (page 627)

1. $\frac{7}{12} + \frac{7}{15} = \frac{35}{60} + \frac{28}{60} = \frac{63}{60} = 1\frac{3}{60} = 1\frac{1}{20}$

2. $15 = 3 \times 5$
$18 = 2 \times 3^2$
$35 = 5 \times 7$
$\text{LCM} = 2 \times 3^2 \times 5 \times 7 = 630$

3.

$$\begin{array}{r} {\scriptstyle 12\ 13} \\ {\scriptstyle 8\not{2}\ \not{3}\ 12} \\ \not{9}\not{3}.\not{4}\not{2} \\ -\ 57.69 \\ \hline 35.73 \end{array}$$

4.

$$\begin{array}{r} {\scriptstyle 1\ \ \ 11} \\ 78.320 \\ 4.089 \\ +\ \ 0.139 \\ \hline 82.548 \end{array}$$

5. $\frac{15}{16} \div \frac{5}{8} = \frac{15}{16} \cdot \frac{8}{5} = \frac{3}{2}$, or $1\frac{1}{2}$

6. $3\frac{5}{7} \cdot 35 = \frac{26}{7} \cdot \frac{35}{1} = \frac{130}{1} = 130$

7. 199 is prime.

8. $0.62\overline{)23.764}$

$\Downarrow$

$$\begin{array}{r} 38.329 \\ 62\overline{)2376.400} \\ \underline{186} \\ 516 \\ \underline{496} \\ 204 \\ \underline{186} \\ 180 \\ \underline{124} \\ 560 \\ \underline{558} \end{array}$$

$38.329 \approx 38.33$

9.

$$\begin{array}{r} 21 \\ -\,13\frac{11}{15} \\ \hline \end{array} \Rightarrow \begin{array}{r} 20\frac{15}{15} \\ -\,13\frac{11}{15} \\ \hline 7\frac{4}{15} \end{array}$$

10. $(0.0945)(10{,}000) = 945$

11. $314.9278 \approx 314.93$

12. Put three decimal places in the product.

$$\begin{array}{r} \scriptstyle 2 \\ \not{1}\,\not{4} \\ 11.6 \\ \times\quad 4.07 \\ \hline 812 \\ 0000 \\ 46400 \\ \hline 47.212 \end{array}$$

13. $36\% = 36 \cdot \dfrac{1}{100} = \dfrac{36}{100} = \dfrac{9}{25}$

14.

$$\begin{array}{r} 9\frac{11}{15} \\ +\quad \frac{5}{6} \\ \hline \end{array} \Rightarrow \begin{array}{r} 9\frac{22}{30} \\ +\quad \frac{25}{30} \\ \hline 9\frac{47}{30} = 10\frac{17}{30} \end{array}$$

15.
$$\frac{5}{16} = \frac{x}{24}$$
$$5(24) = 16x$$
$$120 = 16x$$
$$\frac{120}{16} = \frac{16x}{16}$$
$$\frac{15}{2} = x, \text{ or } x = 7\frac{1}{2}$$

16. ten-thousandths

17.
$$\frac{3}{4} = \frac{3}{4} \cdot \frac{5}{5} = \frac{15}{20}$$
$$\frac{4}{5} = \frac{4}{5} \cdot \frac{4}{4} = \frac{16}{20}$$
$$\frac{7}{10} = \frac{7}{10} \cdot \frac{2}{2} = \frac{14}{20}$$
$$\frac{7}{10} < \frac{3}{4} < \frac{4}{5}$$

18. $\dfrac{23}{25} = \dfrac{92}{100} = 92\%$

19.
$$\begin{array}{r} 19.7391 \\ 46\overline{)908.0000} \\ \underline{46} \\ 448 \\ \underline{414} \\ 340 \\ \underline{322} \\ 180 \\ \underline{138} \\ 420 \\ \underline{414} \\ 60 \\ \underline{46} \end{array}$$
$$19.7391 \approx 19.739$$

20. $\dfrac{17}{33} = 17 \div 33 \approx 0.515$

21. 72,000.05

22.
$$0.043\overline{)0.34787}$$
$$\Downarrow$$
$$\begin{array}{r} 8.09 \\ 43\overline{)347.87} \\ \underline{344} \\ 38 \\ \underline{0} \\ 387 \\ \underline{387} \\ 0 \end{array}$$

23. $57\frac{5}{8}\% = 0.57625$

24.
$$\frac{39}{100} = \frac{19.5}{x}$$
$$39x = 100(19.5)$$
$$39x = 1950$$
$$\frac{39x}{39} = \frac{1950}{39}$$
$$x = 50$$

25. $0.0935 = 9.35\%$

26. $\dfrac{71}{250} = 71 \div 250 = 0.284$

FINAL EXAMINATION

27. $\dfrac{73}{100} = \dfrac{x}{82}$
$73(82) = 100x$
$5986 = 100x$
$\dfrac{5986}{100} = \dfrac{100x}{100}$
$59.86 = x$

28. $(0.26)(4.5)(0.55) = 1.17(0.55) = 0.6435$

29. $\dfrac{7}{24.78} = \dfrac{18}{x}$
$7x = 24.78(18)$
$7x = 446.04$
$\dfrac{7x}{7} = \dfrac{446.04}{7}$
$x = 63.72$
They will cost \$63.72.

30. Discount $= 345 - 280 = 65$
$\dfrac{x}{100} = \dfrac{65}{345}$
$x(345) = 100(65)$
$345x = 6500$
$\dfrac{345x}{345} = \dfrac{6500}{345}$
$x \approx 18.8 \Rightarrow 18.8\%$

31. $51 \cdot 1 = 51;\ 51 \cdot 2 = 102;\ 51 \cdot 3 = 153;$
$51 \cdot 4 = 204;\ 51 \cdot 5 = 255$

32. eight thousand thirty-seven and thirty-seven thousandths

33. $\dfrac{1.9}{22} \stackrel{?}{=} \dfrac{5.8}{59}$
$1.9(59) \stackrel{?}{=} 22(5.8)$
$112.1 \neq 127.6$
false

34. $750 = 75 \cdot 10 = 3 \cdot 25 \cdot 2 \cdot 5 = 2 \cdot 3 \cdot 5^3$

35. $\dfrac{315}{450} = \dfrac{7 \cdot \cancel{45}}{10 \cdot \cancel{45}} = \dfrac{7}{10}$

36. $0.945 = \dfrac{945}{1000} = \dfrac{189}{200}$

37.
$$\begin{array}{r} 48 \\ 9\overline{)436} \\ \underline{36} \\ 76 \\ \underline{72} \\ 4 \end{array} \Rightarrow 48\frac{4}{9}$$

38. $\dfrac{6}{35} \cdot \dfrac{42}{54} = \dfrac{\overset{2}{\cancel{6}}}{\underset{5}{\cancel{35}}} \cdot \dfrac{\overset{6}{\cancel{42}}}{\underset{18}{\cancel{54}}} = \dfrac{2}{5} \cdot \dfrac{\overset{1}{\cancel{6}}}{\underset{3}{\cancel{18}}} = \dfrac{2}{15}$

39. $0.145 = \dfrac{145}{1000} = \dfrac{29}{200}$

40. $\dfrac{27}{50} = 27 \div 50 = 0.54 = 54\%$

41. no

42. $\begin{array}{r} 14\frac{3}{5} \\ -\ \ \frac{11}{15} \\ \hline \end{array} \Rightarrow \begin{array}{r} 14\frac{9}{15} \\ -\ \ \frac{11}{15} \\ \hline \end{array} \Rightarrow \begin{array}{r} 13\frac{24}{15} \\ -\ \ \frac{11}{15} \\ \hline 13\frac{13}{15} \end{array}$

43. $13\frac{7}{18} = \frac{18(13)+7}{18} = \frac{234+7}{18} = \frac{241}{18}$

44. 2.299, 2.32, 2.322, 2.332

45. 408 : 1, 2, 3, 4, 6, 8, 12, 17, 24, 34, 51, 68, 102, 136, 204, 408

46. $45.893 \div 10^5 = 0.00045893$

47. $4\frac{7}{12} \div 1\frac{9}{16} = \frac{55}{12} \div \frac{25}{16} = \frac{55}{12} \cdot \frac{16}{25} = \frac{\overset{11}{\cancel{55}}}{\underset{3}{\cancel{12}}} \cdot \frac{\overset{4}{\cancel{16}}}{\underset{5}{\cancel{25}}} = \frac{44}{15} = 2\frac{14}{15}$

48. $\frac{85¢}{\$5} = \frac{85¢}{500¢} = \frac{17}{100}$

49. $0.67 + 0.30 = 0.97$ for 5 miles

$$\frac{0.97}{5} = \frac{x}{7500}$$
$$0.97(7500) = 5x$$
$$7275 = 5x$$
$$\frac{7275}{5} = \frac{5x}{5}$$
$$1455 = x$$

It will cost $1455.

50. $1500 + 0.035(467{,}800) = 1500 + 16{,}373 = \$17{,}873$

51.

$$A = R \times B$$
$$4.75 = R \times 55$$
$$\frac{4.75}{55} = \frac{R \times 55}{55}$$
$$0.086 \approx R \Rightarrow R \approx 8.6\%$$

The rate is about 8.6%.

52. $0.30(245.50) = 73.65$; $245.50 - 73.65 = 171.85$; $0.065(171.85) \approx 11.17$; $171.85 + 11.17 = \$183.02$

53. $82 - 7 \cdot 6 + 45 \div 5 = 82 - 42 + 9 = 40 + 9 = 49$

54. $79.15 - 5.1(8.3) \div 3.4 = 79.15 - 42.33 \div 3.4 = 79.15 - 12.45 = 66.7$

55.
$$\begin{array}{r} 5\text{ hr } 47\text{ min } 32\text{ sec} \\ +\ 2\text{ hr } 36\text{ min } 48\text{ sec} \\ \hline 7\text{ hr } 83\text{ min } 80\text{ sec} \\ 7\text{ hr } 84\text{ min } 20\text{ sec} \\ 8\text{ hr } 24\text{ min } 20\text{ sec} \end{array}$$

56.
$$\begin{array}{r} 9\text{ m } 45\text{ cm} \\ -\ 5\text{ m } 72\text{ cm} \\ \hline \end{array} \Rightarrow \begin{array}{r} 8\text{ m } 145\text{ cm} \\ -\ 5\text{ m } \ \ 72\text{ cm} \\ \hline 3\text{ m } \ \ 73\text{ cm} \end{array}$$

57. $\dfrac{5¢}{1\text{ g}} = \dfrac{5¢}{1\text{ g}} \cdot \dfrac{\$1}{100¢} \cdot \dfrac{1000\text{ g}}{1\text{ kg}} = \dfrac{\$5000}{100\text{ kg}} = \$50\text{ per kg}$

58. $63.7 + 74.2 + 21.5 + 23.6 = 183\text{ ft}$

59. $A = \dfrac{1}{2}bh = \dfrac{1}{2}(9.3\text{ m})(7.2\text{ m}) = \dfrac{1}{2}\left(66.96\text{ m}^2\right) = 33.48\text{ m}^2$

60. $A = (12\text{ yd})(4\text{ yd}) + \dfrac{1}{2}\pi(3\text{ yd})^2 = 48\text{ yd}^2 + \dfrac{1}{2}(3.14)\left(9\text{ yd}^2\right) = 48\text{ yd}^2 + 14.13\text{ yd}^2 = 62.13\text{ yd}^2$

61. $V = (4.6\text{ ft})(7\text{ in.})(5\text{ in.}) = (55.2\text{ in.})(7\text{ in.})(5\text{ in.}) = 1932\text{ in.}^3$

62. $\sqrt{578.4} \approx 24.05$

63. $c = \sqrt{23^2 + 27^2} = \sqrt{1258} \approx 35.5\text{ cm}$

64. $(-34) + (-23) + 41 = -57 + 41 = -16$

65. $(-73) - (-94) = -73 + 94 = 21$

66. $5(-7)(-1)(-3) = -35(-1)(-3) = 35(-3) = -105$

67. $(-46.5) \div (-15) = 3.1$

68. $(-6 - 4)(-4) \div (-5) - (-10) = (-10)(-4) \div (-5) - (-10) = 40 \div (-5) + 10 = -8 + 10 = 2$

69.
$$\begin{aligned} 15a + 108 &= 33 \\ 15a + 108 - 108 &= 33 - 108 \\ 15a &= -75 \\ \frac{15a}{15} &= \frac{-75}{15} \\ a &= -5 \end{aligned}$$

70.
$$\begin{aligned} 9C &= 5F - 160 \\ 9(-15) &= 5F - 160 \\ -135 &= 5F - 160 \\ -135 + 160 &= 5F - 160 + 160 \\ 25 &= 5F \\ \frac{25}{5} &= \frac{5F}{5} \\ 5 &= F \Rightarrow 5°\text{ F} \end{aligned}$$